Ching-Yuan CHIANG
Department of Mathematics
and Computer Science
James Madison Univ.

Nonparametric
Statistical
Inference

McGraw-Hill Series in Probability and Statistics

David Blackwell and Herbert Solomon, Consulting Editors

Nonparametric
Statistical
Inference

Jean Dickinson Gibbons, Ph.D.
Professor of Statistics
University of Alabama

Formerly Associate Professor
of Statistics
University of Pennsylvania

McGraw-Hill Book Company
New York   St. Louis   San Francisco   Düsseldorf
London   Mexico   Panama   Sydney   Toronto

Nonparametric Statistical Inference

Library of Congress Catalog Card Number 74-126747

07-023166-4

345678910 KPKP 79876

**To my mother and
the memory of my father**

Preface

During the last few years many institutions offering graduate programs in statistics have experienced a demand for a course devoted exclusively to a survey of nonparametric techniques and their justifications. This demand has arisen both from their own majors and from majors in social science or other quantitatively oriented fields such as psychology, sociology, or economics. Although the basic statistics courses often include a brief description of some of the better-known and simpler nonparametric methods, usually the treatment is necessarily perfunctory and perhaps even misleading. Discussion of only a few techniques in a highly condensed fashion may leave the impression that nonparametric statistics consists of a "bundle of tricks" which are simply applied by following a list of instructions dreamed up by some genie as a panacea for all sorts of vague and ill-defined problems.

One of the deterrents to meeting this demand has been the lack of a suitable textbook in nonparametric techniques. Our experience at the University of Pennsylvania has indicated that an appropriate text would provide a theoretical but readable survey. Only a moderate amount of pure mathematical sophistication should be required so that the course would be comprehensible to a wide variety of graduate students and perhaps even some advanced undergraduates. The course should be available to anyone who has completed at least the rather traditional one-year sequence in probability and statistical inference at the level of Parzen, Mood and Graybill, Hogg and Craig, etc. The time allotment should be a full semester, or perhaps two semesters if outside reading in journal publications is desirable.

The texts presently available which are devoted exclusively to non-parametric statistics are few in number and seem to be predominantly either of the handbook style, with few or no justifications, or of the highly rigorous mathematical style. The present book is an attempt to bridge the gap between these extremes. It assumes the reader is well acquainted with statistical inference for the traditional parametric-estimation and hypothesis-testing procedures, basic probability theory, and random-sampling distributions. The survey is not intended to be exhaustive, as the field is so extensive. The purpose of the book is to provide a compendium of some of the better-known nonparametric techniques for each problem situation. Those derivations, proofs, and mathematical details which are relatively easily grasped or which illustrate typical procedures in general nonparametric statistics are included. More advanced results are simply stated with references. For example, some of the asymptotic distribution theory for order statistics is derived since the methods are equally applicable to other statistical problems. However, the Glivenko-Cantelli theorem is given without proof since the mathematics may be too advanced. Generally those proofs given are not mathematically rigorous, ignoring details such as existence of derivatives or regularity conditions. At the end of each chapter, some problems are included which are generally of a theoretical nature but on the same level as the related text material they supplement.

The organization of the material is primarily according to the type of statistical information collected and the type of questions to be answered by the inference procedures or according to the general type of mathematical derivation. For each statistic, the null distribution theory is derived, or when this would be too tedious, the procedure one could follow is outlined, or when this would be overtheoretical, the results are stated without proof. Generally the other relevant mathematical details necessary for nonparametric inference are also included. The purpose is to acquaint the reader with the mathematical logic on which a test is based, those test properties which are essential for understanding the procedures, and the basic tools necessary for comprehending the extensive literature published in the statistics journals. The book is not intended to be a user's manual for the application of nonparametric techniques. As a result, almost no numerical examples or problems are provided to illustrate applications or elicit applied motivation. With this approach, reproduction of an extensive set of tables is not required.

The reader may already be acquainted with many of the nonparametric methods. If not, the foundations obtained from this book should enable anyone to turn to a user's handbook and quickly grasp the application. Once armed with the theoretical background, the user of nonparametric methods is much less likely to apply tests indiscriminately or view the field as a collection of simple prescriptions. The only insurance against misapplication is a thorough understanding. Although some of the strengths and weaknesses of the tests covered are alluded to, no definitive judgments are attempted regarding the relative merits of comparable tests. For each topic covered, some references are given which provide further information about the tests or are specifically related to the approach used in this book. These references are necessarily incomplete, as the literature is vast. The interested reader may consult Savage's "Bibliography" (1962).

I wish to acknowledge the helpful comments of the reviewers and the assistance provided unknowingly by the authors of other textbooks in the area of nonparametric statistics, particularly Gottfried E. Noether and James V. Bradley, for the approach to presentation of several topics, and Maurice G. Kendall, for much of the material on measures of association. The products of their endeavors greatly facilitated this project. It is a pleasure also to acknowledge my indebtedness to Herbert A. David, both as friend and mentor. His training and encouragement helped make this book a reality. Particular gratitude is also due to the Lecture Note Fund of the Wharton School, for typing assistance, and the Department of Statistics and Operations Research at the University of Pennsylvania for providing the opportunity and time to finish this manuscript. Finally, I thank my husband for his enduring patience during the entire writing stage.

Jean Dickinson Gibbons

Contents

1
Introduction, Review, and Notation

1.1 INTRODUCTION

In many elementary statistics courses, the subject matter is somewhat arbitrarily divided into two categories, called descriptive and inductive statistics. *Descriptive statistics* usually relates only to the calculation or presentation of figures (visual or conceptual) to summarize or characterize a set of data. For such procedures, no assumptions are made or implied, and there is no question of legitimacy of techniques. The descriptive figures may be a mean, median, variance, range, histogram, etc. Each of these figures summarizes a set of numbers in its own unique way; each is a distinguishable and well-defined characterization of data. If such data constitute a random sample from a certain population, the sample represents the population in miniature and any set of descriptive statistics provides some information regarding this universe. The term *parameter* is generally employed to connote a characteristic of the population. A parameter is often an unspecified constant appearing in a family of probability distributions, but the word can also be interpreted in a broader

1

sense to include almost all descriptions of population characteristics within a family.

When sample descriptions are used to infer some information about the population, the subject is called *inductive statistics* or *statistical inference*. The two types of problems most frequently encountered here are estimation and tests of hypotheses. The factor which makes inference a scientific method, thereby differentiating it from mere guessing, is the ability to make evaluations or probability statements concerning the accuracy of an estimate or reliability of a decision. Unfortunately, such scientific evaluations cannot be made without some information regarding the probability distribution of the random variable relating to the sample description used in the inference procedure. This means that certain types of sample descriptions will be more popular than others, because of their distribution properties or mathematical tractability. The sample mean is a popular figure for describing the characteristic of central tendency for many reasons but perhaps least of all because it is a mean. The unique position of the mean in inference stems largely from its "almost normal" distribution properties. If some other measure, say the sample median, had a property as useful as the central-limit theorem, surely it would share the spotlight as a favorite description of location.

The entire body of classical statistical-inference techniques is based on fairly specific assumptions regarding the nature of the underlying population distribution; usually its form and some parameter values must be stated. Given the right set of assumptions, certain test statistics can be developed using mathematics which is frequently elegant and beautiful. The derived distribution theory is qualified by certain prerequisite conditions, and therefore all conclusions reached using these techniques are exactly valid only so long as the assumptions themselves can be substantiated. In textbook problems, the requisite postulates are frequently just stated and the student practices applying the appropriate technique. However, in a real-world problem, everything does not come packaged with labels of population of origin. A decision must be made as to what population properties may judiciously be assumed for the model. If the reasonable assumptions are not such that the traditional techniques are applicable, the classical methods may be used and inference conclusions stated only with the appropriate qualifiers, e.g., "If the population is normal, then . . . ."

The mathematical statistician may claim that it is the users' problem to decide on the legitimacy of the postulates. Frequently in practice, those assumptions which are deemed reasonable by empirical evidence or past experience are not the desired ones, i.e., those for which a set of standard statistical techniques has been developed. Alternatively, the sample may be too small or previous experience too limited to determine what is a reasonable assumption. Or, if the researcher is a product of the "cookbook school" of statistics, his particular expertise being in the area of application, he may not understand or even be aware of the pre-conditions implicit in the derivation of the statistical technique. In any of these three situations, the result often is a substitution of blind faith for scientific method, either because of ignorance or with the rationalization that an approximately accurate inference based on recognized and accepted scientific techniques is better than no answer at all or a con-clusion based on common sense or intuition.

There is an alternative set of techniques available, and the mathe-matical bases for these procedures are the subject of this book. They may be classified as distribution-free and nonparametric procedures. In a *distribution-free* inference, whether for testing or estimation, the methods are based on functions of the sample observations whose corresponding random variable has a distribution which does not depend on the specific distribution function of the population from which the sample was drawn. Therefore, assumptions regarding the underlying population are not necessary. On the other hand, strictly speaking, the term *nonparametric test* implies a test for a hypothesis which is not a statement about parameter values. The type of statement permissible then depends on the definition accepted for the term parameter. If parameter is interpreted in the broader sense, the hypothesis can only be concerned with the form of the population, as in goodness-of-fit tests, or with some characteristic of the probability distribution of the sample data, as in tests of randomness and trend. Needless to say, distribution-free tests and nonparametric tests are not synonymous labels or even in the same spirit, since one relates to the distribution of the test statistic and the other to the type of hypothesis to be tested. A distribution-free test may be for a hypothesis concerning the median, which is certainly a population parameter within our broad definition of the term.

In spite of the inconsistency in nomenclature, we shall follow the customary practice and consider both types of tests as procedures in nonparametric inference, making no distinction between the two classi-fications. For the purpose of differentiation, the classical statistical techniques, whose justification in probability is based on specific assump-tions about the population sampled, may be called *parametric methods*. This implies a definition of nonparametric statistics then as the treatment of either nonparametric types of inferences or analogies to standard statistical problems when specific distribution assumptions are replaced by very general assumptions and the analysis is based on some function of the sample observations whose sampling distribution can be deter-mined without knowledge of the specific distribution function of the underlying population. The assumption most frequently required is simply that the population be continuous.   More restrictive assumptions are sometimes made, e.g., that the population is symmetrical, but not to the extent that the distribution is specifically postulated.   The information used in making nonparametric inferences generally relates to some function of the actual magnitudes of the random variables in the sample. For example, if the actual observations are replaced by their relative rankings within the sample and the probability distribution of some function of these sample ranks can be determined by postulating only very general assumptions about the basic population sampled, this function will provide a distribution-free technique for estimation or hypothesis testing.   Inferences based on descriptions of these derived sample data may relate to whatever parameters are relevant and adaptable, such as the median for a location parameter.   The nonparametric and parametric hypotheses are analogous, both relating to location, and identical in the case of a continuous and symmetrical population.

Tests of hypotheses which are not statements about parameter values have no counterpart in parametric statistics, and thus here nonparametric statistics provides techniques for solving new kinds of problems.   On the other hand, a distribution-free test simply relates to a different approach to solving standard statistical problems, and therefore comparisons of the merits of the two types of techniques are relevant. Some of the more obvious general advantages of nonparametric-inference procedures can be appreciated even before our systematic study begins. Nonparametric methods generally are quick and easy to apply, since they involve extremely simple arithmetic.   The theory of nonparametric inference relates to properties of the statistic used in the inductive procedure.   Discussion of these properties requires derivation of the random sampling distribution of the pertinent statistic, but this generally involves much less sophisticated mathematics than classical statistics.   The test statistic in most cases is a discrete random variable with nonzero probabilities assigned to only a finite number of values, and its exact sampling distribution can often be determined by enumeration or simple combinatorial formulas.   The asymptotic distributions are usually normal, chi-square, or other well-known functions.   The derivations are easier to understand, especially for nonmathematically trained users of statistics. A cookbook approach to learning techniques is then not necessary, which lessens the danger of misuse of procedures.   This advantage also minimizes the opportunities for inappropriate and indiscriminate applications, because the assumptions are so general.   When no stringent postulations regarding the basic population are needed, there is little problem of violation of assumptions, with the result that conclusions reached in nonparametric methods usually need not be tempered by qualifiers.   The types of assumption made in nonparametric statistics are generally easily satisfied, and decisions regarding their legitimacy almost obvious. Besides, in many cases the assumptions are sufficient, but not necessary, for the test's validity. Assumptions regarding the sampling process, usually that it is a random sample, are not relaxed with nonparametric methods, but a careful experimenter can generally adopt sampling techniques which render this problem academic. With so-called "dirty data," most nonparametric techniques are, relatively speaking, much more appropriate than parametric methods. The basic data available need not be actual measurements in many cases; if the test is to be based on ranks, for example, only the ranks are needed. The process of collecting and compiling sample data then may be less expensive and time-consuming. Some new types of problems relating to sample-distribution characteristics are soluble with nonparametric tests. The scope of application is also wider because the techniques may be legitimately applied to phenomena for which it is impractical or impossible to obtain quantitative measurements. When information about actual observed sample magnitudes is provided but not used as such in drawing an inference, it might seem that some of the available information is being discarded, for which one usually pays a price in efficiency. This is really not true, however. The information embodied in these actual magnitudes, which is not directly employed in the inference procedure, really relates to the underlying distribution, information which is not relevant for distribution-free tests. On the other hand, if the underlying distribution is known, a classical approach to testing may legitimately be used and so this would not be a situation requiring nonparametric methods. The information of course may be consciously ignored, say for the purpose of speed or simplicity.

This discussion of relative merits has so far been concerned mainly with the application of nonparametric techniques. Performance is certainly a matter of concern to the experimenter, but generalizations about reliability are always difficult because of varying factors like sample size, significance levels or confidence coefficients, evaluation of the importance of speed, simplicity and cost factors, and the nonexistence of a fixed and universally acceptable criterion of good performance. Box and Anderson state that "to fulfill the needs of the experimenter, statistical criteria should (1) be sensitive to change in the specific factors tested, (2) be insensitive to changes, of a magnitude likely to occur in practice, in extraneous factors."[1] These properties, usually called *power* and *robustness*, respectively, are generally agreed upon as the primary requirements of good performance in hypothesis testing. Parametric tests are often

---

[1] G. E. P. Box and S. L. Anderson (1955): Permutation Theory in the Derivation of Robust Criteria and the Study of Departures from Assumption, *J. Roy. Statist. Soc.*, **B17**: 1.

derived in such a way that the first requirement above is satisfied for an assumed specific probability distribution, e.g., using the likelihood-ratio technique of test construction. However, since such tests are, strictly speaking, not even valid unless the assumptions are met, robustness is of great concern in parametric statistics. On the other hand, non-parametric tests are inherently robust because their construction requires only very general assumptions. One would expect some sacrifice in power to result. It is therefore natural to look at robustness as a performance criterion for parametric tests and power for nonparametric tests. How then do we compare analogous tests of the two types?

Power calculations for any test require knowledge of the probability distribution of the test statistic under the alternative, while alternatives in nonparametric problems are often extremely general. When the requisite assumptions are met, many of the classical parametric tests are known to be most powerful. In those cases where comparison studies have been made, however, nonparametric tests are frequently almost as powerful, especially for small samples, and therefore may be considered more desirable whenever there is any doubt about assumptions. No generalizations can be made for moderate-sized samples. The criterion of asymptotic relative efficiency is theoretically relevant only for very large samples. When the classical tests are known to be robust, comparisons may also be desirable for distributions which deviate somewhat from the exact parametric assumptions. However, with inexact assumptions, calculation of power of classical tests is often difficult except by Monte Carlo techniques, and studies of power here have been less extensive. Either type of test may be more reliable, depending on the particular tests compared and type or degree of deviations assumed. The difficulty with all these comparisons is that they can only be made for specific nonnull distribution assumptions, which are closely related to the conditions under which the parametric test is exactly valid and optimal.

Perhaps the chief advantage of nonparametric tests lies in their very generality, and an assessment of their performance under conditions unrestricted by, and different from, the intrinsic postulates in classical tests seems more expedient. A comparison under more nonparametric conditions would seem especially desirable for two or more nonparametric tests which are designed for the same general hypothesis-testing situation. Unlike the body of classical techniques, nonparametric techniques frequently offer a selection from interchangeable methods. With such a choice, some judgments of relative merit would be particularly useful. Power comparisons have been made, predominantly among the many tests designed to detect location differences, but again we must add that even with comparisons of nonparametric tests, power can be determined only with fairly specific distribution assumptions. The relative merits of the different tests depend on the conditions imposed. Comprehensive conclusions are thus still impossible for blanket comparisons of very general tests.

In conclusion, the extreme generality of nonparametric techniques and their wide scope of usefulness, while definite advantages in application, are factors which discourage objective criteria, particularly power, as assessments of performance, relative either to each other or to parametric techniques. The comparison studies so frequently published in the literature are certainly interesting, informative, and valuable, but they do not provide the sought-for comprehensive answers under more nonparametric conditions. Perhaps we can even say that specific comparisons are really contrary to the spirit of nonparametric methods. No definitive rules of choice will be provided in this book. The interested reader will find many pertinent articles in all the statistics journals. This book is a compendium of many of the large number of nonparametric techniques which have been proposed for various inference situations.

Before embarking on a systematic treatment of new concepts, some basic notation and definitions must be agreed upon and the groundwork prepared for development. Therefore, the remainder of this chapter will be devoted to an explanation of the notation adopted here and an abbreviated review of some of those definitions and terms from classical inference which are also relevant to the special world of nonparametric inference. A few new concepts and terms will also be introduced which are uniquely useful in nonparametric theory. The general theory of order statistics will be the subject of Chap. 2, since they play a fundamental role in many nonparametric techniques. A few direct applications of order statistics to inference problems will be specifically covered, and others alluded to. Starting with Chap. 3, the important nonparametric techniques will be discussed in turn, organized according to the type of inference problem (hypothesis to be tested) in the case of hypotheses not involving statements about parameters, or the type of sampling situation, e.g., one sample, two independent samples, etc., in the case of distribution-free techniques, or whichever seems more pertinent. Chapters 3 and 4 will treat tests of randomness and goodness-of-fit tests, respectively, both nonparametric hypotheses which have no counterpart in classical statistics. Chapter 5 is a general introduction to rank-order statistics, which are used in many of the inference situations of the chapters following. Chapters 6 and 7, respectively, deal with distribution-free tests for one-sample (or paired-sample) and two-independent-sample tests of hypotheses involving the value of a location parameter, i.e., analogies of the one-sample (paired-sample) and two-sample tests for the population mean (mean difference) and difference of two means. Chapter 8 is an introduction to a particular form of nonparametric test statistic, called a linear rank statistic, which is especially useful for testing a hypothesis that two independent samples are drawn from identical populations. Those linear rank statistics which are particularly sensitive to differences only in location and only in scale are the subjects of Chaps. 9 and 10, respectively.   Chapter 11 extends this situation to the hypothesis that k independent samples are drawn from identical populations.   Chapters 12 and 13 are concerned with measures of association and tests of independence in bivariate and multivariate sample situations, respectively. For almost all tests the discussion will center on logical justification, null distribution and moments of the test statistic, asymptotic distribution, and other relevant distribution properties.   Whenever possible, related methods of interval estimation of parameters are also included.   During the course of discussion, only the briefest attention will be paid to relative merits of comparable tests.   Chapter 14 presents some theorems relating to calculation of asymptotic relative efficiency, a possible criterion for evaluating large-sample performance of nonparametric tests relative to that parametric test which is optimal for the same sampling situation when certain population-distribution assumptions are met.   These techniques are then applied to evaluating the efficiency of many of the tests presented in the previous chapters.   Following each chapter some problems of a more or less theoretical nature are included to help the reader assess his understanding of the text material and to amplify or solidify the related theory.

1.2 FUNDAMENTAL STATISTICAL CONCEPTS

In this section a few of the basic definitions and concepts of classical statistics are reviewed, but only very briefly since the main purpose is to explain notation and terms taken for granted later on.   A few of the new fundamentals needed for the development of nonparametric inference will also be introduced here.

BASIC DEFINITIONS

A *sample space* is the set of all possible outcomes of a random experiment.

A *random variable* is a set function whose domain is the elements of a sample space on which a probability function has been defined and whose range is the set of all real numbers.   Alternatively, X is a random variable if for every real number x there exists a probability that the value assumed by the random variable does not exceed x, denoted by $P(X \leq x)$ or $F_X(x)$, and called the *cumulative distribution function* (cdf) of X.

The customary practice is to denote the random variable by a capital letter like X and the actual value assumed (value observed in the experiment) by the corresponding letter in lowercase, x.   This practice will generally be adhered to in this book. However, it is not always possible, strictly speaking, to make such a distinction. Occasional inconsistencies will therefore be unavoidable, but the statistically sophisticated reader is no doubt already accustomed to this type of conventional confusion in notation.

The mathematical properties of any function $F_X(\cdot)$ which is a cdf of a random variable X are as follows:

1. $F_X(x_1) \leq F_X(x_2)$ for all $x_1 < x_2$.
2. $\lim\limits_{x \to -\infty} F_X(x) = 0$, and $\lim\limits_{x \to \infty} F_X(x) = 1$.
3. $F_X(x)$ is continuous from the right, or, symbolically, as $\epsilon \to 0$ through positive values, $\lim\limits_{\epsilon \to 0} F_X(x + \epsilon) = F_X(x)$.

A random variable X is called *continuous* if its cdf is continuous. Every continuous cdf in this book will be assumed differentiable everywhere with the possible exception of a finite number of points. The derivative of the cdf will be denoted by $f_X(x)$, a nonnegative function called the *probability density function* (pdf) of X. Thus when X is continuous

$$F_X(x) = \int_{-\infty}^{x} f_X(t) \, dt \qquad f_X(x) = \frac{d}{dx} F_X(x) = F'_X(x)$$

$$\int_{-\infty}^{\infty} f_X(x) \, dx = 1$$

The *probability mass function* (pmf) of a random variable X is defined as

$$f_X(x) = P(X = x) = F_X(x) - \lim_{\epsilon \to 0} F_X(x - \epsilon)$$

where $\epsilon \to 0$ through positive values. A random variable is called *discrete* if $\sum\limits_{\text{all } x} f_X(x) = 1$, where the expression "all x" is to be interpreted as meaning all x at which $F_X(x)$ is not continuous.

The term *probability function* (pf) or *probability distribution* will be used to denote either a pdf or a pmf. For notation, capital letters will always be reserved for the cdf, while the corresponding lowercase letter denotes the pf.

The *expected value* of a function $g(X)$ of a random variable X, denoted by $E[g(X)]$, is

$$E[g(X)] = \begin{cases} \int_{-\infty}^{\infty} g(x)f_X(x) \, dx & \text{if } X \text{ is continuous} \\ \sum\limits_{\text{all } x} g(x)f_X(x) & \text{if } X \text{ is discrete} \end{cases}$$

Joint probability functions and expectations for functions of more than one random variable are similarly defined and denoted by replacing single symbols by vectors, sometimes abbreviated to

$$X^{(n)} = (X_1, X_2, \ldots, X_n)$$

A set of n random variables $X_1, X_2, \ldots, X_n$ is *independent* if and only if their joint probability function equals the product of the n individual marginal probability functions.

A set of n random variables $X_1, X_2, \ldots, X_n$ is called a *random sample* of the random variable X (or from the population f_X) if they are independent and identically distributed according to X, so that

$$f_{X^{(n)}}(x_1, x_2, \ldots, x_n) = f_{X_1, X_2, \ldots, X_n}(x_1, x_2, \ldots, x_n) = \prod_{i=1}^{n} f_X(x_i)$$

A *statistic* is any function of observable or sample random variables which does not involve unknown parameters.

A *moment* is a particular type of population parameter. The kth moment of X is $\mu'_k = E(X^k)$, and the kth central moment is

$$\mu_k = E[(X - \mu)^k]$$

where $\mu = E(X)$. The kth *factorial moment* is $E[X(X - 1) \cdots (X - k + 1)]$. The variance of X is μ_2, or

$$\mu_2 = \text{var}(X) = \sigma^2(X) = E(X^2) - \mu^2$$

For two random variables, their *covariance* and *correlation*, respectively, are

$$\text{cov}(X, Y) = E[(X - \mu_X)(Y - \mu_Y)] = E(XY) - \mu_X \mu_Y$$

$$\text{corr}(X, Y) = \rho(X, Y) = \frac{\text{cov}(X, Y)}{\sigma(X)\sigma(Y)}$$

The *moment-generating function* (mgf) of a function $g(X)$ of X is

$$M_{g(X)}(t) = E\{\exp[tg(X)]\}$$

Some special properties of the mgf are

$$M_{aX+b}(t) = e^{bt} M_X(at) \qquad \text{for } a \text{ and } b \text{ const}$$

$$\mu'_k = \frac{d^k}{dt^k} M_X(t) \Big|_{t=0} = M_X^{(k)}(0)$$

MOMENTS OF LINEAR COMBINATIONS OF RANDOM VARIABLES

Let $X_1, X_2, \ldots, X_n$ be n random variables and $a_i, b_i, i = 1, 2, \ldots, n,$ be any constants. Then

$$E\left(\sum_{i=1}^{n} a_i X_i\right) = \sum_{i=1}^{n} a_i E(X_i)$$

$$\operatorname{var}\left(\sum_{i=1}^{n} a_i X_i\right) = \sum_{i=1}^{n} a_i^2 \operatorname{var}(X_i) + 2 \sum\sum_{1 \le i < j \le n} a_i a_j \operatorname{cov}(X_i, X_j)$$

$$\operatorname{cov}\left(\sum_{i=1}^{n} a_i X_i, \sum_{i=1}^{n} b_i X_i\right) = \sum_{i=1}^{n} a_i b_i \operatorname{var}(X_i)$$

$$+ \sum\sum_{1 \le i < j \le n} (a_i b_j + a_j b_i) \operatorname{cov}(X_i, X_j)$$

PROBABILITY FUNCTIONS

Table 2.1   Some special probability functions

Name	Probability function $f_X(x)$	mgf	$E(X)$	$\operatorname{var}(X)$
Bernoulli	$p^x(1-p)^{1-x}$ $x = 0, 1$ $0 \le p \le 1$	$pe^t + 1 - p$	p	$p(1-p)$
Binomial	$\binom{n}{x} p^x(1-p)^{n-x}$ $x = 0, 1, \ldots, n$ $0 \le p \le 1$	$(pe^t + 1 - p)^n$	np	$np(1-p)$
Hypergeometric	$\dfrac{\binom{Np}{x}\binom{Nq}{n-x}}{\binom{N}{n}}$ $x = 0, 1, \ldots, n$ $p + q = 1$	†	np	$\dfrac{npq(N-n)}{N-1}$
Uniform on (α,β)	$(\beta - \alpha)^{-1}$ $\alpha < x < \beta$ $\alpha < \beta$	$\dfrac{e^{t\beta} - e^{t\alpha}}{t(\beta - \alpha)}$	$\dfrac{\beta + \alpha}{2}$	$\dfrac{(\beta - \alpha)^2}{12}$
Normal	$\dfrac{\exp\left[-\frac{1}{2}\left(\dfrac{x-\mu}{\sigma}\right)^2\right]}{(2\pi\sigma^2)^{1/2}}$ $\sigma > 0$	$\exp(t\mu + \tfrac{1}{2}t^2\sigma^2)$	μ	σ^2
Gamma	$\dfrac{x^{\alpha-1}e^{-x/\beta}}{\beta^\alpha \Gamma(\alpha)}$ $x > 0$ $\alpha > 0, \beta > 0$	$(1 - \beta t)^{-\alpha}$	$\alpha\beta$	$\alpha\beta^2$
Exponential	Same as gamma with $\alpha = 1$	$(1 - \beta t)^{-1}$	β	β^2
Chi square (ν)	Same as gamma with $\alpha = \dfrac{\nu}{2}, \beta = 2$	$(1 - 2t)^{-\nu/2}$	ν	2ν
Beta	$\dfrac{x^{\alpha-1}(1-x)^{\beta-1}}{B(\alpha,\beta)}$ $0 < x < 1$ $\alpha > 0, \beta > 0$	†	$\dfrac{\alpha}{\alpha + \beta}$	$\dfrac{\alpha\beta}{(\alpha + \beta)^2(\alpha + \beta + 1)}$

† The mgf is omitted here because the expression is too complicated.

The term standard normal will designate the particular member of the normal family where $\mu = 0$ and $\sigma = 1$. The symbols $\varphi(x)$ and $\Phi(x)$ will be reserved for the standard normal density and cumulative distribution functions, respectively.

Three other important distributions are:

Student's $t(\nu)$:

$$f_X(x) = \frac{\nu^{-\frac{1}{2}}(1 + x^2/\nu)^{-(\nu+1)/2}}{B(\nu/2, \frac{1}{2})} \qquad \nu > 0$$

Snedecor's $F(\nu_1, \nu_2)$:

$$f_X(x) = \left(\frac{\nu_1}{\nu_2}\right)^{\nu_1/2} x^{(\nu_1/2)-1} \frac{(1 + \nu_1 x/\nu_2)^{-(\nu_1+\nu_2)/2}}{B(\nu_1/2, \nu_2/2)} \qquad \begin{array}{l} x > 0 \\ \nu_1,\, \nu_2 > 0 \end{array}$$

Fisher's $z(\nu_1, \nu_2)$:

$$f_X(x) = 2\left(\frac{\nu_1}{\nu_2}\right)^{\nu_1/2} e^{\nu_1 x} \frac{(1 + \nu_1 e^{2x}/\nu_2)^{-(\nu_1+\nu_2)/2}}{B(\nu_1/2, \nu_2/2)} \qquad \begin{array}{l} x > 1 \\ \nu_1,\, \nu_2 > 0 \end{array}$$

TRANSFORMATIONS USING THE METHOD OF JACOBIANS

Let $X_1, X_2, \ldots, X_n$ be n continuous random variables with joint pdf $f_{X^{(n)}}(x_1, x_2, \ldots, x_n)$, which is nonzero for an n-dimensional region S_x. Define the transformation

$$y_1 = u_1(x_1, x_2, \ldots, x_n),\ y_2 = u_2(x_1, x_2, \ldots, x_n),$$
$$\ldots, y_n = u_n(x_1, x_2, \ldots, x_n)$$

which maps S_x onto S_y, where S_x can be written as the union of a finite number m of disjoint spaces $S_1, S_2, \ldots, S_m$ such that the transformation from S_k onto S_y is one to one, for all $k = 1, 2, \ldots, m$. Then for each k there exists a unique inverse transformation, denoted by

$$x_1 = w_{1k}(y_1, y_2, \ldots, y_n),\ x_2 = w_{2k}(y_1, y_2, \ldots, y_n),$$
$$\ldots, x_n = w_{nk}(y_1, y_2, \ldots, y_n)$$

Assume that for each of these m sets of inverse transformations, the Jacobian

$$J_k(y_1, y_2, \ldots, y_n) = J_k(y^{(n)}) = \frac{\partial(w_{1k}, w_{2k}, \ldots, w_{nk})}{\partial(y_1, y_2, \ldots, y_n)} = \det \frac{\partial w_{ik}}{\partial y_j}$$

exists and is continuous and nonzero in S_y, where $\det(a_{ij})$ denotes the determinant of the $n \times n$ matrix with entry a_{ij} in the ith row and jth column.   Then the joint pdf of the n random variables $Y_1, Y_2, \ldots, Y_n$, where $Y_i = u_i(X_1, X_2, \ldots, X_n)$, is

$$f_{Y^{(n)}}(y_1, y_2, \ldots, y_n)$$
$$= \sum_{k=1}^{m} |J_k(y_1, y_2, \ldots, y_n)| f_{X^{(n)}}[w_{1k}(y_1, y_2, \ldots, y_n),$$
$$w_{2k}(y_1, y_2, \ldots, y_n), \ldots, w_{nk}(y_1, y_2, \ldots, y_n)]$$

for all $(y_1, y_2, \ldots, y_n) \in S_y$ and zero otherwise.   The Jacobian of the inverse transformation is the reciprocal of the Jacobian of the direct transformation,

$$\frac{\partial(w_{1k}, w_{2k}, \ldots, w_{nk})}{\partial(y_1, y_2, \ldots, y_n)} = \left[\frac{\partial(u_1, u_2, \ldots, u_n)}{\partial(x_1, x_2, \ldots, x_n)} \right]^{-1} \Bigg|_{x_i = w_{ik}(y_1, y_2, \ldots, y_n)}$$

or

$$J_k(y^{(n)}) = [J_k(x^{(n)})]^{-1} \Big|_{x_i = w_{ik}(y^{(n)})}$$

Thus the pdf above can also be written as

$$f_{Y^{(n)}}(y_1, y_2, \ldots, y_n) = \sum_{k=1}^{m} |[J_k(x_1, x_2, \ldots, x_n)]^{-1}| f_{X^{(n)}}(x_1, x_2, \ldots, x_n)$$

where the right-hand side is evaluated at $x_i = w_{ik}(y_1, y_2, \ldots, y_n)$ for $i = 1, 2, \ldots, n$.   If $m = 1$ so that the transformation from S_x onto S_y is one to one, the subscript k and the summation sign may be dropped. When $m = 1$ and $n = 1$, this reduces to the familiar result

$$f_Y(y) = \left[f_X(x) \left| \frac{dy}{dx} \right|^{-1} \right] \Bigg|_{x = u^{-1}(y)}$$

POINT AND INTERVAL ESTIMATION

A point estimate of a parameter is any single function of random variables whose observed value is used to estimate the true value.   Let $\hat{\theta}_n = u(X_1, X_2, \ldots, X_n)$ be a point estimate of a parameter θ.   Some desirable properties of $\hat{\theta}_n$ are defined as follows for all θ in the relevant parameter space.

1. *Unbiasedness:* $E(\hat{\theta}_n) = \theta$.
2. *Sufficiency:* $f_{X_1, X_2, \ldots, X_n | \hat{\theta}_n}(x_1, x_2, \ldots, x_n \mid \hat{\theta}_n)$ does not depend on θ, or, equivalently,

$$f_{X_1, X_2, \ldots, X_n}(x_1, x_2, \ldots, x_n; \theta) = g(\hat{\theta}_n; \theta) H(x_1, x_2, \ldots, x_n)$$

where $H(x_1, x_2, \ldots, x_n)$ does not depend on θ.

3. *Consistency* (also called stochastic convergence and convergence in probability): $\lim_{n \to \infty} P(|\hat{\theta}_n - \theta| > \epsilon) = 0$ for every $\epsilon > 0$.

   a. If $\hat{\theta}_n$ is an unbiased estimate of θ and $\lim_{n \to \infty} \text{var}(\hat{\theta}_n) = 0$, $\hat{\theta}_n$ is a consistent estimate of θ, by Chebyshev's inequality.

   b. $\hat{\theta}_n$ is a consistent estimate of θ if the limiting distribution of $\hat{\theta}_n$ is a degenerate distribution with probability 1 at θ.

4. *Minimum mean squared error:* $E[(\hat{\theta}_n - \theta)^2] \leq E[(\theta_n^* - \theta)^2]$, for any other estimate θ_n^*.

An *interval estimate* of a parameter θ with confidence coefficient $1 - \alpha$, or a $100(1 - \alpha)$ percent *confidence interval* for θ, is a random interval whose end points U and V are functions of observable random variables such that the probability statement $P(U < \theta < V) = 1 - \alpha$ is satisfied.

A useful technique for finding point estimates for parameters which appear as unspecified constants (or as functions of such constants) in a family of probability functions, say $f_X(\cdot;\theta)$, is the method of maximum likelihood. The *likelihood function* of a random sample of size n from the population $f_X(\cdot;\theta)$ is the joint probability function of the sample variables regarded as a function of θ, or

$$L(x_1, x_2, \ldots, x_n; \theta) = \prod_{i=1}^{n} f_X(x_i; \theta)$$

A *maximum-likelihood estimate* (MLE) of θ is a value $\hat{\theta}$ such that for all θ

$$L(x_1, x_2, \ldots, x_n; \hat{\theta}) \geq L(x_1, x_2, \ldots, x_n; \theta)$$

Subject to certain regularity conditions, MLEs are sufficient and consistent and asymptotically are unbiased, minimum variance, and normally distributed. Another useful property is invariance, which says that if $g(\theta)$ is a one-to-one function of θ and $\hat{\theta}$ is the MLE of θ, then the MLE of $g(\theta)$ is $g(\hat{\theta})$. If more than one parameter is involved, θ above may be interpreted as a vector.

HYPOTHESIS TESTING

A *statistical hypothesis* is an assertion about the probability function of one or more random variables or a statement about the population(s) from which one or more samples are drawn, e.g., its form, shape, or parameter values. A hypothesis is called *simple* if the statement completely specifies the population. Otherwise it is called *composite*. The *null hypothesis* H_0 is the hypothesis under test. The *alternative hypothesis*, H_1 or H_A, is the conclusion reached if the null hypothesis is rejected.

A *test* of a statistical hypothesis is a rule which enables one to make a decision whether or not H_0 should be rejected on the basis of the observed value of a *test statistic*, which is some function of a set of observable random variables.

A *critical region* or *rejection region* R for a test is that subset of real numbers assumed by the test statistic which, in accordance with a certain test, leads to rejection of H_0. The *critical values* of a test statistic are the bounds of R. For example, if a test statistic T prescribes rejection of H_0 for $t \leq t_\alpha$, t_α is the critical value and R is written symbolically as

$$T \in R \qquad \text{for } t \leq t_\alpha$$

A *type I error* is committed if the null hypothesis is rejected when it is true. A *type II error* is failure to reject a false H_0. For a test statistic T of H_0: $\theta \in \omega$ versus H_1: $\theta \in \Omega - \omega$, the probabilities of these errors are, respectively,

$$\alpha(\theta) = P(T \in R \mid \theta \in \omega) \qquad \text{and} \qquad \beta(\theta) = P(T \notin R \mid \theta \in \Omega - \omega)$$

In classical statistics, the maximum value of $\alpha(\theta)$ for all $\theta \in \omega$ is often called the *size of the test*. The *significance level* is a preselected nominal bound for $\alpha(\theta)$, which may not be attained if the relevant probability function is discrete. Since this is usually the case in nonparametric hypothesis testing, some confusion might arise if these distinctions were adhered to here. The symbol α will be used to denote either the size of the test or the significance level or the probability of a type I error, prefaced by the adjective "exact" whenever $\max_{\theta \in \omega} \alpha(\theta) = \alpha$.

The *power* of a test is the probability that the test statistic will lead to rejection of H_0, denoted by $Pw(\theta) = P(T \in R)$. Power is of interest mainly as the probability of a correct decision, which implies H_0 is false, and then $Pw(\theta) = 1 - \beta(\theta)$. In this case, power depends on the following four variables:

1. The degree of falseness of H_0, that is, the amount of discrepancy between the assertion as stated in H_0 and the true condition
2. The size of the test α
3. The number of observable random variables involved in the test statistic, generally the sample size
4. The subset of real numbers chosen for R

The *power function* of a test is the power when all but one of these variables are held constant, usually item 1.

A test is said to be *most powerful* for a specified alternative hypothesis if no other test of the same size has greater power against the same alternative.

A test is *uniformly most powerful* against a class of alternatives if it is most powerful with respect to each specific alternative within the class of alternative hypotheses.

A "good" test statistic is one which is reasonably successful in distinguishing correctly between the conditions as stated in the null and alternative hypotheses. A method of constructing tests which often have good properties is the *likelihood-ratio principle*. A random sample of size n is drawn from the population $f_X(\cdot;\theta)$ with likelihood function $L(x_1,x_2, \ldots ,x_n;\theta)$, where θ is to be interpreted as a vector if more than one parameter is involved. Suppose that $f_X(x;\theta)$ is a specified family of functions for every $\theta \in \Omega$ and ω is a subset of Ω. The likelihood-ratio test of

$$H_0: \theta \in \omega \qquad \text{versus} \qquad H_1: \theta \in \Omega - \omega$$

has the rejection region

$$T \in R \qquad \text{for } t \leq c,\, 0 \leq c \leq 1$$

where T is the ratio

$$T = \frac{L(\hat{\omega})}{L(\hat{\Omega})}$$

and $L(\hat{\omega})$ and $L(\hat{\Omega})$ are the maximums of the likelihood function with respect to θ for $\theta \in \omega$ and $\theta \in \Omega$, respectively. For an exact size α test of a simple H_0, the number c which defines R is chosen such that $P(T \leq c \mid H_0) = \alpha$. Any monotonic function of T, say $g(T)$, can also be employed for the test statistic as long as the rejection region is stated in terms of the appropriate values of $g(T)$. The likelihood-ratio test is always a function of sufficient statistics, and the principle often produces a uniformly most powerful test when such exists. A particularly useful property of T for constructing tests based on large samples is that, subject to certain regularity conditions, the probability distribution of $-2 \log T$ approaches the chi-square distribution with $k_1 - k_2$ degrees of freedom as $n \to \infty$, where k_1 and k_2 are, respectively, the dimensions of the spaces Ω and ω, $k_2 < k_1$.

All these concepts should have been familiar to the reader, since they are an integral part of any standard introductory probability and inference course. We now turn to a few concepts which are probably new and which are relevant mainly to nonparametric inference.

CONSISTENCY

A test is *consistent* for a specified alternative if the power of the test when that alternative is true approaches 1 as the sample size approaches infinity. A test is consistent for a class (or subclass) of alternatives if the power of the test when any member of the class (subclass) of alternatives is true approaches 1 as the sample size approaches infinity.

Although consistency is a "good" test criterion relevant to both parametric and nonparametric methods, it is seldom even mentioned in classical inference since all the standard test procedures clearly share this property. However, in nonparametric statistics the alternatives are often extremely general, and a wide selection of tests may be available for any one experimental situation. The consistency criterion provides an objective method of choosing among these tests (or at least eliminating some from consideration) when a less general subclass of alternatives is of major interest to the experimenter. A test which is known to be consistent against a specified subclass is said to be especially sensitive to that type of alternative and can generally be recommended for use when the experimenter wishes particularly to detect differences of the type expressed in that subclass.

Consistency can often be shown by investigating convergence in probability for a test statistic or some function of it. An especially useful method of investigating consistency is described as follows. A random sample of size n is drawn from the family $f_X(\cdot;\theta)$, $\theta \in \Omega$. Let T be a test statistic for the general hypothesis $\theta \in \omega$ versus $\theta \in \Omega - \omega$, and let $g(\theta)$ be some function of θ such that

$$g(\theta) = \theta_0 \qquad \text{if } \theta \in \omega$$

and

$$g(\theta) \neq \theta_0 \qquad \text{if } \theta \in \Delta \text{ for } \Delta \subset \Omega - \omega$$

If for all θ we have

$$E(T) = g(\theta) \qquad \text{and} \qquad \lim_{n \to \infty} \text{var}(T) = 0$$

then the size α test T with rejection region

$$T \in R \qquad \text{for } |t - \theta_0| > c_\alpha$$

is consistent for the subclass Δ. Similarly, for a one-sided subclass of alternatives where

$$g(\theta) = \theta_0 \qquad \text{if } \theta \in \omega$$

and

$$g(\theta) > \theta_0 \qquad \text{if } \theta \in \Delta \text{ for } \Delta \subset \Omega - \omega$$

the consistent test of size α has rejection region

$$T \in R \qquad \text{for } t - \theta_0 > c'_\alpha$$

These results follow directly from Chebyshev's inequality. [For a proof, see Fraser (1957), pp. 267–268.]

PITMAN EFFICIENCY

Another sort of objective criterion may be useful in choosing between two or more tests which are comparable in a well-defined way, namely, the concept of Pitman efficiency. In the theory of point estimation, the efficiency of two unbiased estimators for a parameter is defined as the ratio of their variances. In some situations, the limiting value of this ratio may be interpreted as the relative number of additional observations needed using the less efficient estimator to obtain the same accuracy. The idea of efficiency of two test statistics is closely related, where power is regarded as a measure of accuracy, but the tests must be compared under equivalent conditions (as both estimators were specified to be unbiased), and there are many variables in hypothesis testing. The most common way of comparing two tests is to make all factors equivalent except sample size.

The *power efficiency* of a test A relative to a test B, where both tests are for the same simple null and alternative hypotheses, the same type of rejection region, and the same significance level, is the ratio n_b/n_a, where n_a is the number of observations required by test A for the power of test A to equal the power of test B when n_b observations are employed. Since power efficiency generally depends on the selected significance level, hypotheses, and n_b, it is difficult to calculate and interpret. The problem can be avoided in many cases by defining a type of limiting power efficiency.

Let A and B be two consistent tests of a null hypothesis H_0 and alternative hypothesis H_1 at significance level α. The *asymptotic relative efficiency* (ARE) of test A relative to test B is the limiting value of the ratio n_b/n_a, where n_a is the number of observations required by test A for the power of test A to equal the power of test B based on n_b observations while simultaneously $n_b \to \infty$ and $H_1 \to H_0$.

In many applications of this definition, the ratio is the same for all choices of α, so that the ARE is a single number with a well-defined interpretation for large samples. The requirement that both tests be consistent against H_1 is not a limitation in application, since most tests under consideration for a particular type of alternative will be consistent anyway. But with two consistent tests, their powers both approach 1 with increasing sample sizes. Therefore, we must let H_1 approach H_0 so that the power of each test lies on the open interval $(\alpha,1)$ for finite sample sizes and the limiting ratio will generally be some number other than 1. The ARE is sometimes also called local asymptotic efficiency since it relates to large sample power in the vicinity of the null hypothesis. A few studies have been conducted which seem to indicate that in several important cases the ARE is a reasonably close approximation to the exact efficiency for moderate-sized samples and alternatives not too far from the null case. Especially in the case of small samples, however, the implications of the ARE value cannot be considered particularly meaningful. Methods of calculating the ARE for comparisons of particular tests will be treated fully in Chap. 14.

The problem of evaluating the relative merits of two or more comparable test statistics is by no means solved by introducing the criteria of consistency and asymptotic relative efficiency. Both are large-sample properties and may not have much import for small- or even moderate-sized samples. As discussed in Sec. 1, exact power calculations are tedious and often too specific to shed much light on the problem as it relates to nonparametric tests, which may explain the general acceptance of asymptotic criteria in the field of nonparametric inference.

RANDOMIZED TESTS

We now turn to a different problem which, although not limited to non-parametric inference, is of particular concern in this area. For most classical test procedures, the experimenter chooses a "reasonable" significance level α in advance and determines the rejection-region boundary such that the probability of a type I error is exactly α for a simple hypothesis and does not exceed α for a composite hypothesis. When the null probability distribution of the test statistic is continuous, any real number between 0 and 1 may be chosen as the significance level. Let us call this preselected number the nominal α. If the test statistic T can take on only a countable number of values, i.e., if the sampling distribution of T is discrete, the number of possible exact probabilities of a type I error is limited to the number of jump points in the cdf of the test statistic. These exact probabilities will be called *exact α values* or *natural significance levels*. The rejection region can then be chosen such that either (1) the exact α is the largest number which does not exceed the nominal α or (2) the exact α is the number closest to the nominal α. Although most statisticians seem to prefer the first approach, as it is more consistent with classical test procedures for a composite H_0, this has not been universally agreed upon. As a result, two sets of tables of critical values of a test statistic may not be identical for the same nominal α. This is an unfortunate circumstance which occasionally leads to confusion in reading tables, but a carefully presented table will explain which approach was used in construction.

Disregarding that problem now, suppose we wish to compare the performance, as measured by power, of two different discrete test statistics. Their natural significance levels are unlikely to be the same, so that identical nominal α values do not ensure identical exact probabilities of a type I error. Power is certainly affected by exact α, and power comparisons of tests may be quite misleading without identical exact α

values. A method of equalizing exact α values is provided by *randomized test procedures*.

A *randomized decision rule* is one which prescribes rejection of H_0 always for a certain range of values of the test statistic, rejection sometimes for another nonoverlapping range, and acceptance otherwise. A typical rejection region of exact size α might be written: $T \in R$ with probability 1 if $t \geq t_2$ and with probability p if $t_1 \leq t < t_2$ where $t_1 < t_2$ and $0 < p < 1$ are chosen such that

$$P(T \geq t_2 | H_0) + pP(t_1 \leq T < t_2 | H_0) = \alpha$$

Some random device could be used to make the decision in practice, like drawing one card at random from 100, of which $100p$ are labeled reject. Such decision rules may seem an artificial device and are probably seldom employed by experimenters, but the technique is useful in discussions of theoretical properties of tests.

A simple example will suffice to explain the procedure. A random sample of size 5 is drawn from the Bernoulli population. We wish to test $H_0: \theta = 0.5$ versus $H_1: \theta > 0.5$ at significance level 0.05. The test statistic is X, the number of successes in the sample, which has the binomial distribution with parameter θ and $n = 5$. A reasonable rejection region would be large values of X, and thus the six exact significance levels obtainable without using a randomized test are:

c	5	4	3	2	1	0
$P(X \geq c \mid \theta = 0.5)$	$\frac{1}{32}$	$\frac{6}{32}$	$\frac{16}{32}$	$\frac{26}{32}$	$\frac{31}{32}$	1

A nonrandomized test procedure of nominal size 0.05 but exact

$$\alpha = \frac{1}{32} = 0.03125$$

has rejection region

$$X \in R \qquad \text{for } x = 5$$

The randomized decision rule with exact $\alpha = 0.05$ is found as follows:

$$\frac{1}{32} + \frac{5p}{32} = 0.05 \qquad p = 0.12$$

$X \in R$ with probability 1 if $x = 5$ and with probability 0.12 if $x = 4$.

CONTINUITY CORRECTIONS

The exact null distribution of most test statistics used in nonparametric inference is discrete. Tables of rejection regions or cumulative distributions are often available for small sample sizes only. However, in many cases some simple approximations to these null distributions are accurate enough for practical applications with moderate-sized samples. When these asymptotic distributions are continuous (like the normal or chi square), the approximation may be improved by introducing a correction for continuity. This is accomplished by regarding the value of the discrete test statistic as the midpoint of an interval. For example, if the domain of a test statistic T is only integer values, the observed value is considered to be $t \pm 0.5$. If the decision rule is to reject for $T \geq t_{\alpha/2}$ or $T \leq t'_{\alpha/2}$ and the large-sample approximation to the distribution of $[T - E(T)]/\sigma(T)$ is the standard normal, the rejection region with continuity correction incorporated is determined by solving the equations

$$\frac{t_{\alpha/2} - 0.5 - E(T)}{\sigma(T)} = z_{\alpha/2} \qquad \frac{t'_{\alpha/2} + 0.5 - E(T)}{\sigma(T)} = -z_{\alpha/2}$$

where $z_{\alpha/2}$ satisfies $\Phi(z_{\alpha/2}) = 1 - \alpha/2$. If the test is performed by calculating the critical ratio for the observed value t, the number to be compared with $z_{\alpha/2}$ is $[|t - E(T)| - 0.5]/\sigma(T)$. One-sided rejection regions or critical ratios employing continuity corrections are found similarly. Generally t is reduced (increased) by c when it exceeds (is exceeded by) $E(T)$ if the continuity correction is the amount c.

2
Order Statistics

2.1 INTRODUCTION AND PROBABILITY–INTEGRAL TRANSFORMATION

Let $X_1, X_2, \ldots, X_n$ denote a random sample from a population with continuous cumulative distribution function F_X. Since F_X is assumed to be continuous, the probability of any two or more of these random variables assuming equal magnitudes is zero. Therefore, there exists a unique ordered arrangement within the sample. Suppose that $X_{(1)}$ denotes the smallest of the set $X_1, X_2, \ldots, X_n$; $X_{(2)}$ denotes the second smallest, etc.; and $X_{(n)}$ denotes the largest. Then

$$X_{(1)} < X_{(2)} < \cdots < X_{(n)}$$

denotes the original random sample after arrangement in increasing order of magnitude, and these are collectively termed the *order statistics* of the random sample $X_1, X_2, \ldots, X_n$. $X_{(r)}$, for $1 \leq r \leq n$, is called the rth order statistic. The subject of order statistics generally deals with the properties of $X_{(r)}$ itself or functions of some subset of the n order statistics.

Order statistics are particularly useful in nonparametric statistics because the transformation $U_{(r)} = F_X(X_{(r)})$ produces a random variable which is the rth order statistic from the uniform population on the interval (0,1), and therefore $U_{(r)}$ is distribution-free. This property is due to the so-called *probability-integral transformation*, which is proved in the following theorem.

Theorem 1.1 Probability-integral transformation *Let the random variable X have the cumulative distribution function F_X. If F_X is continuous, the random variable Y produced by the transformation $Y = F_X(X)$ has the uniform probability distribution over the interval (0,1).*

Proof Since $0 \leq F_X(x) \leq 1$ for all x, we have $F_Y(y) = 0$ for $y \leq 0$ and $F_Y(y) = 1$ for $y \geq 1$. For $0 < y < 1$, define u to be the largest number satisfying $F_X(u) = y$. Then $F_X(X) \leq y$ if and only if $X \leq u$, and it follows that

$$F_Y(y) = P[F_X(X) \leq y] = P(X \leq u) = F_X(u) = y$$

which is the uniform distribution.

The theorem can also be proved using moment-generating functions, but this approach will be left as an exercise for the reader.

As a result of this theorem, we can conclude that if $X_1, X_2, \ldots, X_n$ is a random sample from *any* population with continuous distribution F_X, then $F_X(X_1), F_X(X_2), \ldots, F_X(X_n)$ constitutes a random sample from the uniform population. Similarly, if $X_{(1)} < X_{(2)} < \cdots < X_{(n)}$ are the order statistics for the original sample, then

$$F_X(X_{(1)}) < F_X(X_{(2)}) < \cdots < F_X(X_{(n)})$$

are order statistics from the uniform distribution on (0,1). These order statistics may be termed distribution-free, in the sense that their probability distribution is known to be uniform regardless of the original distribution F_X as long as it is continuous.

Some familiar applications of order statistics which are obvious on reflection follow:

1. $X_{(n)}$, the maximum value in the sample, is of interest in the study of floods and other extreme meteorological phenomena.
2. $X_{(1)}$, the minimum value, is useful for phenomena where, for example, the strength of a chain depends on the weakest link.
3. The sample median, defined as $X_{[(n+1)/2]}$ for n odd, and any number between $X_{(n/2)}$ and $X_{(n/2+1)}$ for n even, is a measure of location and an estimate of the population central tendency.

4. The sample midrange, defined as $(X_{(1)} + X_{(n)})/2$, is also a measure of central tendency.
5. The sample range $X_{(n)} - X_{(1)}$ is a measure of dispersion.
6. The sample interquartile range $(Q_3 - Q_1)/2$ is also a measure of dispersion.
7. Normal scores are used to test the hypothesis that two independent samples are drawn from identical populations. These will be discussed in Chaps. 9 and 10.
8. In censored samples, the sampling process ceases after completing r observations out of n. For example, in life testing of electric light bulbs, one may start with a group of n bulbs but stop taking observations after the rth bulb burns out. Then information is available only on $X_{(1)}, X_{(2)}, \ldots, X_{(r)}$ where $r \leq n$.
9. Order statistics are used to study outliers or extreme observations, e.g., when so-called dirty data are suspected.

Many other applications of order statistics will readily come to mind. When the observation $X_{(r)}$ is replaced by its rank r, the subject is called rank-order statistics, which also has important uses in nonparametric statistics. This subject will be treated in a later chapter.

The study of order statistics in this chapter will be limited to their mathematical and statistical properties, including the joint probability distribution, marginal probability distributions, exact moments, asymptotic moments, and asymptotic marginal distributions. Two general uses of order statistics in distribution-free inference, however, will be discussed, namely, interval estimation of population quantiles and tolerance limits for distributions.

2.2 JOINT DISTRIBUTION OF n ORDER STATISTICS

Since the original random sample from a continuous population with probability density function f_X is composed of independent, identically distributed random variables, the joint probability density function is

$$f_{X_1, X_2, \ldots, X_n}(x_1, x_2, \ldots, x_n) = \prod_{i=1}^{n} f_X(x_i)$$

The joint distribution of the n order statistics for this random sample is not the same since the order statistics are obviously neither independent nor identically distributed. Their distribution is easily derived, however, using the method of Jacobians for transformations.

The set of n order statistics is produced by the transformation

$Y_1 = $ smallest of $(X_1, X_2, \ldots, X_n)$
$Y_2 = $ second smallest of $(X_1, X_2, \ldots, X_n)$
$\cdot \cdot \cdot \cdot \cdot \cdot \cdot \cdot \cdot \cdot \cdot \cdot \cdot \cdot \cdot \cdot \cdot \cdot \cdot \cdot \cdot \cdot \cdot \cdot$
$Y_r = r$th smallest of $(X_1, X_2, \ldots, X_n)$
$\cdot \cdot \cdot \cdot \cdot \cdot \cdot \cdot \cdot \cdot \cdot \cdot \cdot \cdot \cdot \cdot \cdot \cdot \cdot \cdot \cdot$
$Y_n = $ largest of $(X_1, X_2, \ldots, X_n)$

This transformation is not one to one. In fact, since there are a total of $n!$ possible arrangements of the original random variables in increasing order of magnitude, there exist $n!$ inverses to the transformation.

One of these $n!$ permutations might be

$$X_5 < X_1 < X_{n-1} < \cdots < X_n < X_2$$

and the corresponding inverse transformation is

$X_5 = Y_1$
$X_1 = Y_2$
$X_{n-1} = Y_3$
$\cdot \cdot \cdot \cdot \cdot \cdot \cdot \cdot$
$X_n = Y_{n-1}$
$X_2 = Y_n$

The Jacobian of this transformation would be the determinant of an $n \times n$ identity matrix with rows rearranged, since each new Y_i equals one and only one of the original $X_1, X_2, \ldots, X_n$, and therefore equals ± 1. The joint density function of the random variables in this particular transformation is thus

$$f_{X_1, X_2, \ldots, X_n}(y_2, y_n, \ldots, y_3, y_{n-1})|J| = \prod_{i=1}^{n} f_X(y_i)$$

$$\text{for } y_1 < y_2 < \cdots < y_n$$

The same expression results for each of the $n!$ arrangements, since each Jacobian has absolute value 1 and multiplication is commutative. Therefore, applying the general Jacobian technique described in Chap. 1, the result is

$$f_{X_{(1)}, X_{(2)}, \ldots, X_{(n)}}(y_1, y_2, \ldots, y_n) = \sum_{\substack{\text{over all} \\ n! \text{ inverse} \\ \text{transformations}}} \prod_{i=1}^{n} f_X(y_i)$$

$$= n! \prod_{i=1}^{n} f_X(y_i)$$

$$\text{for } y_1 < y_2 < \cdots < y_n \quad (2.1)$$

In other words, the joint probability density function of the n order statistics is $n!$ times the joint distribution of the original sample. For example, for a random sample from the normal distribution, we have

$$f_{X_{(1)},X_{(2)},\ldots,X_{(n)}}(x_1,x_2,\ldots,x_n) = \frac{n!}{(2\pi\sigma^2)^{n/2}} \exp\left[-\frac{1}{2\sigma^2} \sum_{i=1}^{n} (x_i - \mu)^2 \right]$$

2.3　MARGINAL DISTRIBUTIONS OF ORDER STATISTICS

The usual method of finding the marginal distribution of any random variable can be applied to the rth order statistic by integrating out the remaining $n - 1$ variables in the joint distribution found in (2.1).

For the largest element in the sample, $X_{(n)}$, we have

$$f_{X_{(n)}}(y_n) = n!f_X(y_n) \int_{-\infty}^{y_n} \int_{-\infty}^{y_{n-1}} \cdots \int_{-\infty}^{y_3} \int_{-\infty}^{y_2} \prod_{i=1}^{n-1} f_X(y_i)\, dy_i$$

$$= n!f_X(y_n) \int_{-\infty}^{y_n} \int_{-\infty}^{y_{n-1}} \cdots \int_{-\infty}^{y_3} [F_X(y_2)f_X(y_2)]$$
$$\prod_{i=3}^{n-1} f_X(y_i)\, dy_2 \cdots dy_{n-1}$$

$$= n!f_X(y_n) \int_{-\infty}^{y_n} \int_{-\infty}^{y_{n-1}} \cdots \int_{-\infty}^{y_4} \frac{[F_X(y_3)]^2}{2(1)} f_X(y_3)$$
$$\prod_{i=4}^{n-1} f_X(y_i)\, dy_3 \cdots dy_{n-1}$$

$$\cdots \cdots \cdots \cdots \cdots \cdots \cdots \cdots \cdots \cdots \cdots$$

$$= n!f_X(y_n) \frac{[F_X(y_n)]^{n-1}}{(n - 1)!}$$

$$= n[F_X(y_n)]^{n-1}f_X(y_n) \tag{3.1}$$

Similarly, for the smallest element, $X_{(1)}$,

$$f_{X_{(1)}}(y_1) = n!f_X(y_1) \int_{y_1}^{\infty} \int_{y_2}^{\infty} \cdots \int_{y_{n-2}}^{\infty} \int_{y_{n-1}}^{\infty}$$
$$\prod_{i=2}^{n} f_X(y_i)\, dy_n\, dy_{n-1} \cdots dy_3\, dy_2$$

$$= n!f_X(y_1) \int_{y_1}^{\infty} \int_{y_2}^{\infty} \cdots \int_{y_{n-2}}^{\infty} [1 - F_X(y_{n-1})]f_X(y_{n-1})$$
$$\prod_{i=2}^{n-2} f_X(y_i)\, dy_{n-1}\, dy_{n-2} \cdots dy_2$$

$$= n! f_X(y_1) \int_{y_1}^{\infty} \int_{y_2}^{\infty} \cdots \int_{y_{n-3}}^{\infty} \frac{[1 - F_X(y_{n-2})]^2}{2(1)} f_X(y_{n-2})$$

$$\prod_{i=2}^{n-3} f_X(y_i) \, dy_{n-2} \cdots dy_2$$

$$\cdots \cdots \cdots \cdots \cdots \cdots \cdots \cdots \cdots$$

$$= n! f_X(y_1) \frac{[1 - F_X(y_1)]^{n-1}}{(n - 1)!}$$

$$= n[1 - F_X(y_1)]^{n-1} f_X(y_1) \tag{3.2}$$

For the rth order statistic, the order of integration which is easiest to handle would be $\infty > y_n > y_{n-1} > \cdots > y_r$ followed by $-\infty < y_1 < y_2 < \cdots < y_r$, so that we have the following combination of techniques used for $X_{(n)}$ and $X_{(1)}$:

$$f_{X_{(r)}}(y_r) = n! f_X(y_r) \int_{-\infty}^{y_r} \int_{-\infty}^{y_{r-1}} \cdots \int_{-\infty}^{y_2} \int_{y_r}^{\infty} \int_{y_{r+1}}^{\infty} \cdots \int_{y_{n-1}}^{\infty}$$

$$\prod_{\substack{i=1 \\ i \neq r}}^{n} f_X(y_i) \, dy_n \cdots dy_{r+2} \, dy_{r+1} \, dy_1 \cdots dy_{r-1}$$

$$= n! f_X(y_r) \frac{[1 - F_X(y_r)]^{n-r}}{(n - r)!} \int_{-\infty}^{y_r} \int_{-\infty}^{y_{r-1}} \cdots \int_{-\infty}^{y_2}$$

$$\prod_{i=1}^{r-1} f_X(y_i) \, dy_1 \cdots dy_{r-2} \, dy_{r-1}$$

$$\cdots \cdots \cdots \cdots \cdots \cdots \cdots \cdots \cdots$$

$$= n! f_X(y_r) \frac{[1 - F_X(y_r)]^{n-r}}{(n - r)!} \frac{[F_X(y_r)]^{r-1}}{(r - 1)!}$$

$$= \frac{n!}{(r - 1)!(n - r)!} [F_X(y_r)]^{r-1}[1 - F_X(y_r)]^{n-r} f_X(y_r) \tag{3.3}$$

This method could be applied to find the marginal distribution of a subset of two or more order statistics. The approach, although direct, involves tiresome integration.

A much simpler method can be used which appeals to probability theory instead of pure mathematics. The technique will be illustrated first for the single order statistic $X_{(r)}$. By the definition of a derivative, we have

$$f_{X_{(r)}}(x) = \lim_{h \to 0} \frac{F_{X_{(r)}}(x + h) - F_{X_{(r)}}(x)}{h}$$

$$= \lim_{h \to 0} \frac{P(x < X_{(r)} \leq x + h)}{h} \tag{3.4}$$

Suppose that the x axis is divided into the following three disjoint intervals:

$$I_1 = (-\infty, x]$$
$$I_2 = (x, x + h]$$
$$I_3 = (x + h, \infty)$$

For the population F_X, the probabilities that X lies in each of these intervals are

$$p_1 = P(X \in I_1) = F_X(x)$$
$$p_2 = P(X \in I_2) = F_X(x + h) - F_X(x)$$
$$p_3 = P(X \in I_3) = 1 - F_X(x + h)$$

Now, $X_{(r)}$ is the rth order statistic of the set $X_1, X_2, \ldots, X_n$ *and* lies in the interval I_2 if and only if exactly $r - 1$ of the original X random variables lie in the interval I_1, exactly $n - r$ of the original X's lie in the interval I_3 and $X_{(r)}$ lies in the interval I_2. Since the original X values are independent and the intervals are disjoint, the multinomial probability distribution with parameters p_1, p_2, and p_3 can be used to evaluate the probability in (3.4). The result is

$$
\begin{aligned}
f_{X_{(r)}}(x) &= \lim_{h \to 0} \binom{n}{r-1, 1, n-r} p_1^{r-1} p_2 p_3^{n-r} \\
&= \frac{n!}{(r-1)!(n-r)!} [F_X(x)]^{r-1} \\
&\quad \lim_{h \to 0} \left\{ \frac{F_X(x+h) - F_X(x)}{h} [1 - F_X(x+h)]^{n-r} \right\} \\
&= \frac{n!}{(r-1)!(n-r)!} [F_X(x)]^{r-1} f_X(x) [1 - F_X(x)]^{n-r} \qquad (3.5)
\end{aligned}
$$

which agrees with the result previously obtained in (3.3).

Now let $X_{(r)}$ and $X_{(s)}$ be any two order statistics from the set $X_{(1)} < X_{(2)} < \cdots < X_{(n)}$. By the definition of partial derivatives, the joint density function can be written

$$
\begin{aligned}
f_{X_{(r)}, X_{(s)}}(x, y) &= \lim_{\substack{h \to 0 \\ t \to 0}} \frac{\begin{aligned}F_{X_{(r)}, X_{(s)}}(x+h, y+t) - F_{X_{(r)}, X_{(s)}}(x, y+t)\\ - F_{X_{(r)}, X_{(s)}}(x+h, y) + F_{X_{(r)}, X_{(s)}}(x, y)\end{aligned}}{ht} \\
&= \lim_{\substack{h \to 0 \\ t \to 0}} \frac{\begin{aligned}P(x < X_{(r)} \leq x+h, X_{(s)} \leq y+t)\\ - P(x < X_{(r)} \leq x+h, X_{(s)} \leq y)\end{aligned}}{ht} \\
&= \lim_{\substack{h \to 0 \\ t \to 0}} \frac{P(x < X_{(r)} \leq x+h, y < X_{(s)} \leq y+t)}{ht} \qquad (3.6)
\end{aligned}
$$

For any $x < y$, the x axis can be divided into the following five disjoint intervals with the corresponding probabilities that an original X observation lies in that interval:

Interval I	$P(X \in I)$
$I_1 = (-\infty, x]$	$p_1 = F_X(x)$
$I_2 = (x, x + h]$	$p_2 = F_X(x + h) - F_X(x)$
$I_3 = (x + h, y]$	$p_3 = F_X(y) - F_X(x + h)$
$I_4 = (y, y + t]$	$p_4 = F_X(y + t) - F_X(y)$
$I_5 = (y + t, \infty)$	$p_5 = 1 - F_X(y + t)$

With this interval separation and assuming without loss of generality that $r < s$, $X_{(r)}$ and $X_{(s)}$ are the rth and sth order statistics, respectively, *and* lie in the respective intervals I_2 and I_4 if and only if the n X values are distributed along the x axis in such a way that exactly $r - 1$ lie in I_1 and 1 in I_2, 1 in I_4, and $n - s$ in I_5 since the one in I_4 is the sth in magnitude, and the remaining $s - r - 1$ must therefore lie in I_3. Applying the multinomial probability distribution to these five types of outcomes with the corresponding probabilities, we obtain

$$\binom{n}{r - 1, 1, s - r - 1, 1, n - s} p_1^{r-1} p_2 p_3^{s-r-1} p_4 p_5^{n-s}$$

Substituting this for the probability expression in (3.6) gives

$$f_{X_{(r)}, X_{(s)}}(x,y) = \binom{n}{r - 1, 1, s - r - 1, 1, n - s} [F(x)]^{r-1}$$

$$\lim_{\substack{h \to 0 \\ t \to 0}} \left\{ \frac{F_X(x + h) - F_X(x)}{h} [F_X(y) - F_X(x + h)]^{s-r-1} \right\}$$

$$\lim_{\substack{h \to 0 \\ t \to 0}} \left\{ \frac{F_X(y + t) - F_X(y)}{t} [1 - F_X(y + t)]^{n-s} \right\}$$

$$= \frac{n!}{(r - 1)!(s - r - 1)!(n - s)!} [F_X(x)]^{r-1} [F_X(y)$$

$$- F_X(x)]^{s-r-1} [1 - F_X(y)]^{n-s} f_X(x) f_X(y) \qquad \text{for all } x < y \quad (3.7)$$

This method could be extended in a similar manner to find the joint marginal distribution of any subset of the n order statistics. In general, for any $k \leq n$, to find the marginal distribution of k order statistics, the x axis must be divided into $k + (k - 1) + 2 = 2k + 1$ disjoint intervals and the multinomial probability law applied.

Of particular interest in distribution-free techniques is the case where the set $X_{(1)} < X_{(2)} < \cdots < X_{(n)}$ is the order statistics for the uniform distribution over the interval $(0,1)$.   Then the marginal distributions of $X_{(r)}$ and $X_{(r)}$ and $X_{(s)}$ for $r < s$ are, respectively,

$$f_{X_{(r)}}(x) = \frac{n!}{(r-1)!(n-r)!} x^{r-1}(1-x)^{n-r} \qquad 0 < x < 1 \qquad (3.8)$$

$$f_{X_{(r)},X_{(s)}}(x,y) = \frac{n!}{(r-1)!(s-r-1)!(n-s)!} x^{r-1}$$
$$(y-x)^{s-r-1}(1-y)^{n-s} \qquad 0 < x < y < 1 \quad (3.9)$$

from (3.3) and (3.7).   The density function in (3.8) will be recognized as that of the beta distribution with parameters r and $n - r + 1$.

2.4 DISTRIBUTION OF THE MEDIAN AND RANGE

As indicated in Sec. 2.1, the median and range of a random sample are measures based on order statistics which are descriptive of the central tendency and dispersion of the population, respectively.   Their distributions are easily obtained from the results found in Sec. 2.3.

For n odd, the median of a sample has the distribution of (3.3) with $r = (n+1)/2$.   If n is even and a unique value is desired for the sample median U, the usual definition is

$$U = \frac{X_{(n/2)} + X_{[(n+2)/2]}}{2}$$

so that the distribution of U must be derived from the joint density of these two order statistics.   Letting $n = 2m$, from (3.7) we have for $x < y$

$$f_{X_{(m)},X_{(m+1)}}(x,y) = \frac{(2m)!}{[(m-1)!]^2} [F_X(x)]^{m-1}[1 - F_X(y)]^{m-1}f_X(x)f_X(y)$$

Making the transformation

$$u = \frac{x+y}{2}$$
$$v = y$$

and using the method of Jacobians, the distribution of the median U is

$$f_U(u) = \frac{(2m)!2}{[(m-1)!]^2} \int_u^\infty [F_X(2u-v)]^{m-1}$$
$$[1 - F_X(v)]^{m-1}f_X(2u-v)f_X(v) \, dv \quad (4.1)$$

As an example, consider the uniform distribution over (0,1). The integrand in (4.1) is nonzero for the intersection of the regions

$$0 < 2u - v < 1 \quad \text{and} \quad 0 < v < 1$$

The region of integration then is the intersection of the three regions

$$u < v \qquad \frac{v}{2} < u < \frac{v+1}{2} \qquad \text{and} \qquad 0 < v < 1$$

which is depicted graphically in Fig. 4.1. We see that the limits on the integral in (4.1) must be $u < v < 2u$ for $0 < u \le \frac{1}{2}$ and $u < v < 1$ for $\frac{1}{2} < u < 1$. Thus if $m = 2$, say, the distribution of the median of a sample of size 4 is

$$f_U(u) = \begin{cases} 8u^2(3 - 4u) & \text{for } 0 < u \le \frac{1}{2} \\ 8(4u^3 - 9u^2 + 6u - 1) & \text{for } \frac{1}{2} < u < 1 \end{cases} \qquad (4.2)$$

Verification of this result will be left as an exercise for the reader.

A similar procedure can be used to obtain the distribution of the range, defined as

$$R = X_{(n)} - X_{(1)}$$

In the joint distribution for $x < y$

$$f_{X_{(1)}, X_{(n)}}(x,y) = n(n - 1)[F_X(y) - F_X(x)]^{n-2} f_X(x) f_X(y)$$

we make the transformation

$$u = y - x$$
$$v = y$$

and obtain for all $u > 0$

$$f_R(u) = \int_{-\infty}^{\infty} n(n - 1)[F_X(v) - F_X(v - u)]^{n-2} f_X(v - u) f_X(v) \, dv$$

$$(4.3)$$

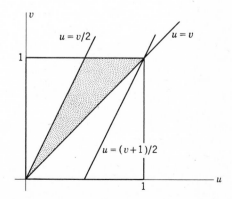

Fig. 4.1

For the uniform distribution, the integrand in (4.3) is nonzero for the intersection of the regions

$$0 < v - u < 1 \quad \text{and} \quad 0 < v < 1$$

but this is simply $0 < u < v < 1$. Therefore, the probability function of the range is

$$f_R(u) = n(n - 1)u^{n-2}(1 - u) \qquad \text{for } 0 < u < 1$$

This result is quite easy to handle. For many distributions F_X, the integral in (4.3) is difficult to evaluate. In the case of a standard normal population, Hartley (1942) has tabulated the cumulative distribution of the range for sample sizes not exceeding 20. The asymptotic distribution of the range is discussed in Gumbel (1944).

2.5 EXACT MOMENTS OF ORDER STATISTICS

Any individual or joint moments of order statistics can be found using (3.3) and (3.7), respectively, for a specified continuous population F_X. The only practical limitation is the complexity of integration involved. Any distribution for which $F_X(x)$ is not easily expressible in closed form is particularly difficult.

The easiest distribution to handle is the uniform population over $(0,1)$, since then the simpler probability densities given in (3.8) and (3.9) are applicable. For the kth moment about the origin of the rth order statistic from the uniform distribution, we have

$$
\begin{aligned}
E(X_{(r)}{}^k) &= \frac{n!}{(r - 1)!(n - r)!} \int_0^1 x^{r+k-1}(1 - x)^{n-r}\,dx \\
&= \frac{n!}{(r - 1)!(n - r)!} B(r + k, n - r + 1) \\
&= \frac{n!(r + k - 1)!}{(n + k)!(r - 1)!} \\
&= \frac{(r + k - 1)(r + k - 2) \cdots (r + 1)r}{(n + k)(n + k - 1) \cdots (n + 2)(n + 1)}
\end{aligned}
\tag{5.1}
$$

for any $1 \leq r \leq n$ and integer k. In particular, the mean is

$$E(X_{(r)}) = \frac{r}{n + 1} \tag{5.2}$$

and the variance is easily obtained as follows:

$$\text{var}(X_{(r)}) = E(X_{(r)}{}^2) - [E(X_{(r)})]^2 = \frac{r(r+1)}{(n+2)(n+1)} - \frac{r^2}{(n+1)^2}$$

$$= \frac{r(n-r+1)}{(n+2)(n+1)^2} \tag{5.3}$$

In order to determine the covariance of the two order statistics $X_{(r)}$ and $X_{(s)}$, from (3.9) the following integral must be evaluated:

$$E(X_{(r)}X_{(s)}) = \frac{n!}{(r-1)!(s-r-1)!(n-s)!}$$

$$\int_0^1 \int_0^y x^r y (y-x)^{s-r-1} (1-y)^{n-s} \, dx \, dy$$

$$= \frac{n!}{(r-1)!(s-r-1)!(n-s)!}$$

$$\int_0^1 y(1-y)^{n-s} [y^s B(r+1, s-r)] \, dy$$

$$= \frac{n!}{(r-1)!(s-r-1)!(n-s)!}$$

$$B(s+2, n-s+1)B(r+1, s-r)$$

$$= \frac{n!(s+1)!(n-s)!r!(s-r-1)!}{(r-1)!(s-r-1)!(n-s)!(n+2)!s!}$$

$$= \frac{r(s+1)}{(n+1)(n+2)}$$

Now the covariance is found using

$$\text{cov}(X_{(r)},X_{(s)}) = E(X_{(r)}X_{(s)}) - E(X_{(r)})E(X_{(s)})$$

$$= \frac{r(s+1)}{(n+1)(n+2)} - \frac{rs}{(n+1)^2}$$

$$= \frac{r(n-s+1)}{(n+1)^2(n+2)} \qquad \text{for } r < s \tag{5.4}$$

The correlation coefficient then is

$$\text{corr}(X_{(r)},X_{(s)}) = \left[\frac{r(n-s+1)}{s(n-r+1)}\right]^{\frac{1}{2}} \qquad \text{for } r < s \tag{5.5}$$

In particular, the correlation between the minimum and maximum value in a sample of n is

$$\text{corr}(X_{(1)},X_{(n)}) = \frac{1}{n}$$

so that correlation is inversely proportional to sample size.

When the population is such that $F_X(x)$ cannot be expressed in closed form, evaluation of moments is often tedious or even impossible without the aid of a computer for numerical integration. Since the expected values of the order statistics from a normal probability distribution have especially useful applications, these results have been tabulated, and are available, for example, in Harter (1961). For n small, even these normal moments can be evaluated with appropriate techniques of integration. For example, if $n = 2$ and F_X is the standard normal, the mean of the first order statistic is

$$E(X_{(1)}) = 2 \int_{-\infty}^{\infty} x \left(1 - \int_{-\infty}^{x} \frac{1}{\sqrt{2\pi}} e^{-\frac{1}{2}t^2} dt \right) \frac{1}{\sqrt{2\pi}} e^{-\frac{1}{2}x^2} dx$$

$$= \frac{1}{\pi} \int_{-\infty}^{\infty} \int_{x}^{\infty} x e^{-\frac{1}{2}(t^2 + x^2)} dt\, dx$$

Introducing a change to polar coordinates with

$$x = r \cos \theta \qquad t = r \sin \theta$$

the integral above becomes

$$E(X_{(1)}) = \frac{1}{\pi} \int_{\pi/4}^{5\pi/4} \int_{0}^{\infty} r^2 \cos \theta\, e^{-\frac{1}{2}r^2} dr\, d\theta$$

$$= \frac{\sqrt{2}}{\sqrt{\pi}} \int_{\pi/4}^{5\pi/4} \cos \theta\, \frac{1}{2} \int_{-\infty}^{\infty} \frac{r^2}{\sqrt{2\pi}} e^{-\frac{1}{2}r^2} dr\, d\theta$$

$$= \frac{1}{\sqrt{2\pi}} \int_{\pi/4}^{5\pi/4} \cos \theta\, d\theta = \frac{1}{\sqrt{2\pi}} \left(-\frac{1}{\sqrt{2}} - \frac{1}{\sqrt{2}} \right) = -\frac{1}{\sqrt{\pi}}$$

By symmetry, of course, $E(X_{(2)}) = 1/\sqrt{\pi}$.

Other examples of these techniques will be found in the problems.

2.6 LARGE-SAMPLE APPROXIMATION TO THE MEAN AND VARIANCE OF THE rTH ORDER STATISTIC

Evaluation of the exact moments of $X_{(r)}$ directly from the probability density function in (3.3) requires numerical integration for many F_X of interest. However, the rth order statistic from any continuous distribution is a function of the rth order statistic from the uniform distribution because of the probability-integral transformation. Letting $U_{(r)}$ denote the rth order statistic from a uniform distribution over the interval $(0,1)$, this functional relationship can be denoted

$$X_{(r)} = F_X^{-1}(U_{(r)}) \tag{6.1}$$

Since the moments of $U_{(r)}$ were easily evaluated, an approximation to the moments of a function in terms of some function of these moments would enable us to approximate the moments of $X_{(r)}$ for any specified continuous F_X. Such an approximation to the moments of any function in terms of the moments of the argument can be found as follows.

The Taylor's series expansion of a function $g(z)$ about a point μ is

$$g(z) = g(\mu) + \sum_{i=1}^{\infty} \frac{(z-\mu)^i}{i!} g^{(i)}(\mu) \tag{6.2}$$

where $g^{(i)}(\mu) = d^i g(z)/dz^i \big|_{z=\mu}$, and this series converges if

$$\lim_{n \to \infty} \frac{(z-\mu)^n}{n!} g^{(n)}(z_1) = 0 \qquad \text{for } \mu < z_1 < z$$

Writing (6.2) for any random variable Z which has mean μ and taking the expected value of both sides, we obtain

$$E[g(Z)] = g(\mu) + \frac{\text{var}(Z)}{2!} g^{(2)}(\mu) + \sum_{i=3}^{\infty} \frac{E[(Z-\mu)^i]}{i!} g^{(i)}(\mu) \tag{6.3}$$

From this result, it is clear that:

1. A first approximation to $E[g(Z)]$ is $g(\mu)$.
2. A second approximation to $E[g(Z)]$ is $g(\mu) + \{[\text{var}(Z)]/2!\}g^{(2)}(\mu)$.

To find similar approximations to $\text{var}(Z)$, we can form the difference between Eqs. (6.2) and (6.3), square it, and then take expected values as follows:

$$g(z) - E[g(Z)] = (z-\mu)g^{(1)}(\mu) + g^{(2)}(\mu)\frac{1}{2!}[(z-\mu)^2 - \text{var}(Z)]$$

$$+ \sum_{i=3}^{\infty} \frac{g^{(i)}(\mu)}{i!} \{(z-\mu)^i - E[(Z-\mu)^i]\}$$

$$\{g(z) - E[g(Z)]\}^2 = (z-\mu)^2[g^{(1)}(\mu)]^2 + \tfrac{1}{4}[g^{(2)}(\mu)]^2[\text{var}^2(Z)$$
$$- 2\,\text{var}(Z)(z-\mu)^2] - g^{(1)}(\mu)g^{(2)}(\mu)\,\text{var}(Z)(z-\mu) + h(z)$$

$$\text{var}[g(Z)] = \text{var}(Z)[g^{(1)}(\mu)]^2$$
$$- \tfrac{1}{4}[g^{(2)}(\mu)]^2\,\text{var}^2(Z) + E[h(Z)] \tag{6.4}$$

where $E[h(Z)]$ involves third or higher central moments of Z. The first two terms provide first and second approximations to $\text{var}[g(Z)]$ in terms of $\text{var}(Z)$. The goodness of any of these approximations of course depends upon the magnitude of the terms ignored, i.e., the order of the higher central moments of Z.

In order to apply these generally useful results to the rth order statistic $X_{(r)}$ of a sample of n from the continuous distribution F_X, we note that the functional relationship in (6.1) implies our g function must be

$$g(u_{(r)}) = x_{(r)} = F_X^{-1}(u_{(r)}) \quad \text{and} \quad u_{(r)} = F_X(x_{(r)})$$

and the appropriate moments

$$\mu = E(U_{(r)}) = \frac{r}{n+1} \quad \text{and} \quad \text{var}(U_{(r)}) = \frac{r(n-r+1)}{(n+1)^2(n+2)}$$

were found in (5.2) and (5.3). The derivatives involved are

$$g^{(1)}(\mu) = \frac{dx_{(r)}}{du_{(r)}}\Big|_{u_{(r)}=\mu} = \left(\frac{du_{(r)}}{dx_{(r)}}\right)^{-1}\Big|_{u_{(r)}=\mu}$$

$$= [f_X(x_{(r)})]^{-1}\Big|_{x_{(r)}=F_X^{-1}(\mu)}$$

$$= \left\{f_X\left[F_X^{-1}\left(\frac{r}{n+1}\right)\right]\right\}^{-1}$$

and similarly

$$g^{(2)}(\mu) = \frac{d}{du_{(r)}}[f_X(x_{(r)})]^{-1}\Big|_{x_{(r)}=F_X^{-1}(\mu)}$$

$$= -[f_X(x_{(r)})]^{-2}f_X'(x_{(r)})\frac{dx_{(r)}}{du_{(r)}}\Big|_{x_{(r)}=F_X^{-1}(\mu)}$$

$$= -f_X'\left[F_X^{-1}\left(\frac{r}{n+1}\right)\right]\left\{f_X\left[F_X^{-1}\left(\frac{r}{n+1}\right)\right]\right\}^{-3}$$

Substituting these results above, we can obtain first and second approximations to the mean and variance of $X_{(r)}$. The first approximations are

$$E(X_{(r)}) \approx F_X^{-1}\left(\frac{r}{n+1}\right) \tag{6.5}$$

$$\text{var}(X_{(r)}) \approx \frac{r(n-r+1)}{(n+1)^2(n+2)}\left\{f_X\left[F_X^{-1}\left(\frac{r}{n+1}\right)\right]\right\}^{-2} \tag{6.6}$$

Using (5.1), the third central moment of $U_{(r)}$ is found to be

$$E[(U_{(r)} - \mu)^3] = \frac{r(2n^2 - 6nr + 4n + 4r^2 - 6r + 2)}{(n+1)^3(n+2)(n+3)} \tag{6.7}$$

so that for large n and finite r or r/n fixed, the terms from (6.3) and (6.4) which were ignored in reaching these approximations are of small order. For greater accuracy, the second- or higher-order approximations can be found. This will be left as an exercise for the reader.

The use of (6.5) and (6.6) is particularly simple when f_X and F_X are tabulated. For example, to approximate the mean and variance of the fourth order statistic of a sample of 19 from the standard normal population, we have

$$E(X_{(4)}) \approx \Phi^{-1}(0.20) = -0.84$$

$$\text{var}(X_{(4)}) \approx \frac{4(16)}{20^2(21)}\, [\varphi(-0.84)]^{-2} = \frac{0.16}{21}\, 0.2803^{-2} = 0.097$$

The exact values of the means and variances of the normal order statistics are widely available; for example, in Ruben (1954) and Sarhan and Greenberg (1962). For comparison with the results in this example, the exact mean and variance of $X_{(4)}$ when $n = 19$ from these tables are -0.8859 and 0.107406, respectively.

2.7 ASYMPTOTIC DISTRIBUTION OF ORDER STATISTICS

Since the exact probability density function of the rth order statistic from any continuous population F_X is rather irksome to deal with for large n, information concerning the form of the asymptotic distribution would increase the usefulness of order statistics in applications for large samples. In speaking of a general asymptotic distribution for any r, however, two distinct cases must be considered:

Case 1: As n approaches infinity, r/n remains fixed.
Case 2: As n approaches infinity, r or $n - r$ remains finite.

Case 1 would be of interest, for example, in the distribution of quantiles, whereas case 2 would be mainly appropriate for the distribution of extreme values. Case 2 will not be considered here. The reader is referred to Wilks (1948) for a discussion of the asymptotic distribution of $X_{(r)}$ for fixed r under various conditions and to Gumbel (1958) for asymptotic distributions of extremes.

Under the assumptions of case 1, we shall show in this section that the distribution of the standardized rth order statistic from the uniform distribution approaches the standard normal distribution. This result can be shown in either of two ways. The most direct approach is to show that the probability density function of a standardized $U_{(r)}$ approaches the function $\varphi(u)$. In the density for $U_{(r)}$,

$$f_{U_{(r)}}(u) = \frac{n!}{(r-1)!(n-r)!}\, u^{r-1}(1-u)^{n-r} \qquad 0 < u < 1$$

we make the transformation

$$Z_{(r)} = \frac{U_{(r)} - \mu}{\sigma}$$

and obtain for all z

$$
\begin{aligned}
f_{Z_{(r)}}(z) &= \frac{n!}{(r-1)!(n-r)!} (\sigma z + \mu)^{r-1}(1 - \sigma z - \mu)^{n-r}\sigma \\
&= n\binom{n-1}{r-1} \sigma\mu^{r-1}(1-\mu)^{n-r}\left(1 + \frac{\sigma z}{\mu}\right)^{r-1}\left(1 - \frac{\sigma z}{1-\mu}\right)^{n-r} \\
&= n\binom{n-1}{r-1} \sigma\mu^{r-1}(1-\mu)^{n-r}e^{v}
\end{aligned}
\tag{7.1}
$$

where

$$v = (r-1)\log\left(1 + \frac{\sigma z}{\mu}\right) + (n-r)\log\left(1 - \frac{\sigma z}{1-\mu}\right) \tag{7.2}$$

Now using the Taylor's series expansion

$$\log(1 + x) = \sum_{i=1}^{\infty} (-1)^{i-1}\frac{x^i}{i}$$

which converges for $-1 < x \le 1$, and with the notation

$$\frac{\sigma}{\mu} = c_1 \qquad \frac{\sigma}{1-\mu} = c_2$$

we have

$$
\begin{aligned}
v &= (r-1)\left(c_1 z - c_1^2\frac{z^2}{2} + c_1^3\frac{z^3}{3} - \cdots\right) \\
&\qquad\qquad - (n-r)\left(c_2 z + c_2^2\frac{z^2}{2} + c_2^3\frac{z^3}{3} + \cdots\right) \\
&= z[c_1(r-1) - c_2(n-r)] - \frac{z^2}{2}[c_1^2(r-1) + c_2^2(n-r)] \\
&\qquad + \frac{z^3}{3}[c_1^3(r-1) - c_2^3(n-r)] - \cdots
\end{aligned}
\tag{7.3}
$$

Since we are going to take the limit of v as $n \to \infty$, $p = r/n$ remaining fixed, c_1 and c_2 can be approximated as

$$c_1 = \left[\frac{(n-r+1)}{r(n+2)}\right]^{1/2} \approx \left(\frac{1-p}{pn}\right)^{1/2}$$

$$c_2 = \left[\frac{r}{(n-r+1)(n+2)}\right]^{1/2} \approx \left[\frac{p}{(1-p)n}\right]^{1/2}$$

Substitution of these values in (7.3) gives the coefficient of z

$$\frac{(r-1)\sqrt{1-p}}{\sqrt{np}} - \frac{(n-r)\sqrt{p}}{\sqrt{n(1-p)}} = \frac{r-np-(1-p)}{\sqrt{np(1-p)}}$$

$$= -\frac{\sqrt{1-p}}{\sqrt{np}} \to 0 \qquad \text{as } n \to \infty$$

and the coefficient of $-z^2/2$

$$\frac{(r-1)(1-p)}{np} + \frac{(n-r)p}{n(1-p)} = (1-p) - \frac{(1-p)}{np} + p$$

$$= 1 - \frac{(1-p)}{np} \to 1 \qquad \text{as } n \to \infty$$

and the coefficient of $z^3/3$

$$\frac{(r-1)(1-p)^{3/2}}{(np)^{3/2}} - \frac{(n-r)p^{3/2}}{[n(1-p)]^{3/2}} = \frac{(np-1)}{n^{3/2}}\left(\frac{1-p}{p}\right)^{3/2}$$

$$- \frac{p^{3/2}}{[n(1-p)]^{1/2}} \to 0 \qquad \text{as } n \to \infty$$

Substituting these results back in (7.3) and ignoring terms of order $n^{-1/2}$ and higher, the limiting value is

$$\lim_{n\to\infty} v = -\frac{z^2}{2}$$

For the limiting value of the constant term in (7.1), we must use Stirling's formula

$$k! \approx \sqrt{2\pi}\, e^{-k}k^{k+1/2}$$

for the factorials, which are to be multiplied by

$$\sigma\mu^{r-1}(1-\mu)^{n-r} = \frac{r^{r-1/2}(n-r+1)^{n-r+1/2}}{(n+1)^n(n+2)^{1/2}} \approx \frac{r^{r-1/2}(n-r+1)^{n-r+1/2}}{(n+1)^{n+1/2}}$$

The entire constant of (7.1) is written as

$$n\binom{n-1}{r-1}\sigma\mu^{r-1}(1-\mu)^{n-r} = \frac{(n+1)!}{r!(n-r+1)!}$$

$$\frac{r(n-r+1)}{n+1}\sigma\mu^{r-1}(1-\mu)^{n-r}$$

$$\approx \frac{\sqrt{2\pi}\,e^{-(n+1)}(n+1)^{n+3/2}}{2\pi e^{-r}r^{r+1/2}e^{-(n-r+1)}(n-r+1)^{n-r+3/2}}$$

$$\frac{r^{r+1/2}(n-r+1)^{n-r+3/2}}{(n+1)^{n+3/2}} = \frac{1}{\sqrt{2\pi}}$$

Thus we have the desired result

$$\lim_{n \to \infty} f_{Z_{(r)}}(z) = \frac{1}{\sqrt{2\pi}} e^{-\frac{1}{2}z^2}$$

and

$$\lim_{n \to \infty} P(U_{(r)} \leq t) = \Phi\left(\frac{t - \mu}{\sigma}\right)$$

If $X_{(r)}$ is the rth order statistic from any continuous distribution F_X, the probability-integral transformation allows us to conclude that the same asymptotic distribution holds for $X_{(r)}$ as long as the appropriate μ and σ are substituted. Using the approximate mean and variance found in (6.5) and (6.6) for $r/n = p$ fixed and n large, we can substitute

$$\mu = F_X^{-1}(p) \qquad \text{and} \qquad \sigma = \frac{[p(1 - p)]^{\frac{1}{2}}[f_X(\mu)]^{-1}}{n^{\frac{1}{2}}}$$

and state the following theorem.

Theorem 7.1 *If $X(r)$ denotes the rth order statistic of n from any continuous cumulative distribution F_X, then as n approaches infinity and r/n remains fixed, the distribution of*

$$\left[\frac{n}{p(1 - p)}\right]^{\frac{1}{2}} f_X(\mu)[X_{(r)} - \mu]$$

tends to the standard normal, where μ satisfies $F_X(\mu) = p$, for $p = r/n$.

A similar result, attributed to Smirnov (1935), can be stated for the asymptotic joint distribution of two order statistics $X_{(r)}$ and $X_{(s)}$, where n approaches infinity in such a way that $r/n = p_1$ and $s/n = p_2$, $0 < p_1 < p_2 < 1$, remain fixed. In this case, $X_{(r)}$ and $X_{(s)}$ are asymptotically bivariate normal with means μ_i, variances $p_i(1 - p_i)[f_X(\mu_i)]^{-2}/n$, for $i = 1, 2$, and covariance $p_1(1 - p_2)[nf_X(\mu_1)f_X(\mu_2)]^{-1}$, where μ_i satisfies $F_X(\mu_i) = p_i$.

2.8 CONFIDENCE-INTERVAL ESTIMATES FOR POPULATION QUANTILES

A quantile of a continuous distribution $f_X(x)$ of a random variable X is a real number which divides the area under the probability density function into two parts of specified amounts. Only the area to the left of the number need be specified since the entire area is equal to 1. Let the pth quantile, or quantile of order p, be denoted by κ_p for all $0 \leq p \leq 1$. In terms of the cumulative distribution function, then, κ_p is defined to be any real number which is a solution to the equation

$$F_X(\kappa_p) = p$$

We shall assume here that a unique solution exists, as would be the case for a strictly increasing function F_X. κ_p for any value $0 \le p \le 1$ is a parameter of the population F_X. For example, $\kappa_{0.50}$ is the median of the distribution, a measure of central tendency.

Suppose that a confidence-interval estimate of this parameter κ_p is desired for some specified value of p. A logical point estimate of κ_p would be related to order statistics, say the sample quantile of order p. We know from Theorem 7.1 that the rth order statistic is a consistent estimator of the pth quantile of the distribution where $r/n = p$ remains fixed. A definition of the pth sample quantile which provides a unique number $x_{(r)}$ which is an order statistic of the sample for any n and p is to choose

$$r = \begin{cases} np & \text{if } np \text{ is an integer} \\ [np + 1] & \text{if } np \text{ is not an integer} \end{cases}$$

where $[x]$ denotes the largest integer not exceeding x. Other conventions are sometimes adopted to define sample quantiles.

However, the value of the point estimate will not concern us here. Consistency is, of course, only a large-sample property. We are seeking a procedure for interval estimation of κ_p which will enable us to attach a confidence coefficient to our estimate for the given sample size. The logical choices for the confidence-interval end points are two order statistics for a random sample of n drawn from the population F_X. In other words, we must find two numbers, say r and s, where $r < s$, such that

$$P(X_{(r)} < \kappa_p < X_{(s)}) = 1 - \alpha$$

for some chosen number $0 < \alpha < 1$. Now the event $X_{(r)} < \kappa_p$ occurs if and only if either $X_{(r)} < \kappa_p < X_{(s)}$ or $\kappa_p > X_{(s)}$, and these latter two events are clearly mutually exclusive. Therefore, for all $r < s$,

$$P(X_{(r)} < \kappa_p) = P(X_{(r)} < \kappa_p < X_{(s)}) + P(\kappa_p > X_{(s)})$$

or, equivalently,

$$P(X_{(r)} < \kappa_p < X_{(s)}) = P(X_{(r)} < \kappa_p) - P(X_{(s)} < \kappa_p) \tag{8.1}$$

Since F_X is a strictly increasing function,

$$X_{(r)} < \kappa_p \quad \text{if and only if} \quad F_X(X_{(r)}) < F_X(\kappa_p)$$

But the probability distribution of the random variable $F_X(X_{(r)})$ is known to be that of the rth order statistic from the uniform distribution over the interval (0,1) because of the probability-integral transformation, and further $F_X(\kappa_p) = p$ by the definition of κ_p. Therefore, for $X_{(r)}$ the rth order statistic from any continuous population with pth quantile κ_p, we have

$$P(X_{(r)} < \kappa_p) = P[F_X(X_{(r)}) < p]$$
$$= \int_0^p \frac{n!}{(r-1)!(n-r)!} x^{r-1}(1-x)^{n-r}\,dx \qquad (8.2)$$

Thus, while the distribution of the rth order statistic depends on the population distribution F_X, the probability in (8.2) does not. A confidence-interval procedure based on (8.1) is then distribution-free.

In order to find the interval estimate of κ_p, substitution of (8.2) back into (8.1) indicates that r and s should be chosen such that

$$\int_0^p n \binom{n-1}{r-1} x^{r-1}(1-x)^{n-r}\,dx$$
$$- \int_0^p n \binom{n-1}{s-1} x^{s-1}(1-x)^{n-s}\,dx = 1 - \alpha \qquad (8.3)$$

This will not give a unique solution for r and s. If we want the narrowest possible interval for a fixed confidence coefficient, r and s should be chosen such that (8.3) is satisfied *and* $X_{(s)} - X_{(r)}$ is as small as possible. Alternatively, we could minimize $s - r$.

The integrals in (8.2) or (8.3) can be evaluated by parts, or tables of the incomplete beta function can be used. However, (8.2) can be expressed in another form after integration by parts as follows:

$$P(X_{(r)} < \kappa_p) = \int_0^p n \binom{n-1}{r-1} x^{r-1}(1-x)^{n-r}\,dx$$
$$= n \binom{n-1}{r-1} \left[\frac{x^r}{r}(1-x)^{n-r} \Big|_0^p \right.$$
$$\left. + \frac{n-r}{r} \int_0^p x^r(1-x)^{n-r-1}\,dx \right]$$
$$= \binom{n}{r} p^r(1-p)^{n-r}$$
$$+ n \binom{n-1}{r} \left[\frac{x^{r+1}}{r+1}(1-x)^{n-r-1} \Big|_0^p \right.$$
$$\left. + \left(\frac{n-r-1}{r+1} \right) \int_0^p x^{r+1}(1-x)^{n-r-2}\,dx \right]$$
$$= \binom{n}{r} p^r(1-p)^{n-r} + \binom{n}{r+1} p^{r+1}(1-p)^{n-r-1}$$
$$+ n \binom{n-1}{r+1} \int_0^p x^{r+1}(1-x)^{n-r-2}\,dx$$

After repeating this integration by parts $n - r$ times, the result will be

$$\binom{n}{r} p^r (1 - p)^{n-r} + \binom{n}{r+1} p^{r+1} (1 - p)^{n-r-1} + \cdots$$

$$+ \binom{n}{n-1} p^{n-1} (1 - p) + n \binom{n-1}{n-1} \int_0^p x^{n-1} (1 - x)^0 \, dx$$

$$= \sum_{j=0}^{n-r} \binom{n}{r+j} p^{r+j} (1 - p)^{n-r-j}$$

or

$$P(X_{(r)} < \kappa_p) = \sum_{i=r}^{n} \binom{n}{i} p^i (1 - p)^{n-i} \tag{8.4}$$

In this final form, the integral in (8.2) is expressed as the sum of the last $n - r + 1$ terms of the binomial probability distribution with parameter p, and tables of those cumulative probabilities are widely available. The confidence coefficient of (8.1) can thus also be written as

$$P(X_{(r)} < \kappa_p < X_{(s)}) = \sum_{i=r}^{n} \binom{n}{i} p^i (1 - p)^{n-i} - \sum_{i=s}^{n} \binom{n}{i} p^i (1 - p)^{n-i}$$

$$= \sum_{i=r}^{s-1} \binom{n}{i} p^i (1 - p)^{n-i} \tag{8.5}$$

This form is probably the easiest to use in choosing r and s such that $s - r$ is a minimum for fixed α.

The binomial form in (8.4) found by integration of (8.2) can also be argued directly as follows. For any p, we have $X_{(r)} < \kappa_p$ if and only if at least r of the sample values $X_1, X_2, \ldots, X_n$ are less than κ_p. These sample values are independent, and if each is classified according to whether it is less than the number κ_p, the n random variables can be considered the result of n repeated, independent trials of a Bernoulli variable with parameter

$$P(X < \kappa_p) = p$$

Thus the number of observations which are less than κ_p follows the binomial probability distribution with parameter p.

HYPOTHESIS TESTING FOR POPULATION QUANTILES

In a hypothesis-testing type of inference problem concerned with distribution quantiles, a distribution-free procedure is also possible. Given the order statistics $X_{(1)} < X_{(2)} < \cdots < X_{(n)}$ from any unspecified but continuous distribution F_X, a null hypothesis concerning the value of the

pth quantile is written

$$H_0: \kappa_p = \kappa_p{}^0$$

where $\kappa_p{}^0$ and p are both specified numbers. Then under H_0,

$$P(X < \kappa_p{}^0) = p$$

If the alternative is one-sided, say

$$H_1: \kappa_p > \kappa_p{}^0$$

and the decision rule is to be based on the magnitude of some order statistic, we might choose a rejection region of the form

$$X_{(r)} \in R \qquad \text{for } x_{(r)} > \kappa_p{}^0 \tag{8.6}$$

For a specified significance level α, r should be chosen such that

$$\alpha = P(X_{(r)} > \kappa_p{}^0 \mid H_0) = 1 - P(X_{(r)} \le \kappa_p{}^0 \mid H_0)$$

or, using (8.4), r satisfies

$$\alpha = 1 - \sum_{i=r}^{n} \binom{n}{i} p^i (1 - p)^{n-i} = \sum_{i=0}^{r-1} \binom{n}{i} p^i (1 - p)^{n-i} \tag{8.7}$$

However, $X_{(r)} > \kappa_p{}^0$ if and only if there are at least $n - r + 1$ plus signs among the n differences $X_{(i)} - \kappa_p{}^0$, so that this test in (8.6) is equivalent to a rejection region of the form

$$K \in R \qquad \text{for } k \ge n - r + 1$$

where the random variable K represents the number of positive differences $X_i - \kappa_p{}^0$, for $i = 1, 2, \ldots, n$. These differences are independent, and the probability of a plus sign under H_0 is

$$P(X > \kappa_p{}^0) = 1 - p$$

and thus r must be chosen to satisfy

$$\sum_{k=n-r+1}^{n} \binom{n}{k} (1 - p)^k p^{n-k} = \alpha$$

which agrees with the statement in (8.7).

    A similar approach would be used for a one-sided alternative reversing the order of magnitude or a two-sided alternative. The test will not be discussed further, since it is equivalent to the ordinary one-sample sign test discussed at length in Chap. 6.

2.9 TOLERANCE LIMITS FOR DISTRIBUTIONS AND COVERAGES

In setting confidence intervals for parameters which are population quantiles, we found that the procedure did not depend in any way upon the particular population as long as its distribution function was continuous. A similar type of application of order statistics which is also distribution-free is setting *tolerance limits* for distributions.

A tolerance interval for a continuous distribution with tolerance coefficient γ is a random interval such that the probability is γ that the area between the end points of the interval and under the probability density function is at least a certain preassigned number p. In other words, the probability is γ that this random interval covers or includes at least a specified percentage $(100p)$ of the distribution. If the end points of the tolerance interval are two order statistics of a random sample of size n, the tolerance interval $(X_{(r)}, X_{(s)})$, $r < s$, symbolically satisfies the condition

$$P[P(X_{(r)} < X < X_{(s)}) \geq p] = \gamma \tag{9.1}$$

But if X has the cumulative distribution F_X, which is continuous,

$$P(X_{(r)} < X < X_{(s)}) = P(X < X_{(s)}) - P(X < X_{(r)})$$
$$= F_X(X_{(s)}) - F_X(X_{(r)})$$

By the probability-integral transformation,

$$F_X(X_{(r)}) = U_{(r)} \qquad \text{and} \qquad F_X(X_{(s)}) = U_{(s)}$$

are the rth and sth order statistics of a random sample of n from the uniform distribution over the interval $(0,1)$. Substituting these results back in (9.1), the tolerance interval satisfies

$$P(U_{(s)} - U_{(r)} \geq p) = \gamma \tag{9.2}$$

Since the joint distribution of $U_{(r)}$ and $U_{(s)}$ was found in (3.9), the required result for all $0 < p < 1$, $r < s$, is

$$\gamma = \int_p^1 \int_0^{y-p} \frac{n!}{(r-1)!(s-r-1)!(n-s)!} x^{r-1}(y-x)^{s-r-1}(1-y)^{n-s} \, dx \, dy$$

This double integral may be difficult to evaluate for any r, s, and n. Thus, let us instead find the probability distribution of the difference of the two order statistics, and then only a single integral will be required for (9.1). We shall use the method of Jacobians for the transformation

$$U = U_{(s)} - U_{(r)} \qquad \text{and} \qquad V = U_{(s)}$$

The joint distribution is then

$$f_{U,V}(u,v) = \frac{n!}{(r-1)!(s-r-1)!(n-s)!}$$
$$(v-u)^{r-1}u^{s-r-1}(1-v)^{n-s} \qquad 0 < u < v < 1$$

and

$$f_U(u) = \frac{n!}{(r-1)!(s-r-1)!(n-s)!} u^{s-r-1}$$
$$\int_u^1 (v-u)^{r-1}(1-v)^{n-s}\,dv$$

Under the integral sign, we make the change of variable $v - u = t(1-u)$ and obtain

$$f_U(u) = \frac{n!}{(r-1)!(s-r-1)!(n-s)!} u^{s-r-1}(1-u)^{r-1}$$
$$\int_0^1 t^{r-1}[(1-u)-t(1-u)]^{n-s}(1-u)\,dt$$

$$= \frac{n!}{(r-1)!(s-r-1)!(n-s)!}$$
$$u^{s-r-1}(1-u)^{n-s+r}B(r, n-s+1)$$

$$= \frac{n!}{(s-r-1)!(n-s+r)!} u^{s-r-1}(1-u)^{n-s+r}$$
$$0 < u < 1 \quad (9.3)$$

which is another beta distribution. The required result in (9.2) can now be written

$$\gamma = P(U \geq p) = \int_p^1 n\binom{n-1}{s-r-1} u^{s-r-1}(1-u)^{n-s+r}\,du$$

from which the tolerance coefficient is easily found for a given r and s.

COVERAGES

The difference $U_{(s)} - U_{(r)} = F_X(X_{(s)}) - F_X(X_{(r)})$ is called the *coverage* of the random interval $(X_{(r)}, X_{(s)})$ or sometimes simply an $s - r$ *cover*. Coverages are generally important in nonparametric statistics because of their distribution-free character. We define a set of elementary coverages as the differences

$$C_1 = F_X(X_{(1)}) = U_{(1)}$$
$$C_2 = F_X(X_{(2)}) - F_X(X_{(1)}) = U_{(2)} - U_{(1)}$$
$$C_3 = F_X(X_{(3)}) - F_X(X_{(2)}) = U_{(3)} - U_{(2)}$$
$$\cdots\cdots\cdots\cdots\cdots\cdots\cdots\cdots\cdots\cdots \qquad (9.4)$$
$$C_n = F_X(X_{(n)}) - F_X(X_{(n-1)}) = U_{(n)} - U_{(n-1)}$$
$$C_{n+1} = 1 - F_X(X_{(n)}) = 1 - U_{(n)}$$

or, generally,

$$C_i = F_X(X_{(i)}) - F_X(X_{(i-1)}) = U_{(i)} - U_{(i-1)}$$

for $i = 1, 2, \ldots, n + 1$, where $X_{(0)} = -\infty$, $X_{(n+1)} = \infty$, and $U_{(0)} = 0$, $U_{(n+1)} = 1$. Although the distribution of $X_{(i)}$ depends on F_X, the distribution of C_i does not, since it is the difference of two uniform order statistics whose joint distribution is

$$f_{U_{(i-1)}, U_{(i)}}(x,y) = \frac{n!}{(i-2)!(n-i)!} \, x^{i-2}(1-y)^{n-i} \qquad 0 < x < y < 1$$

Using (9.3) with $r = i - 1$, $s = i$, the marginal distribution of C_i is

$$f_{C_i}(u) = n(1-u)^{n-1} \qquad 0 < u < 1 \tag{9.5}$$

and

$$E(C_i) = \int_0^1 nu(1-u)^{n-1}\, du = nB(2,n) = \frac{1}{n+1}$$

From this result, we can draw the interpretation that the n order statistics partition the area under the probability density function into $n + 1$ parts, each of which has the same expected area or, equivalently, the same expected proportion of the probability.

Since the Jacobian of the transformation defined in (9.4) mapping $U_{(1)}, U_{(2)}, \ldots, U_{(n)}$ into $C_1, C_2, \ldots, C_n$ is equal to 1, the joint distribution of the n coverages is

$$f_{C_1, C_2, \ldots, C_n}(c_1, c_2, \ldots, c_n) = n! \qquad \text{for } c_i \geq 0,$$

$$i = 1, 2, \ldots, n, \ \sum_{i=1}^{n} c_i = 1$$

A sum of any r successive elementary coverages is called an r coverage. We have the sum $C_i + C_{i+1} + \cdots + C_{i+r} = U_{(i+r)} - U_{(i)}$, $i + r \leq n$. Since the distribution of $C_1, C_2, \ldots, C_n$ is symmetric in $c_1, c_2, \ldots, c_n$, the marginal distribution of the sum of any r of the coverages must be the same for each fixed value of r, in particular equal to that of

$$C_1 + C_2 + \cdots + C_r = U_{(r)}$$

which is given in (3.8). The expected value of an r coverage then is $r/(n + 1)$, with the same interpretation as before.

PROBLEMS

2.1. Prove the probability-integral transformation (Theorem 1.1) by finding the moment-generating function of the random variable $Y = F_X(X)$, where X has the continuous cumulative distribution F_X.

2.2. If X is a continuous random variable with probability density function $f_X(x) = 2(1 - x), 0 < x < 1$, find that transformation $Y = g(X)$ such that the random variable Y has the uniform distribution over $(0,2)$.

2.3. The order statistics for a random sample of size n from a discrete distribution are defined as in the continuous case except that now we have $X_{(1)} \leq X_{(2)} \leq \cdots \leq X_{(n)}$. Suppose a random sample of size 5 is taken with replacement from the discrete distribution $f_X(x) = \frac{1}{6}$ for $x = 1, 2, \ldots, 6$. Find the probability mass function of $X_{(1)}$, the smallest order statistic.

2.4. A random sample of size 3 is drawn from the population $f_X(x) = \exp[-(x-\theta)]$ for $x > \theta$. We wish to find a 95 percent confidence-interval estimate for the parameter θ. Since the maximum-likelihood estimate for θ is $X_{(1)}$, the smallest order statistic, a logical choice for the limits of the confidence interval would be some functions of $X_{(1)}$. If the upper limit is $X_{(1)}$, find the corresponding lower limit $g(X_{(1)})$ such that the confidence coefficient is 0.95.

2.5. For the n order statistics of a sample from the uniform distribution over $(0,\theta)$, show that the interval $(X_{(n)}, X_{(n)}/\alpha^{1/n})$ is a $100(1-\alpha)$ percent confidence-interval estimate of the parameter θ.

2.6. Ten points are chosen randomly and independently on the interval $(0,1)$.

    (a) Find the probability that the point nearest 1 exceeds 0.90.

    (b) Find the number c such that the probability is 0.5 that the point nearest zero will exceed c.

2.7. Find the expected value of the largest order statistic in a random sample of size 3 from:

    (a) The exponential distribution $f_X(x) = \exp(-x)$ for $x \geq 0$.

    (b) The standard normal distribution.

2.8. Verify the result given in (4.2) for the distribution of the median of a sample of size $2m$ from the uniform distribution over $(0,1)$ when $m = 2$. Show that this distribution is symmetric about $\frac{1}{2}$ by writing (4.2) in the form

$$f_U(u) = 8(\tfrac{1}{2} - |u - \tfrac{1}{2}|)^2(1 + 4|u - \tfrac{1}{2}|) \qquad 0 < u < 1$$

2.9. Find the mean and variance of the median of a random sample of n from the uniform distribution over $(0,1)$:

    (a) When n is odd.

    (b) When n is even and U is defined as in Sec. 4.

2.10. Find the probability that the range of a random sample of size n from the population $f_X(x) = 2e^{-2x}$ for $x \geq 0$ does not exceed 4.

2.11. Find the distribution of the range of a random sample of size n from the exponential distribution $f_X(x) = 4\exp(-4x)$ for $x \geq 0$.

2.12. Give an expression similar to (4.3) for the probability density function of the midrange for any continuous distribution and use it to find the density function in the case of a uniform population over $(0,1)$.

2.13. By making the transformation $U = nF_X(X_{(1)})$, $V = n[1 - F_X(X_{(n)})]$ in (3.7) with $r = 1, s = n$, for any continuous F_X, show that U and V are independent random variables in the limiting case as $n \to \infty$, so that the two extreme values of a random sample are asymptotically independent.

2.14. Use (6.5) and (6.6) to approximate the mean and variance of:

    (a) The median of a sample of size $2m + 1$ from a normal distribution with mean μ and variance σ^2.

    (b) The fifth order statistic of a random sample of size 19 from the exponential distribution $f_X(x) = \exp(-x)$ for $x \geq 0$.

2.15. Let $X_{(n)}$ be the largest value of a sample of size n from the population f_X.

    (a) Show that $\lim\limits_{n \to \infty} P(n^{-1}X_{(n)} \leq x) = \exp(-\alpha/\pi x)$ if $f_X(x) = \alpha/[\pi(\alpha^2 + x^2)]$

(Cauchy).

(b) Show that $\lim_{n \to \infty} P(n^{-2}X_{(n)} \leq x) = \exp(-\alpha\sqrt{2/\pi x})$ if $f_X(x) = (\alpha/\sqrt{2\pi})$ $x^{-3/2}\exp(-\alpha^2/2x)$ for $x \geq 0$.

2.16. Let $X_{(r)}$ be the rth order statistic of a random sample of size n from a continuous distribution F_X.

(a) Show that $P(X_{(r)} \leq t) = \sum_{k=r}^{n} \binom{n}{k} [F_X(t)]^k [1 - F_X(t)]^{n-k}$.

(b) Verify the probability density function of $X_{(r)}$ given in (3.5) by differentiation of the result in (a).

(c) By considering $P(X_{(r)} > t/n)$ in the form of (a), find the asymptotic distribution of $X_{(r)}$ for r fixed and $n \to \infty$ if $F_X(x)$ is the uniform distribution over (0,1).

2.17. Let $X_{(1)} < X_{(2)} < \cdots < X_{(n)}$ be order statistics for a random sample from the exponential distribution $f_X(x) = \exp(-x)$ for $x \geq 0$.

(a) Show that $X_{(r)}$ and $X_{(s)} - X_{(r)}$ are independent for any $s > r$.

(b) Find the distribution of $X_{(r+1)} - X_{(r)}$.

(c) Interpret the significance of these results if the sample arose from a life test on n light bulbs with exponential lifetimes.

2.18. Let $X_{(1)} < X_{(2)} < \cdots < X_{(n)}$ denote the order statistics of a sample from a continuous unspecified distribution F_X. Define the n random variables

$$V_i = \frac{F_X(X_{(i)})}{F_X(X_{(i+1)})} \quad \text{for } 1 \leq i \leq n-1 \quad \text{and} \quad V_n = F_X(X_{(n)})$$

(a) Find the marginal distribution of V_r, $1 \leq r \leq n$.

(b) Find the joint distribution of V_r and $F_X(X_{(r+1)})$, $1 \leq r \leq n-1$, and show they are independent.

(c) Find the joint distribution of $V_1, V_2, \ldots, V_n$.

(d) Show that $V_1, V_2, \ldots, V_n$ are independent.

(e) Show that $V_1, V_2^2, V_3^3, \ldots, V_n^n$ are independent and identically distributed with the uniform distribution over (0,1).

2.19. Let $X_{(r)}$ denote the rth order statistic of a random sample of size 5 from any continuous population and κ_p denote the pth quantile of this population. Find:

(a) $P(X_{(1)} < \kappa_{0.5} < X_{(5)})$

(b) $P(X_{(1)} < \kappa_{0.25} < X_{(3)})$

(c) $P(X_{(4)} < \kappa_{0.80} < X_{(5)})$

2.20. For order statistics of a random sample of size n from any continuous population F_X, show that the interval $(X_{(r)}, X_{(n-r+1)})$, $r < n/2$, is a $100(1-\alpha)$ percent confidence-interval estimate for the median of F_X, where

$$1 - \alpha = 1 - 2n\binom{n-1}{r-1}\int_0^{0.5} x^{n-r}(1-x)^{r-1}\,dx$$

2.21. If $X_{(1)}$ and $X_{(n)}$ are the smallest and largest values, respectively, of a sample of size n from any continuous population F_X with median $\kappa_{0.50}$, find the smallest value of n such that:

(a) $P(X_{(1)} < \kappa_{0.50} < X_{(n)}) \geq 0.99$

(b) $P[F_X(X_{(n)}) - F_X(X_{(1)}) \geq 0.5] \geq 0.95$

2.22. Let the random variable U denote the proportion of the population lying between the two extreme values of a sample of n from some unspecified continuous population. Find the mean and variance of U.

3
Tests Based on Runs

3.1 INTRODUCTION

Passing a line of ten persons waiting to buy a ticket at a movie theater on a Saturday afternoon, suppose we observe the arrangement of five males and five females in the line to be M, F, M, F, M, F, M, F, M, F. Would this be considered a random arrangement by sex? Intuitively, the answer is no, since the alternation of the two types of symbols suggests intentional mixing by pairs. This arrangement is an extreme case, as is the configuration M, M, M, M, M, F, F, F, F, F, with intentional clustering. In the less extreme situations, the randomness of an arrangement can be tested statistically using the theory of runs.

Given an ordered sequence of two or more types of symbols, a *run* is defined to be a succession of one or more identical symbols which are followed and preceded by a different symbol or no symbol at all. Clues to lack of randomness are provided by any tendency of the symbols to exhibit a definite pattern in the sequence. Both the number of runs and the lengths of the runs, which are of course interrelated, should reflect the existence of some sort of pattern. Tests for randomness can therefore be based on either criterion or some combination thereof. Too few runs, too many runs, a run of excessive length, or too many runs of excessive length, etc., can be used as statistical criteria for rejection of the null hypothesis of randomness, since these situations should occur rarely in a truly random sequence.

The alternative to randomness is often simply nonrandomness. In a test based on the total number of runs, both too few and too many runs suggest lack of randomness. A null hypothesis of randomness would consequently be rejected if the total number of runs is either too large or too small. However, the two situations may indicate different types of lack of randomness. In the movie theater example, a sequence with too many runs, tending toward the sexes alternating, might suggest that the movie is popular with teenagers and young adults, whereas the other extreme arrangement may result if the movie is more popular with younger children.

Tests of randomness are an important addition to statistical theory, because the theoretical bases for almost all the classical techniques, as well as distribution-free procedures, begin with the assumption of a random sample. If this assumption is valid, every arrangement of the observations is equally likely, so that the particular sequential order is of no consequence. However, if the randomness of the observations is suspect, the information about order, which is almost always available, can be used to test a hypothesis of randomness. Runs analysis has also been found particularly useful in time-series and quality-control studies.

The symbols studied for pattern may arise naturally, as with the theater example, or may be artificially imposed according to some dichotomizing criterion. Thus the runs tests are applicable to either qualitative or quantitative data. In the latter case, the dichotomy is usually effected by comparison of the magnitude of each number with a focal point, commonly the median or mean of the sample, and noting whether each observation exceeds or is exceeded by this value. When the data consist of numerical observations, another type of runs analysis can be applied to reach a conclusion about randomness. This technique, called *runs up and down*, uses more of the available information and is especially effective when the alternative to randomness is a trend.

3.2 TESTS BASED ON THE TOTAL NUMBER OF RUNS

Assume an ordered sequence of n elements of two types, n_1 of the first type and n_2 of the second type, where $n_1 + n_2 = n$. If r_1 is the number of runs of type 1 elements and r_2 is the number of runs of type 2, the total number of runs in the sequence is $r = r_1 + r_2$. In order to derive a test for randomness based on the random variable R, we need the probability distribution of R when the null hypothesis of randomness is true.

EXACT NULL DISTRIBUTION OF R

The distribution of R will be found by first determining the joint probability distribution of R_1 and R_2 and then the distribution of their sum. Since under the null hypothesis every arrangement of the $n_1 + n_2$ objects is equiprobable, the probability that $R_1 = r_1$ and $R_2 = r_2$ is the number of distinguishable arrangements of $n_1 + n_2$ objects with r_1 runs of type 1 and r_2 runs of type 2 objects divided by the total number of distinguishable arrangements, which is $n!/n_1!n_2!$. For the numerator quantity, the following counting lemma can be used.

Lemma 1 *The number of distinguishable ways of distributing n like objects into r distinguishable cells with no cell empty is* $\binom{n-1}{r-1}, n \geq r.$

Proof Suppose that the n-like objects are all white balls. Place these n balls in a row and effect the division into r cells by inserting each of $r - 1$ black balls between any two white balls in the line. Since there are $n - 1$ positions in which each black ball can be placed, the total number of arrangements is $\binom{n-1}{r-1}$.

In order to obtain a sequence with r_1 runs of the objects of type 1, the n_1-like objects must be placed into r_1 cells, which can be done in $\binom{n_1-1}{r_1-1}$ different ways. The same reasoning applies to obtain r_2 runs of the other n_2 objects. The total number of distinguishable arrangements starting with a run of type 1 then is the product $\binom{n_1-1}{r_1-1}\binom{n_2-1}{r_2-1}$. Similarly for a sequence starting with a run of type 2. The blocks of objects of type 1 and type 2 must alternate, and consequently either $r_1 = r_2 \pm 1$, or $r_1 = r_2$. If $r_1 = r_2 + 1$, the sequence must begin with a run of type 1; if $r_1 = r_2 - 1$, a type 2 run must come first. But if $r_1 = r_2$, the sequence can begin with a run of either type, so that the number of distinguishable arrangements must be doubled. We have thus proved the following result.

Theorem 2.1 *Let R_1 and R_2 denote the respective numbers of runs of n_1 objects of type 1 and n_2 objects of type 2 in a random sample of size $n = n_1 + n_2$. The joint probability distribution of R_1 and R_2 is*

$$f_{R_1,R_2}(r_1,r_2) = \frac{c\binom{n_1-1}{r_1-1}\binom{n_2-1}{r_2-1}}{\binom{n_1+n_2}{n_1}} \quad \begin{array}{l} r_1 = 1, 2, \ldots, n_1 \\ r_2 = 1, 2, \ldots, n_2 \\ r_1 = r_2 \text{ or} \\ r_1 = r_2 \pm 1 \end{array} \quad (2.1)$$

where $c = 2$ if $r_1 = r_2$ and $c = 1$ if $r_1 = r_2 \pm 1$.

Corollary 2.1 *The marginal probability distribution of R_1 is*

$$f_{R_1}(r_1) = \frac{\binom{n_1 - 1}{r_1 - 1}\binom{n_2 + 1}{r_1}}{\binom{n_1 + n_2}{n_1}} \qquad r_1 = 1, 2, \ldots, n_1 \qquad (2.2)$$

Similarly for R_2 with n_1 and n_2 interchanged.

Proof From (2.1), the only possible values of r_2 are $r_2 = r_1, r_1 - 1$, and $r_1 + 1$, for any r_1. Therefore we have

$$f_{R_1}(r_1) = \sum_{r_2} f_{R_1, R_2}(r_1, r_2)$$

$$\binom{n_1 + n_2}{n_1} f_{R_1}(r_1) = 2\binom{n_1 - 1}{r_1 - 1}\binom{n_2 - 1}{r_1 - 1} + \binom{n_1 - 1}{r_1 - 1}\binom{n_2 - 1}{r_1 - 2}$$

$$+ \binom{n_1 - 1}{r_1 - 1}\binom{n_2 - 1}{r_1}$$

$$= \binom{n_1 - 1}{r_1 - 1}\left[\binom{n_2 - 1}{r_1 - 1} + \binom{n_2 - 1}{r_1 - 2} + \binom{n_2 - 1}{r_1 - 1}\right.$$

$$\left. + \binom{n_2 - 1}{r_1}\right]$$

$$= \binom{n_1 - 1}{r_1 - 1}\left[\binom{n_2}{r_1 - 1} + \binom{n_2}{r_1}\right]$$

$$= \binom{n_1 - 1}{r_1 - 1}\binom{n_2 + 1}{r_1}$$

Theorem 2.2 *The probability distribution of R, the total number of runs of $n = n_1 + n_2$ objects, n_1 of type 1 and n_2 of type 2, in a random sample is*

$$f_R(r) = \begin{cases} \dfrac{2\binom{n_1 - 1}{r/2 - 1}\binom{n_2 - 1}{r/2 - 1}}{\binom{n_1 + n_2}{n_1}} & \text{if } r \text{ is even} \\[3em] \dfrac{\binom{n_1 - 1}{(r-1)/2}\binom{n_2 - 1}{(r-3)/2} + \binom{n_1 - 1}{(r-3)/2}\binom{n_2 - 1}{(r-1)/2}}{\binom{n_1 + n_2}{n_1}} & \\[2em] & \text{if } r \text{ is odd} \\ & \qquad\qquad (2.3) \end{cases}$$

for $r = 2, 3, \ldots, n_1 + n_2$.

Proof For r even, there must be the same number of runs of both types. Thus the only possible values of r_1 and r_2 are $r_1 = r_2 = r/2$, and (2.1) is summed over this pair. If $r_1 = r_2 \pm 1$, r is odd. In this case, (2.1) is summed over the two pairs of values $r_1 = (r - 1)/2$ and $r_2 = (r + 1)/2$, $r_1 = (r + 1)/2$ and $r_2 = (r - 1)/2$, obtaining the given result.†

Using the result of Theorem 2.2, tables can be prepared for tests of significance of the null hypothesis of randomness. For example, if $n_1 = 5$ and $n_2 = 4$, we have

$$f_R(9) = \frac{\binom{4}{4}\binom{3}{3}}{\binom{9}{4}} = \frac{1}{126} = 0.008$$

$$f_R(8) = \frac{2\binom{4}{3}\binom{3}{3}}{\binom{9}{4}} = \frac{8}{126} = 0.063$$

$$f_R(2) = \frac{2\binom{4}{0}\binom{3}{0}}{\binom{9}{4}} = \frac{2}{126} = 0.016$$

$$f_R(3) = \frac{\binom{4}{1}\binom{3}{0} + \binom{4}{0}\binom{3}{1}}{\binom{9}{4}} = \frac{7}{126} = 0.056$$

For a two-sided test which rejects the null hypothesis for $R \leq 2$ or $R \geq 9$, the exact significance level α would be $3/126 = 0.024$. For the critical region defined by $R \leq 3$ or $R \geq 8$, $\alpha = 18/126 = 0.143$. As for all tests based on discrete probability distributions, there are a finite number of possible values of α. For a test with significance level at most 0.05, say, the first critical region above would be used even though the actual value of α is only 0.024. Tables which can be used to find rejection regions for the runs test are available in many sources. Swed and Eisenhart (1943) give the probability distribution of R for $n_1 \leq n_2 \leq 20$. Some of the books reproducing parts of these tables are Hoel (1962), Siegel (1956), Tate and Clelland (1957), Owen (1962), and Bradley (1968).

† The binomial coefficient $\binom{a}{b}$ is defined to be zero if $a < b$.

MOMENTS OF THE NULL DISTRIBUTION OF R

The kth moment of R is

$$E(R^k) = \sum_r r^k f_R(r)$$

$$= \sum_{r \text{ even}} 2r^k \binom{n_1 - 1}{r/2 - 1}\binom{n_2 - 1}{r/2 - 1}$$

$$+ \sum_{r \text{ odd}} r^k \left[\binom{n_1 - 1}{(r-1)/2}\binom{n_2 - 1}{(r-3)/2} \right.$$

$$\left. \frac{+ \binom{n_1 - 1}{(r-3)/2}\binom{n_2 - 1}{(r-1)/2} \right]}{\binom{n_1 + n_2}{n_1}}$$

$$\tag{2.4}$$

The smallest value of r is always 2. If $n_1 = n_2$, the largest number of runs occurs when the symbols alternate, in which case $r = 2n_1$. If $n_1 < n_2$, the maximum value of r is $2n_1 + 1$, since the sequence can both begin and end with a type 2 symbol. Assuming without loss of generality that $n_1 \leq n_2$, the range of summation for r is $2 \leq r \leq 2n_1 + 1$. Letting $r = 2i$ for r even and $r = 2i + 1$ for r odd, the range of i is $1 \leq i \leq n_1$. For example, the mean of R is expressed as follows using (2.4):

$$\binom{n_1 + n_2}{n_1} E(R) = \sum_{i=1}^{n_1} 4i \binom{n_1 - 1}{i - 1}\binom{n_2 - 1}{i - 1}$$

$$+ \sum_{i=1}^{n_1} (2i + 1) \binom{n_1 - 1}{i}\binom{n_2 - 1}{i - 1}$$

$$+ \sum_{i=1}^{n_1} (2i + 1) \binom{n_1 - 1}{i - 1}\binom{n_2 - 1}{i} \tag{2.5}$$

To evaluate these three sums, the following lemmas are useful.

Lemma 2 $\displaystyle\sum_{r=0}^{c} \binom{m}{r}\binom{n}{r} = \binom{m + n}{m}$ $\qquad$ *where* $c = \min(m,n)$

Proof $\quad (1 + x)^{m+n} = (1 + x)^m (1 + x)^n \qquad$ for all x

$$\sum_{i=0}^{m+n} \binom{m + n}{i} x^i = \sum_{j=0}^{m} \binom{m}{j} x^j \sum_{k=0}^{n} \binom{n}{k} x^k$$

Assuming without loss of generality that $c = m$ and equating the coefficients of x^m on both sides of this equation, we obtain

$$\binom{m+n}{m} = \sum_{r=0}^{m} \binom{m}{m-r}\binom{n}{r}$$

Lemma 3 $\quad \sum_{r=0}^{c} \binom{m}{r}\binom{n}{r+1} = \binom{m+n}{m+1} \quad$ *where $c = \min(m, n-1)$*

Proof The proof follows as in Lemma 2, equating coefficients of x^{m+1}.

The algebraic process of obtaining $E(R)$ from (2.5) is tedious and will be left as an exercise for the reader. The variance of R can be found by evaluating the factorial moment $E[R(R-1)]$ in a similar manner.

A much simpler approach to finding the moments of R is provided by considering R as the sum of indicator variables as follows for $n = n_1 + n_2$. Let

$$R = 1 + I_2 + I_3 + \cdots + I_n$$

where in an ordered sequence of the two types of symbols, we define

$$I_k = \begin{cases} 1 & \text{if the } k\text{th element} \neq \text{the } (k-1)\text{st element} \\ 0 & \text{otherwise} \end{cases}$$

Then I_k is a Bernoulli random variable with parameter $p = n_1 n_2 / \binom{n}{2}$, so that

$$E(I_k) = E(I_k^2) = \frac{2n_1 n_2}{n(n-1)}$$

Since R is a linear combination of these I_k, we have

$$E(R) = 1 + \sum_{k=2}^{n} E(I_k) = 1 + \frac{2n_1 n_2}{n_1 + n_2} \tag{2.6}$$

$$\text{var}(R) = \text{var}\left(\sum_{k=2}^{n} I_k\right) = (n-1)\,\text{var}(I_k) + \sum_{2 \le j \neq k \le n}\sum \text{cov}(I_j, I_k)$$

$$= (n-1)E(I_k^2) + \sum_{2 \le j \neq k \le n}\sum E(I_j I_k) - (n-1)^2[E(I_k)]^2$$

$$\tag{2.7}$$

To evaluate the $(n-1)(n-2)$ joint moments of the type $E(I_j I_k)$ for $j \neq k$, the subscript choices can be classified as follows:

1. For the $2(n-2)$ selections where $j = k - 1$ or $j = k + 1$,

$$E(I_j I_k) = \frac{n_1 n_2 (n_1 - 1) + n_2 n_1 (n_2 - 1)}{n(n-1)(n-2)} = \frac{n_1 n_2}{n(n-1)}$$

2. For the remaining $(n-1)(n-2) - 2(n-2) = (n-2)(n-3)$ selections of $j \neq k$,

$$E(I_j I_k) = \frac{4 n_1 n_2 (n_1 - 1)(n_2 - 1)}{n(n-1)(n-2)(n-3)}$$

Substitution of these moments in the appropriate parts of (2.7) gives

$$\mathrm{var}(R) = \frac{2 n_1 n_2}{n} + \frac{2(n-2) n_1 n_2}{n(n-1)}$$

$$+ \frac{4 n_1 n_2 (n_1 - 1)(n_2 - 1)}{n(n-1)} - \frac{4 n_1^2 n_2^2}{n^2}$$

$$= \frac{2 n_1 n_2 (2 n_1 n_2 - n_1 - n_2)}{(n_1 + n_2)^2 (n_1 + n_2 - 1)} \tag{2.8}$$

ASYMPTOTIC NULL DISTRIBUTION

Although Eq. (2.3) can be used to find the exact distribution of R for any values of n_1 and n_2, the calculations are laborious unless n_1 and n_2 are both small. For large samples an approximation to the null distribution can be used, however, which gives reasonably good results as long as n_1 and n_2 are both larger than 10.

In order to find the asymptotic distribution, we assume that the total sample size n tends to infinity in such a way that $\lambda = n_1/n$ and $1 - \lambda = n_2/n$ remain constant. For large samples then, the mean and variance of R from (2.6) and (2.8) are

$$\lim_{n \to \infty} E\left(\frac{R}{n}\right) = 2\lambda(1 - \lambda) \qquad \lim_{n \to \infty} \mathrm{var}\left(\frac{R}{n^{1/2}}\right) = 4\lambda^2 (1 - \lambda)^2$$

Forming the standardized random variable

$$Z = \frac{R - 2n\lambda(1 - \lambda)}{2 n^{1/2} \lambda(1 - \lambda)} \tag{2.9}$$

and substituting for r in terms of z in Eq. (2.3), we obtain the standardized probability distribution of R, or $f_Z(z)$. If the factorials in the resulting expression are evaluated by Stirling's formula, the limit [Wald and Wolfowitz (1940)] is

$$\lim_{n \to \infty} \log f_Z(z) = -\log \sqrt{2\pi} - \tfrac{1}{2}z^2$$

which shows that the limiting probability function of Z is the standard normal density.

For a two-sided test of size α using the normal approximation, the null hypothesis of randomness would be rejected when

$$\left| \frac{R - 2n\lambda(1 - \lambda)}{2n^{1/2}\lambda(1 - \lambda)} \right| \geq z_{\alpha/2} \tag{2.10}$$

where z_γ is that number which satisfies $\Phi(z_\gamma) = 1 - \gamma$ or, equivalently, z_γ is the $(1 - \gamma)$th quantile point of the standard normal probability distribution. The exact mean and variance of R given in (2.6) and (2.8) can also be used in forming the standardized random variable, as the asymptotic distribution is unchanged. These approximations are generally improved by using a continuity correction of 0.5, as explained in Chap. 1.

DISCUSSION

This runs test is one of the best known and easiest to apply of the tests for randomness in a sequence of observations. The data may be dichotomous as collected, or if actual measurements are collected, the data may be classified into a dichotomous sequence according as each observation is above or below some fixed number, often the calculated sample median or mean. In this latter case, any observations equal to this fixed number are ignored in the analysis and n_1, n_2, and n reduced accordingly. The runs test can be used with either one- or two-sided alternatives. If the alternative hypothesis is simply nonrandomness, a two-sided test should be used. Since the presence of a trend would usually be indicated by a clustering of like objects, which is reflected by an unusually small number of runs, a one-sided test is more appropriate for trend alternatives.

Because of the generality of alternatives to randomness, no statement can be made concerning the overall performance of this runs test. However, its versatility should not be underrated. Other tests for randomness have been proposed which are especially useful for trend alternatives. The best known of these are tests based on the length of the longest run and tests based on the runs-up-and-down theory. These two types of test criteria will be discussed in the following sections.

3.3 TESTS BASED ON THE LENGTH OF THE LONGEST RUN

A test based on the total number of runs in an ordered sequence of n_1 objects of type 1 and n_2 objects of type 2 is only one way of using information about runs to detect patterns in the arrangement. Other statistics of interest are provided by considering the lengths of these runs. Since a run which is unusually long reflects a tendency for like objects to cluster and therefore possibly a trend, Mosteller (1941) has suggested a test for randomness based on the length of the longest run. Exact and asymptotic probability distributions of the numbers of runs of given length are discussed in Mood (1940).

The joint probability distribution of R_1 and R_2 derived in the previous section, for the numbers of runs of the two types of objects, disregards the individual lengths and is therefore of no use in this problem. We now need the joint probability distribution for a total of $n_1 + n_2$ random variables, representing the numbers of runs of all possible lengths for each of the two types of elements in the dichotomy. Let the random variables R_{ij}, $i = 1, 2; j = 1, 2, \ldots, n_i$, denote, respectively, the numbers of runs of objects of type i which are of length j. Then the following obvious relationships hold:

$$\sum_{j=1}^{n_i} jr_{ij} = n_i \quad \text{and} \quad \sum_{j=1}^{n_i} r_{ij} = r_i \quad \text{for } i = 1, 2$$

The total number of arrangements of the $n_1 + n_2$ symbols is still $\binom{n_1 + n_2}{n_1}$, and each is equally likely under the null hypothesis of randomness. We must compute the number of arrangements in which there are exactly r_{ij} runs of type i and length j, for all i and j. Assume that r_1 and r_2 are held fixed. The number of arrangements of the r_1 runs of type 1 which are composed of r_{1j} runs of length j for $j = 1, 2, \ldots, n_1$ is simply the number of permutations of the r_1 runs with r_{11} runs of length 1, r_{12} runs of length 2, $\ldots$, r_{1n_1} of length n_1, where within each length category the runs cannot be distinguished. This number is $r_1! \Big/ \prod_{j=1}^{n_1} r_{1j}!$. The number of arrangements for the r_2 runs of type 2 objects is similarly $r_2! \Big/ \prod_{j=1}^{n_2} r_{2j}!$. If $r_1 = r_2 \pm 1$, the total number of permutations of the runs of both types of objects is the product of these two expressions, but if $r_1 = r_2$, since the sequence can begin with either type of object, this number must be doubled. Therefore the following theorem is proved.

Theorem 3.1 *Under the null hypothesis of randomness, the probability that the r_1 runs of n_1 objects of type 1 and r_2 runs of n_2 objects of type 2*

consist of exactly $r_{1j}, j = 1, 2, \ldots, n_1,$ *and* $r_{2j}, j = 1, 2, \ldots, n_2,$
runs of length j, respectively, is

$$f(r_{11}, \ldots, r_{1n_1}, r_{21}, \ldots, r_{2n_2}) = \frac{cr_1! r_2!}{\prod\limits_{i=1}^{2} \prod\limits_{j=1}^{n_i} r_{ij}! \binom{n_1 + n_2}{n_1}} \tag{3.1}$$

where $c = 2$ *if* $r_1 = r_2$ *and* $c = 1$ *if* $r_1 = r_2 \pm 1.$

Combining the reasoning for Theorem 2.1 with that of Theorem 3.1, the joint distribution of the $n_1 + 1$ random variables R_2 and $R_{1j}, j = 1,$ $2, \ldots, n_1,$ can be obtained as follows:

$$f(r_{11}, \ldots, r_{1n_1}, r_2) = \frac{cr_1! \binom{n_2 - 1}{r_2 - 1}}{\prod\limits_{j=1}^{n_1} r_{1j}! \binom{n_1 + n_2}{n_1}} \tag{3.2}$$

The result is useful when only the total number, not the lengths, of runs of type 2 objects is of interest. This joint distribution, when summed over the values for r_2, gives the marginal probability distribution of the lengths of the r_1 runs of objects of type 1.

Theorem 3.2 *The probability that the* r_1 *runs of* n_1 *objects of type 1 consist of exactly* $r_{1j}, j = 1, 2, \ldots, n_1,$ *runs of length j, respectively, is*

$$f(r_{11}, \ldots, r_{1n_1}) = \frac{r_1! \binom{n_2 + 1}{r_1}}{\prod\limits_{j=1}^{n_1} r_{1j}! \binom{n_1 + n_2}{n_1}} \tag{3.3}$$

The proof is exactly the same as for the corollary to Theorem 2.1. The distribution of lengths of runs of type 2 objects is similar.

In the probability distributions given in Theorems 3.1 and 3.2, both r_1 and r_2, and r_1, respectively, were assumed to be fixed numbers. In other words, these are conditional probability distributions given the fixed values. If these are not to be considered as fixed, the conditional distributions are simply summed over the possible fixed values, since these are mutually exclusive.

Theorem 3.2 can be used to find the null probability distribution of a test for randomness based on K, the length of the longest run of type 1. For example, the probability that the longest in any number of runs of type 1 is of length k is

$$\sum_{r_1} \sum_{r_{11}, \ldots, r_{1k}} \frac{r_1! \binom{n_2 + 1}{r_1}}{\prod\limits_{j=1}^{k} r_{1j}! \binom{n_1 + n_2}{n_1}} \qquad (3.4)$$

where the sums are extended over all sets of nonnegative integers satisfying $\sum\limits_{j=1}^{k} r_{1j} = r_1$, $\sum\limits_{j=1}^{k} jr_{1j} = n_1$, $r_{1k} \geq 1$, $r_1 \leq n_1 - k + 1$, and $r_1 \leq n_2 + 1$.

For example, if $n_1 = 5$, $n_2 = 6$, the longest possible run is of length 5. There can be no other runs, so that $r_1 = 1$ and $r_{11} = r_{12} = r_{13} = r_{14} = 0$, and

$$P(K = 5) = \frac{\binom{7}{1}}{\binom{11}{5}} = \frac{7}{462}$$

Similarly, we can obtain

$$P(K = 4) = \frac{2!}{1!1!} \frac{\binom{7}{2}}{\binom{11}{5}} = \frac{42}{462}$$

$$P(K = 3) = \frac{\frac{3!}{2!1!} \binom{7}{3} + \frac{2!}{1!1!} \binom{7}{2}}{\binom{11}{5}} = \frac{147}{462}$$

$$P(K = 2) = \frac{\frac{4!}{3!1!} \binom{7}{4} + \frac{3!}{1!2!} \binom{7}{3}}{\binom{11}{5}} = \frac{245}{462}$$

$$P(K = 1) = \frac{5!}{5!} \frac{\binom{7}{5}}{\binom{11}{5}} = \frac{21}{462}$$

For a significance level of at most 0.05 when $n_1 = 5$, $n_2 = 6$, the null hypothesis of randomness is rejected when there is a run of type 1 elements of length 5. In general, the critical region would be those arrangements with at least one run of length t or more.

Theorem 3.1 must be used if the test is to be based on the length of the longest run of either type of element in the dichotomy. These two theorems are tedious to apply unless n_1 and n_2 are both quite small. Tables are available in Bateman (1948) and Mosteller (1941).

Tests based on the length of the longest run may or may not be more powerful than a test based on the total number of runs, depending on the basis for comparison. Both tests use only a portion of the information available, since the total number of runs, although affected by the lengths of the runs, does not directly make use of information regarding these lengths, and the length of the longest run only partially reflects both the lengths of other runs and the total number of runs. Power functions are discussed in Bateman (1948) and David (1947).

3.4 RUNS UP AND DOWN

When numerical observations are available and the sequence of numbers is analyzed for randomness according to the number or lengths of runs of elements above and below the median, some information is lost which might be useful in identifying a pattern in the time-ordered observations. With runs up and down, instead of using a single focal point for the entire series, the magnitude of each element is compared with that of the immediately preceding element in the time sequence. If the next element is larger, a run up is started; if smaller, a run down. We can observe when the sequence increases, and for how long, when it decreases, and for how long. A decision concerning randomness then might be based on the number and lengths of these runs, whether up or down, since a large number of long runs should not occur in a truly random set of numerical observations. Since an excessive number of long runs is usually indicative of some sort of trend exhibited by the sequence, this type of analysis should be most sensitive to trend alternatives.

If the time-ordered observations were 8, 13, 1, 3, 4, 7, there is a run up of length 1, followed by a run down of length 1, followed by a run up of length 3. The sequence of six observations can be represented by the five plus and minus signs, $+$, $-$, $+$, $+$, $+$, indicating their relative magnitudes. More generally, suppose there are n numbers, no two alike, say $a_1 < a_2 < \cdots < a_n$ when arranged in order of magnitude. The time-ordered sequence of observations $S_n = (x_1, x_2, \ldots, x_n)$ represents some permutation of these n numbers. There are $n!$ permutations, each representing a possible set of sample observations. Under the null hypothesis of randomness, each of these $n!$ arrangements is equally likely to occur. The test for randomness using runs up and down for the sequence S_n of dimension n is based on the derived sequence D_{n-1} of dimension $n - 1$, whose ith element is the sign of the difference $x_{i+1} - x_i$, for $i = 1, 2, \ldots, n - 1$. Let R_i denote the number of runs, either up or down, of length exactly i in the sequence D_{n-1} or S_n. We have the obvious restrictions $1 \leq i \leq n - 1$, and $\sum_{i=1}^{n-1} ir_i = n - 1$. The test for randomness will reject the null hypothesis when there are at least r runs of length t or more, where r and t are determined by the desired significance level. Therefore we must find the joint distribution of R_1, R_2, $\ldots$, R_{n-1} under the null hypothesis when every arrangement S_n is equally likely. Let $f_n(r_{n-1}, r_{n-2}, \ldots, r_1)$ denote the probability that there are exactly r_{n-1} runs of length $n - 1$, $\ldots$, r_i runs of length i, $\ldots$, r_1 runs of length 1. If $u_n(r_{n-1}, \ldots, r_1)$ represents the corresponding frequency, then

$$f_n = \frac{u_n}{n!}$$

since there are $n!$ possible arrangements of S_n. Since the probability distribution will be derived as a recursive relation, let us first consider the particular case where $n = 3$ and see how the distribution for $n = 4$ can be generated from it.

Given three numbers $a_1 < a_2 < a_3$, only runs of lengths 1 and 2 are possible. The $3! = 6$ arrangements and their corresponding values of r_2 and r_1 are given in Table 4.1. Since the probability of at least one run of length 2 or more is 2/6, if this defines the critical region the significance level is 0.33. With this size sample, a smaller significance level cannot be obtained without resorting to a randomized decision rule.

Now consider the addition of a fourth number a_4, larger than all the others. For each of the arrangements in S_3, a_4 can be inserted in four different places. In the particular arrangement (a_1, a_2, a_3), for example, insertion of a_4 at the extreme left or between a_2 and a_3 would leave r_2 unchanged but add a run of length 1. If a_4 is placed at the extreme right, a run of length 2 is increased to a run of length 3. If a_4 is inserted between a_1 and a_2, the one run of length 2 is split into three runs, each of length 1.

Extending this analysis to the general case, the extra observation must either split an existing run, lengthen an existing run, or introduce

Table 4.1

S_3	D_2	r_2	r_1	Probability distribution
(a_1, a_2, a_3)	$(+, +)$	1	0	$f_3(1,0) = 2/6$
(a_1, a_3, a_2)	$(+, -)$	0	2	
(a_2, a_1, a_3)	$(-, +)$	0	2	$f_3(0,2) = 4/6$
(a_2, a_3, a_1)	$(+, -)$	0	2	
(a_3, a_1, a_2)	$(-, +)$	0	2	$f_3(r_2, r_1) = 0$ otherwise
(a_3, a_2, a_1)	$(-, -)$	1	0	

Table 4.2

S_3	r_2	r_1	S_4	r_3	r_2	r_1	Case illustrated
(a_1,a_2,a_3)	1	0	(a_4,a_1,a_2,a_3)	0	1	1	1
			(a_1,a_4,a_2,a_3)	0	0	3	3
			(a_1,a_2,a_4,a_3)	0	1	1	1
			(a_1,a_2,a_3,a_4)	1	0	0	2
(a_1,a_3,a_2)	0	2	(a_4,a_1,a_3,a_2)	0	0	3	1
			(a_1,a_4,a_3,a_2)	0	1	1	2
			(a_1,a_3,a_4,a_2)	0	1	1	2
			(a_1,a_3,a_2,a_4)	0	0	3	1

a new run of length 1. The ways in which the run lengths in S_{n-1} are affected by the insertion of an additional observation a_n to make an arrangement S_n can be classified into the following four mutually exclusive and exhaustive cases:

1. An additional run of length 1 can be added in the arrangement S_n.
2. A run of length $i - 1$ in S_{n-1} can be changed into a run of length i in S_n for $i = 2, 3, \ldots, n - 1$.
3. A run of length $h = 2i$ in S_{n-1} can be split into a run of length i, followed by a run of length 1, followed by a run of length i, for $1 \leq i \leq [(n - 2)/2]$, where $[x]$ denotes the largest integer not exceeding x.
4. A run of length $h = i + j$ in S_{n-1} can be split up into either
   a. A run of length i, followed by a run of length 1, followed by a run of length j, or
   b. A run of length j, followed by a run of length 1, followed by a run of length i,
   where $h > i > j$, $3 \leq h \leq n - 2$.

For $n = 4$, the arrangements can be enumerated systematically in a table like Table 4.2 to show how these cases arise. Table 4.2 gives a partial listing. When the table is completed, the number of cases which result in any particular set (r_3,r_2,r_1) can then be counted and divided by 24 to obtain the complete probability distribution. This will be left as an exercise for the reader. The results are:

(r_3,r_2,r_1)	$(1,0,0)$	$(0,1,1)$	$(0,0,3)$	Other values
$f_4(r_3,r_2,r_1)$	2/24	12/24	10/24	0

There is no illustration for case 4 in the completed table of enumerated arrangements since here n is not large enough to permit $h \geq 3$. For $n = 5$, insertion of a_5 in the second position of (a_1,a_2,a_3,a_4) produces the sequence (a_1,a_5,a_2,a_3,a_4). The one run of length 3 has been split into a run of length 1, followed by another run of length 1, followed by a run of length 2, illustrating case 4b with $h = 3$, $j = 1$, $i = 2$. Similarly, case 4a is illustrated by inserting a_5 in the third position.

More generally, the frequency u_n of cases in S_n having exactly r_1 runs of length 1, r_2 runs of length 2, . . . , r_{n-1} runs of length $n - 1$, can be generated from the frequencies for S_{n-1} by the following recursive relation:

$$
\begin{aligned}
u_n(r_{n-1}&,r_{n-2}, \; . \; . \; . \;,r_h, \; . \; . \; . \;,r_i, \; . \; . \; . r_j, \; . \; . \; . \;,r_1) \\
&= 2u_{n-1}(r_{n-2}, \; . \; . \; . \;,r_1 - 1) \\
&\quad + \sum_{i=2}^{n-1} (r_{i-1} + 1)u_{n-1}(r_{n-2}, \; . \; . \; . \;,r_i - 1,r_{i-1} + 1, \; . \; . \; . \;,r_1) \\
&\quad + \sum_{\substack{i=1 \\ (h=2i)}}^{[(n-2)/2]} (r_h + 1)u_{n-1} \\
&\qquad (r_{n-2}, \; . \; . \; . \;,r_h + 1, \; . \; . \; . \;,r_i - 2, \; . \; . \; . \;,r_1 - 1) \\
&\quad + 2 \sum_{\substack{i=2 \\ (h=i+j) \\ h \leq n-2}}^{n-3} \sum_{j=1}^{i-1} (r_h + 1)u_{n-1} \\
&\qquad (r_{n-2}, \; . \; . \; . \;,r_h + 1, \; . \; . \; . \;,r_i - 1, \; . \; . \; . \;,r_j - 1, \; . \; . \; . \;,r_1 - 1)
\end{aligned}
\tag{4.1}
$$

The terms in this sum represent cases 1 to 4 in that order. For case 1, u_{n-1} is multiplied by 2 since for every arrangement in S_{n-1} there are exactly two places in which a_n can be inserted to add a run of length 1. These positions are always at an end or next to the end. If the first run is a run up (down), insertion of a_n at the extreme left (next to extreme left) position adds a run of length 1. A new run of length 1 is also created by inserting a_n at the extreme right (next to extreme right) in S_{n-1} if the last run in S_{n-1} was a run down (up). In case 4, we multiply by 2 because of the (a) and (b) possibilities.

The result is tricky and tedious to use but is much easier than enumeration. The process will be illustrated for $n = 5$, given

$$
u_4(1,0,0) = 2 \qquad u_4(0,1,1) = 12 \qquad u_4(0,0,3) = 10
$$
$$
u_4(r_3,r_2,r_1) = 0 \text{ otherwise}
$$

Using (4.1), we have

$$u_5(r_4,r_3,r_2,r_1) = 2u_4(r_3, r_2, r_1 - 1) + [(r_1 + 1)u_4(r_3, r_2 - 1, r_1 + 1)$$
$$+ (r_2 + 1)u_4(r_3 - 1, r_2 + 1, r_1) + (r_3 + 1)u_4(r_3 + 1, r_2, r_1)]$$
$$+ (r_2 + 1)u_4(r_3, r_2 + 1, r_1 - 3)$$
$$+ 2(r_3 + 1)u_4(r_3 + 1, r_2 - 1, r_1 - 2)$$

$$u_5(1,0,0,0) = 1u_4(1,0,0) = 2$$
$$u_5(0,1,0,1) = 2u_4(1,0,0) + 1u_4(0,1,1) + 2u_4(2,0,1)$$
$$= 4 + 12 + 0 = 16$$
$$u_5(0,0,2,0) = 1u_4(0,1,1) + 1u_4(1,2,0) = 12 + 0 = 12$$
$$u_5(0,0,1,2) = 2u_4(0,1,1) + 3u_4(0,0,3) + 1u_4(1,1,2) + 2u_4(1,0,0)$$
$$= 24 + 30 + 0 + 4 = 58$$
$$u_5(0,0,0,4) = 2u_4(0,0,3) + 1u_4(1,0,4) + 1u_4(0,1,1)$$
$$= 20 + 0 + 12 = 32$$

The means, variances, and covariances of the numbers of runs of length t (or more) are found in Levene and Wolfowitz (1944). Tables of the exact probabilities of at least r runs of length t or more are given in Olmstead (1946) and Owen (1962) for $n \leq 14$, from which appropriate critical regions can be found. Olmstead gives approximate probabilities for larger sample sizes. See Wolfowitz (1944a,b) regarding the asymptotic distribution which is Poisson.

A test for randomness can also be based on the total number of runs, whether up or down, irrespective of their lengths. Since the total number of runs R is related to the R_i by

$$R = \sum_{i=1}^{n-1} R_i$$

the same recursive relation given in (4.1) can be used to find the probability distribution of R. The hypothesis of randomness should be rejected for small values of R. The cumulative probability distribution of R for $n \leq 25$ is given in Edgington (1961) and Bradley (1968). Levene (1952) has shown that the asymptotic distribution of the standardized random variable

$$Z = \frac{R - (2n - 1)/3}{[(16n - 29)/90]^{\frac{1}{2}}} \tag{4.2}$$

is the standard normal distribution, so that tests of significance for large n are easily performed.

PROBLEMS

3.1. Prove Corollary 2.1 using a direct combinatorial argument based on Lemma 1.

3.2. Find the mean and variance of the number of runs R_1 of type 1 elements, using the probability distribution given in (2.2). Since $E(R) = E(R_1) + E(R_2)$, use your result to verify (2.6).

3.3. Use Lemmas 2 and 3 to evaluate the sums in (2.5), obtaining the result given in (2.6) for $E(R)$.

3.4. Show that the asymptotic distribution of the standardized random variable $[R_1 - E(R_1)]/\sigma(R_1)$ is the standard normal distribution, using the distribution of R_1 given in (2.2) and your answer to Prob. 3.2.

3.5. Verify that the asymptotic distribution of the random variable given in (2.9) is the standard normal distribution.

3.6. By considering the ratios $f_R(r)/f_R(r-2)$ and $f_R(r+2)/f_R(r)$, where r is an even positive integer and $f_R(r)$ is given in (2.3), show that if the most probable number of runs is an even integer k, then k satisfies the inequality

$$\frac{2n_1n_2}{n} < k < \frac{2n_1n_2}{n} + 2$$

3.7. Show that the probability that a sequence of n_1 elements of type 1 and n_2 elements of type 2 begins with a type 1 run of length exactly k is

$$\frac{(n_1)_k n_2}{(n_1 + n_2)_{k+1}} \qquad \text{where} \qquad (n)_r = \frac{n!}{(n-r)!}$$

3.8. Find the rejection region with significance level not exceeding 0.10 for a test of randomness based on the length of the longest run when $n_1 = n_2 = 6$.

3.9. Find the complete probability distribution of the number of runs up and down of various lengths when $n = 6$ using (4.1) and the results given for $u_5(r_4, r_3, r_2, r_1)$.

3.10. Use your answers to Prob. 3.9 to obtain the complete probability distribution of the total number of runs up and down when $n = 6$.

4

Tests of Goodness of Fit

4.1 INTRODUCTION

An important problem in statistics relates to obtaining information about the form of the population from which a sample is drawn. The shape of this distribution might be the focus of the investigation. Alternatively, some inference concerning a particular aspect of the population may be of primary interest. In this latter case, in classical statistics, information about the form generally must be postulated or incorporated in the null hypothesis to perform an exact parametric type of inference. For example, suppose we have a small number of observations from an unknown population with unknown variance and the hypothesis of interest concerns the value of the population mean. The traditional parametric test, based on Student's t distribution, is derived under the assumption of a normal population. The exact distribution theory and probabilities of both types of errors depend on this population form. Therefore it might be desirable to check on the reasonableness of the normal assumption before forming any conclusions based on the t distribution. If the

normality assumption appears not to be justified, some type of nonparametric inference for location might be more appropriate with a small
sample size.

The compatibility of a set of observed sample values with a normal
distribution or any other distribution can be checked by a goodness-of-fit
type of test.   These are tests designed for a null hypothesis which is a
statement about the form of the cumulative distribution function or
probability function of the parent population from which the sample is
drawn.   Ideally, the hypothesized distribution is completely specified,
including all parameters.   If the hypothesis states simply some family of
distributions, the unknown parameters must be estimated from the sample data in order to perform any test.   Since the alternative in either
case is necessarily quite broad, including differences only in location,
scale, other parameters, form, or any combination thereof, rejection of
the null hypothesis does not provide much specific information.   Goodness-of-fit tests are customarily used when only the form of the population is in question, with the hope that the null hypothesis will be found
acceptable.

In this chapter we shall consider two types of goodness-of-fit tests:
the chi-square test, proposed by Karl Pearson early in the history of
statistics, and the more recent Kolmogorov-Smirnov tests.

4.2 THE CHI-SQUARE GOODNESS-OF-FIT TEST

A single random sample of size n is drawn from a population with unknown
cumulative distribution function F_X.   We wish to test the null hypothesis

$H_0: F_X(x) = F_0(x)$ for all x

where $F_0(x)$ is completely specified, against the general alternative

$H_1: F_X(x) \neq F_0(x)$ for some x

In order to apply the chi-square test in this situation, the sample
data must first be grouped according to some scheme in order to form a
frequency distribution.   In the case of count or qualitative data, where
the hypothesized distribution would probably be discrete anyway, the
categories would arise naturally in terms of the relevant verbal or numerical classifications.   For example, in tossing a die, the categories would
be the numbers of spots; in tossing a coin, the categories would be the
numbers of heads; in surveys of brand preferences, the categories would
be the brand names considered.   When the sample observations are
quantitative, the categories would be numerical classes chosen by the
experimenter.   In this case, the frequency distribution is not unique, and
some information is necessarily lost.   Even though the hypothesized

distribution is most likely continuous with measurement data, the data must be categorized for analysis.

Assuming that the population distribution is completely specified by the null hypothesis, one can calculate the probability that a random observation will be classified into each of the chosen or fixed categories. These probabilities multiplied by n give the frequencies for each category which would be expected if the null hypothesis were true. Except for sampling variation, there should be close agreement between these expected and observed frequencies if the sample data are compatible with the specified $F_0(x)$. The corresponding observed and expected frequencies can be compared visually using a histogram, frequency polygon, or bar chart. The chi-square goodness-of-fit test provides a probability basis for effecting the comparison and deciding whether the lack of agreement is too great to have occurred by chance.

Assume that the n observations have been grouped into k mutually exclusive categories, and denote by f_i and e_i the observed and expected frequencies, respectively, for the ith group, $i = 1, 2, \ldots, k$. The decision regarding fit is to be based on the deviations $f_i - e_i$. The sum of these k deviations is zero except for rounding. The test criterion suggested by Pearson (1900) is the statistic

$$q = \sum_{i=1}^{k} \frac{(f_i - e_i)^2}{e_i} \tag{2.1}$$

A large value of q would reflect an incompatibility between the observed and expected relative frequencies, and therefore the null hypothesis on which the e_i were calculated should be rejected for q large.

The exact probability distribution of the random variable Q is quite complicated, but for large samples its distribution is approximately chi square with $k - 1$ degrees of freedom. The theoretical basis for this can be argued briefly as follows.

Once the sample is classified into the form of a frequency distribution, the only random variables of concern are the class frequencies $F_1, F_2, \ldots, F_k$. These constitute a set of random variables from the k-variate multinomial distribution with k possible outcomes, the ith outcome being the ith category in the classification system. With $\theta_1, \theta_2, \ldots, \theta_k$ denoting the probabilities of the respective outcomes, the likelihood function of the sample then is

$$L(\theta_1, \theta_2, \ldots, \theta_k) = \prod_{i=1}^{k} \theta_i{}^{f_i} \qquad f_i = 0, 1, \ldots, n;$$

$$\sum_{i=1}^{k} f_i = n; \ \sum_{i=1}^{k} \theta_i = 1 \tag{2.2}$$

The null hypothesis was assumed to specify the population distribution completely, from which the θ_i can be calculated. This hypothesis then is actually concerned only with the values of these parameters and can be equivalently stated as

$$H_0: \theta_i = \theta_i{}^0 = \frac{e_i}{n} \quad \text{for } i = 1, 2, \ldots, k$$

It is easily shown that the maximum-likelihood estimates of the parameters in (2.2) are $\hat{\theta}_i = f_i/n$. The likelihood-ratio statistic for this hypothesis then is

$$T = \frac{L(\hat{\omega})}{L(\hat{\Omega})} = \frac{L(\theta_1{}^0, \theta_2{}^0, \ldots, \theta_k{}^0)}{L(\hat{\theta}_1, \hat{\theta}_2, \ldots, \hat{\theta}_k)} = \prod_{i=1}^{k} \left(\frac{\theta_i{}^0}{\hat{\theta}_i}\right)^{f_i}$$

As was stated in Chap. 1, the distribution of the quantity $-2 \log T$ approximates the chi-square distribution. The degrees of freedom are $k - 1$, since the restriction $\sum_{i=1}^{k} \theta_i = 1$ leaves only $k - 1$ parameters in Ω to be estimated independently. We have here

$$-2 \log t = -2 \sum_{i=1}^{k} f_i \left(\log \theta_i{}^0 - \log \frac{f_i}{n}\right) \tag{2.3}$$

Some statisticians advocate using the expression in (2.3) as a test criterion for goodness of fit. We shall now show that it is asymptotically equivalent to the expression for q given in (2.1). The Taylor's series expansion of $\log \theta_i$ about $f_i/n = \hat{\theta}_i$ is

$$\log \theta_i = \log \hat{\theta}_i + (\theta_i - \hat{\theta}_i)\frac{1}{\hat{\theta}_i} + \frac{(\theta_i - \hat{\theta}_i)^2}{2!}\left(-\frac{1}{\hat{\theta}_i{}^2}\right) + \epsilon$$

or

$$\log \theta_i - \log \frac{f_i}{n} = \left(\theta_i - \frac{f_i}{n}\right)\frac{n}{f_i} - \left(\theta_i - \frac{f_i}{n}\right)^2 \frac{n^2}{2f_i{}^2} + \epsilon$$

$$= \frac{(n\theta_i - f_i)}{f_i} - \frac{(n\theta_i - f_i)^2}{2f_i{}^2} + \epsilon \tag{2.4}$$

where ϵ represents the sum of terms alternating in sign

$$\sum_{j=3}^{\infty} (-1)^{j+1} \left(\theta_i - \frac{f_i}{n}\right)^j \frac{n^j}{j! f_i{}^j}$$

Substituting (2.4) in (2.3), we have

$$-2 \log t = -2 \sum_{i=1}^{k} (n\theta_i{}^0 - f_i) + \sum_{i=1}^{k} \frac{(n\theta_i{}^0 - f_i)^2}{f_i} + \sum_{i=1}^{k} \epsilon'$$

$$= 0 + \sum_{i=1}^{k} \frac{(n\theta_i{}^0 - f_i)^2}{f_i} + \epsilon''$$

By the law of large numbers the random variable F_i/n is known to be a consistent estimator of θ_i, or

$$\lim_{n \to \infty} \left[\frac{1}{n} P(|F_i - n\theta_i| > \epsilon) \right] = 0 \qquad \text{for every } \epsilon > 0$$

Thus we see that $-2 \log T$ approaches the statistic Q defined before, and the probability distribution of Q therefore converges to that of $-2 \log T$, which is chi square with $k - 1$ degrees of freedom. An approximate α-level test then is obtained by rejecting H_0 when Q exceeds the $(1 - \alpha)$th quantile point of the chi-square distribution, denoted by $\chi^2_{k-1,\alpha}$. This approximation can be used with confidence as long as every expected frequency is at least equal to 5. For any e_i smaller than 5, the usual procedure is to combine adjacent groups in the frequency distribution until this restriction is satisfied. The number of degrees of freedom then must be reduced to correspond to the actual number of categories used in the analysis. This rule of 5 should not be considered inflexible, however. It is a conservative value, and the chi-square approximation is often reasonably accurate for expected cell frequencies as small as 1.5.

Any case where the θ_i are completely specified by the null hypothesis is thus easily handled. The more typical situation, however, is where the null hypothesis is composite. Usually the form of the distribution is stated but not all the relevant parameters. For example, when we wish to test whether a sample is drawn from some normal population, μ and σ would not be given. However, in order to calculate the expected frequencies under H_0, μ and σ must be known. If the expected frequencies are estimated from the data as $n_i\hat{\theta}_i{}^0$ for $i = 1, 2, \ldots, k$, the goodness-of-fit test statistic in (2.1) becomes

$$q = \sum_{i=1}^{k} \frac{(f_i - n_i\hat{\theta}_i{}^0)^2}{n_i\hat{\theta}_i{}^0} \tag{2.5}$$

The asymptotic distribution of Q then may depend on the method employed for estimation. When the estimates are found by the method of maximum likelihood for the *grouped* data, the $L(\hat{\omega})$ in the likelihood-ratio test statistic is $L(\hat{\theta}_1{}^0, \hat{\theta}_2{}^0, \ldots, \hat{\theta}_k{}^0)$, where the $\hat{\theta}_i{}^0$ are the MLEs of the θ_i^0 under H_0. The derivation of the distribution of T and therefore Q goes through exactly as before except that the dimension of the space ω is increased. The degrees of freedom for Q then are $k - 1 - s$, where s is the number of independent parameters in $F_0(x)$ which had to be estimated from the grouped data in order to estimate all the θ_i^0. In the normal goodness-of-fit test, for example, the μ and σ parameter estimates would be calculated from the grouped data and used with tables of the normal distribution to find the $n_i\hat{\theta}_i^0$, and the degrees of freedom for k categories would be $k - 3$. When the original data are ungrouped and the MLEs are based on the likelihood function of all the observations, the theory is different. Chernoff and Lehmann (1954) have shown that the limiting distribution of Q is not the chi square in this case and that $P(Q > \chi_\alpha^2) > \alpha$. The test is then anticonservative. Their investigation showed that the error is considerably more serious for the normal distribution than the Poisson. A possible adjustment is discussed in their paper. In practice, however, the statistic in (2.5) is often treated as a chi-square variable anyway.

4.3 THE EMPIRICAL DISTRIBUTION FUNCTION

In the chi-square goodness-of-fit test, the comparison between observed and expected class frequencies is made for a set of k groups. Only k comparisons are made even though there are n observations, where $k \leq n$. If the n sample observations are values of a random variable, as opposed to strictly categorical data, comparisons can be made between observed and expected cumulative relative frequencies for each of the different observed values. The cumulative distribution function of the sample, called the *empirical distribution function*, may be considered an estimate of the population cdf. Several goodness-of-fit test statistics are functions of the deviations between the observed cumulative distribution and the corresponding cumulative probabilities expected under the null hypothesis. The function of these deviations used to perform a test might be the sum of squares, or absolute values, or the maximum deviation, to name only a few. The best-known test is the Kolmogorov-Smirnov one-sample statistic, which will be covered in the next section. Before studying this particular test, we shall discuss some of the properties of this estimated cdf, which is a generally useful tool in nonparametric inference.

The empirical (sample) distribution function of a random sample of size n, denoted by $S_n(x)$ and defined for all real numbers x, is the proportion of sample values which do not exceed the number x. Thus $S_n(x)$ is a step function which increases by the amount $1/n$ at its jump points, which are the order statistics of the sample. Letting $X_{(1)}, X_{(2)}, \ldots ,$

$X_{(n)}$ denote the order statistics of a random sample, its empirical distribution function is defined symbolically as

$$S_n(x) = \begin{cases} 0 & \text{if } x < X_{(1)} \\ \dfrac{k}{n} & \text{if } X_{(k)} \leq x < X_{(k+1)} \text{ for } k = 1, 2, \ldots, n-1 \\ 1 & \text{if } x \geq X_{(n)} \end{cases}$$

$$(3.1)$$

$S_n(x)$ is also sometimes called the statistical image of the population.

For a fixed but otherwise arbitrary value of x, $S_n(x)$ is itself a random variable. Therefore it has a probability distribution, which is given in the following theorem.

Theorem 3.1 *For the random variable $S_n(x)$, which is the empirical distribution function of a random sample $X_1, X_2, \ldots, X_n$ from a distribution F_X, we have*

$$P\left[S_n(x) = \frac{j}{n}\right] = \binom{n}{j} [F_X(x)]^j [1 - F_X(x)]^{n-j} \qquad j = 0, 1, \ldots, n$$

Proof Define the indicator random variables

$$\delta_i(t) = \begin{cases} 1 & \text{if } X_i \leq t \\ 0 & \text{otherwise} \end{cases}$$

The $\delta_1(t), \delta_2(t), \ldots, \delta_n(t)$ constitute a set of n independent random variables from the Bernoulli distribution with parameter θ, where $\theta = P[\delta_i(t) = 1] = P(X_i \leq t) = F_X(t)$. Since we can write

$$S_n(x) = \sum_{i=1}^{n} \frac{\delta_i(x)}{n}$$

the random variable $nS_n(x)$ is the sum of n independent Bernoulli random variables, which follows the binomial distribution with parameter $\theta = F_X(x)$.

Corollary 3.1.1 *The mean and variance of $S_n(x)$ are*

$$E[S_n(x)] = F_X(x) \qquad \text{var}[S_n(x)] = \frac{F_X(x)[1 - F_X(x)]}{n}$$

Proof Since $nS_n(x)$ has the binomial distribution with parameter $F_X(x)$,

$$E[nS_n(x)] = nF_X(x) \qquad \text{and} \qquad \text{var}[nS_n(x)] = nF_X(x)[1 - F_X(x)]$$

Corollary 3.1.2 $S_n(x)$ *is a consistent estimator of $F_X(x)$, or, in other words, $S_n(x)$ converges in probability to $F_X(x)$.*

This result follows from the law of large numbers.

The convergence in Corollary 3.1.2 is for each value of x individually, whereas we are generally interested in all values of x collectively instead of one at a time.   A probability statement can be made simultaneously for all x, as a result of the following important theorem.   The reader is referred to Fisz (1963) for a proof.

Theorem 3.2   Glivenko-Cantelli theorem $S_n(x)$ *converges uniformly to $F_X(x)$; that is for every $\epsilon > 0$,*

$$\lim_{n \to \infty} P[\sup^\dagger_{-\infty < x < \infty} |S_n(x) - F_X(x)| > \epsilon] = 0$$

Another useful property of the empirical distribution function is its asymptotic normality, as given in the following theorem.

Theorem 3.3   *The limiting probability distribution of a standardized $S_n(x)$ is the standard normal, or*

$$\lim_{n \to \infty} P \left\{ \frac{\sqrt{n}[S_n(x) - F_X(x)]}{\sqrt{F_X(x)[1 - F_X(x)]}} \leq t \right\} = \Phi(t)$$

Proof   From the central-limit theorem and Theorem 3.1 and Corollary 3.1.1, the standard normal is the limiting distribution of

$$\frac{nS_n(x) - nF_X(x)}{\sqrt{nF_X(x)[1 - F_X(x)]}} = \frac{\sqrt{n}[S_n(x) - F_X(x)]}{\sqrt{F_X(x)[1 - F_X(x)]}}$$

4.4   THE KOLMOGOROV–SMIRNOV ONE–SAMPLE STATISTIC

A random sample $X_1, X_2, \ldots , X_n$ is drawn from a population with unknown cumulative distribution function $F_X(x)$.   For any value of x, the empirical distribution function of the sample, $S_n(x)$, provides a consistent point estimate for $F_X(x)$ (Corollary 3.1.2).   The Glivenko-Cantelli theorem (Theorem 3.2) states that the step function $S_n(x)$, with

---

† The abbreviation sup stands for supremum, which may be defined simply as the least upper bound.   Similarly, the infimum (inf) is the greatest lower bound.   In general, the supremum and infimum always exist for any set of numbers, but either or both may be infinite.

jumps occurring at the values of the order statistics $X_{(1)}, X_{(2)}, \ldots, X_{(n)}$ for the sample, approaches the true distribution function for all x. Therefore, for large n, the deviations between the true function and its statistical image, $|S_n(x) - F_X(x)|$, should be small for all values of x. This result suggests that the statistic

$$D_n = \sup_x |S_n(x) - F_X(x)| \tag{4.1}$$

is, for any n, a reasonable measure of the accuracy of our estimate.

This D_n statistic, called the *Kolmogorov-Smirnov one-sample statistic*, is particularly useful in nonparametric statistical inference because the probability distribution of D_n does not depend upon $F_X(x)$ as long as F_X is continuous. Therefore, D_n may be called a distribution-free statistic.

The directional deviations defined as

$$D_n^+ = \sup_x [S_n(x) - F_X(x)] \qquad D_n^- = \sup_x [F_X(x) - S_n(x)] \tag{4.2}$$

are called the one-sided Kolmogorov-Smirnov statistics. These measures are also distribution-free, as is proved in the following theorem.

Theorem 4.1 *The statistics D_n, D_n^+, and D_n^- are completely distribution-free for any continuous F_X.*

Proof $D_n = \sup_x |S_n(x) - F_X(x)| = \max_x (D_n^+, D_n^-)$

Defining the additional order statistics $X_{(0)} = -\infty$ and $X_{(n+1)} = \infty$, we can write

$$S_n(x) = \frac{i}{n} \qquad \begin{array}{l} \text{for } X_{(i)} \leq x < X_{(i+1)} \\ i = 0, 1, \ldots, n \end{array}$$

Therefore, we have

$$
\begin{aligned}
D_n^+ = \sup_x [S_n(x) - F_X(x)] &= \max_{0 \leq i \leq n} \sup_{X_{(i)} \leq x < X_{(i+1)}} [S_n(x) - F_X(x)] \\
&= \max_{0 \leq i \leq n} \sup_{X_{(i)} \leq x < X_{(i+1)}} \left[\frac{i}{n} - F_X(x) \right] \\
&= \max_{0 \leq i \leq n} \left[\frac{i}{n} - \inf_{X_{(i)} \leq x < X_{(i+1)}} F_X(x) \right] \\
&= \max_{0 \leq i \leq n} \left[\frac{i}{n} - F_X(X_{(i)}) \right] \\
&= \max \left\{ \max_{1 \leq i \leq n} \left[\frac{i}{n} - F_X(X_{(i)}) \right], 0 \right\} \tag{4.3}
\end{aligned}
$$

Similarly

$$D_n^- = \max \left\{ \max_{1 \leq i \leq n} \left[F_X(X_{(i)}) - \frac{i-1}{n} \right], 0 \right\}$$

$$D_n = \max \left\{ 0, \max_{1 \leq i \leq n} \left[\frac{i}{n} - F_X(X_{(i)}) \right], \max_{1 \leq i \leq n} \left[F_X(X_{(i)}) - \frac{i-1}{n} \right] \right\}$$

(4.4)

The probability distributions of D_n, D_n^+, and D_n^- therefore are seen to depend only on the random variables $F_X(X_{(i)})$, $i = 1, 2, \ldots, n$. These are the order statistics from the uniform distribution on $(0,1)$, regardless of the original F_X as long as it is continuous, because of the probability-integral transformation (Theorem 2.1). Thus D_n, D_n^+, and D_n^- have distributions which are independent of the particular F_X.

A simpler proof can be given by making the transformation $u = F_X(x)$ in D_n, D_n^+, or D_n^-. This will be left to the reader as an exercise. The above proof has the advantage of giving definitions of the Kolmogorov-Smirnov statistics in terms of order statistics.

In order to use the Kolmogorov statistics for inference, their sampling distributions must be known. Since the distributions are independent of F_X, we can assume without loss of generality that F_X is the uniform distribution on $(0,1)$. The derivation of the distribution of D_n is rather tedious. However, the approach below illustrates a number of properties of order statistics and is therefore included here. For an interesting alternative derivation, see Massey (1950).

Theorem 4.2 *For $D_n = \sup_x |S_n(x) - F_X(x)|$ where $F_X(x)$ is any continuous cdf, we have*

$$P\left(D_n < \frac{1}{2n} + v \right) = \begin{cases} 0 & \text{for } v \leq 0 \\[2ex] \int_{1/2n-v}^{1/2n+v} \int_{3/2n-v}^{3/2n+v} \cdots \int_{(2n-1)/2n-v}^{(2n-1)/2n+v} \\ \qquad f(u_1, u_2, \ldots, u_n) \, du_n \cdots du_1 \\ \qquad\qquad\qquad \text{for } 0 < v < \dfrac{2n-1}{2n} \\[3ex] 1 & \text{for } v \geq \dfrac{2n-1}{2n} \end{cases}$$

where

$$f(u_1, u_2, \ldots, u_n) = \begin{cases} n! & \text{for } 0 < u_1 < u_2 < \cdots < u_n < 1 \\ 0 & \text{otherwise} \end{cases}$$

Proof As explained above, $F_X(x)$ can be assumed to be the uniform distribution on $(0,1)$.   We shall first determine the relevant domain of v.   Since both $S_n(x)$ and $F_X(x)$ are between 0 and 1, $0 \leq D_n \leq 1$ always.   Therefore we must determine $P(D_n < c)$ only for $0 < c < 1$, which here requires

$$0 < \frac{1}{2n} + v < 1$$

or

$$-\frac{1}{2n} < v < \frac{2n - 1}{2n}$$

Now, for all $-1/2n < v < (2n - 1)/2n$, where $X_{(0)} = 0$ and $X_{(n+1)} = 1$,

$$P\left(D_n < \frac{1}{2n} + v\right) = P\left[\sup_x|S_n(x) - x| < \frac{1}{2n} + v\right]$$

$$= P\left[|S_n(x) - x| < \frac{1}{2n} + v \text{ for all } x\right]$$

$$= P\left[\left|\frac{i}{n} - x\right| < \frac{1}{2n} + v \text{ for } X_{(i)} \leq x < X_{(i+1)}\right.$$

$$\left. \text{for all } i = 0, 1, \ldots, n\right]$$

$$= P\left[\frac{i}{n} - \frac{1}{2n} - v < x < \frac{i}{n} + \frac{1}{2n} + v\right.$$

$$\left. \text{for } X_{(i)} \leq x < X_{(i+1)} \text{ for all } i = 0, 1, \ldots, n\right]$$

$$= P\left[\frac{2i - 1}{2n} - v < x < \frac{2i + 1}{2n} + v\right.$$

$$\left. \text{for } X_{(i)} \leq x \leq X_{(i+1)} \text{ for all } i = 0, 1, \ldots, n\right]$$

Consider any two consecutive values of i.   We must have, for any $0 \leq i \leq n - 1$, both

$$A_i: \left\{\frac{2i - 1}{2n} - v < x < \frac{2i + 1}{2n} + v \text{ for } X_{(i)} \leq x \leq X_{(i+1)}\right\}$$

and

$$A_{i+1}: \left\{\frac{2i + 1}{2n} - v < x < \frac{2i + 3}{2n} + v \text{ for } X_{(i+1)} \leq x \leq X_{(i+2)}\right\}$$

Since $X_{(i+1)}$ is the random variable common to both events and the common set of x is $(2i + 1)/2n - v < x < (2i + 1)/2n + v$ for $v \geq 0$, the event $A_i \cap A_{i+1}$ for any $0 \leq i \leq n - 1$ is

$$\frac{2i + 1}{2n} - v < X_{(i+1)} < \frac{2i + 1}{2n} + v \qquad \text{for all } v \geq 0$$

In other words,

$$\frac{2i - 1}{2n} - v < x < \frac{2i + 1}{2n} + v \qquad \begin{array}{l} \text{for } X_{(i)} \leq x \leq X_{(i+1)} \\ \text{for all } i = 0, 1, \ldots, n \end{array}$$

if and only if

$$\frac{2i + 1}{2n} - v < X_{(i+1)} < \frac{2i + 1}{2n} + v \qquad \begin{array}{l} \text{for all } i = 0, 1, \ldots, n - 1 \\ v \geq 0 \end{array}$$

The joint probability distribution of the order statistics is

$$f_{X_{(1)}, X_{(2)}, \ldots, X_{(n)}}(x_1, x_2, \ldots, x_n) = n!$$
$$\text{for } 0 < x_1 < x_2 < \cdots < x_n < 1$$

Putting all this together now, we have

$$P\left(D_n < \frac{1}{2n} + v\right) \qquad \text{for all } -\frac{1}{2n} < v < \frac{2n - 1}{2n}$$

$$= P\left(\frac{2i + 1}{2n} - v < X_{(i+1)} < \frac{2i + 1}{2n} + v \text{ for all}\right.$$

$$\left. i = 0, 1, \ldots, n - 1\right) \qquad \text{for all } 0 \leq v < \frac{2n - 1}{2n}$$

$$= P\left[\left(\frac{1}{2n} - v < X_{(1)} < \frac{1}{2n} + v\right) \cap \left(\frac{3}{2n} - v < X_{(2)} < \frac{3}{2n} + v\right)\right.$$

$$\left. \cap \cdots \cap \left(\frac{2n - 1}{2n} - v < X_{(n)} < \frac{2n - 1}{2n} + v\right)\right]$$

$$\text{for all } 0 \leq v < \frac{2n - 1}{2n}$$

which is equivalent to the stated integral.

This result is troublesome to evaluate as it must be used with care. For the sake of illustration, consider $n = 2$. For all $0 \leq v < \frac{3}{4}$,

$$P(D_2 < \frac{1}{4} + v) = 2! \int_{\frac{1}{4}-v}^{\frac{1}{4}+v} \int_{\frac{3}{4}-v}^{\frac{3}{4}+v} du_2 \, du_1$$
$$\scriptstyle 0 < u_1 < u_2 < 1$$

The limits overlap when $\frac{1}{4} + v \geq \frac{3}{4} - v$, or $v \geq \frac{1}{4}$. When $0 \leq v < \frac{1}{4}$, we have $u_1 < u_2$ automatically. Therefore, for $0 \leq v < \frac{1}{4}$,

$$P(D_2 < \tfrac{1}{4} + v) = 2 \int_{\frac{1}{4}-v}^{\frac{1}{4}+v} \int_{\frac{3}{4}-v}^{\frac{3}{4}+v} du_2\, du_1 = 2(2v)^2$$

But for $\frac{1}{4} \leq v < \frac{3}{4}$, the region of integration is as illustrated in Fig. 4.1. Dividing the integral into two pieces, we have for $\frac{1}{4} \leq v < \frac{3}{4}$,

$$P(D_2 < \tfrac{1}{4} + v) = 2 \left[\int_{\frac{3}{4}-v}^{\frac{1}{4}+v} \int_{u_1}^{1} du_2\, du_1 + \int_0^{\frac{3}{4}-v} \int_{\frac{3}{4}-v}^{1} du_2\, du_1 \right]$$
$$= -2v^2 + 3v - \tfrac{1}{8}$$

Collecting the results for all v,

$$P(D_2 < \tfrac{1}{4} + v) = \begin{cases} 0 & \text{for } v \leq 0 \\ 2(2v)^2 & \text{for } 0 < v < \frac{1}{4} \\ -2v^2 + 3v - 0.125 & \text{for } \frac{1}{4} \leq v < \frac{3}{4} \\ 1 & \text{for } v \geq \frac{3}{4} \end{cases}$$

For any given v and n, we can evaluate $P(D_n < 1/2n + v)$ or use table 1 of Birnbaum (1952). The inverse procedure is to find that number $D_{n,\alpha}$ such that $P(D_n > D_{n,\alpha}) = \alpha$. In our numerical example with $n = 2$, $\alpha = 0.05$, we find v such that

$$P(D_2 > \tfrac{1}{4} + v) = 0.05 \qquad \text{or} \qquad P(D_2 < \tfrac{1}{4} + v) = 0.95$$

and then set $D_{2,0.05} = \frac{1}{4} + v$. From the previous evaluation of the D_2 sampling distribution, either

$$2(2v)^2 = 0.95 \qquad \text{and} \qquad 0 < v < \tfrac{1}{4}$$

or

$$-2v^2 + 3v - 0.125 = 0.95 \qquad \text{and} \qquad \tfrac{1}{4} \leq v < \tfrac{3}{4}$$

The first result has no solution, but the second yields the solution $v = 0.5919$. Therefore, $D_{2,0.05} = 0.8419$.

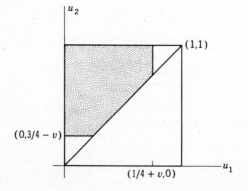

Fig. 4.1

Numerical values of $D_{n,\alpha}$ for $\alpha = 0.01$ and $\alpha = 0.05$ have been tabulated for selected values of n [see, for example, Owen (1962) or Birnbaum (1952), Table 2, p. 431, a portion of which is reproduced here in Table 4.1].

For large samples, Kolmogorov (1933) derived the following convenient approximation to the sampling distribution of D_n, and Smirnov (1939) gave a simpler proof. The result is given here without proof.

Theorem 4.3 *If F_X is any continuous distribution function, then for every $z \geq 0$,*

$$\lim_{n \to \infty} P\left(D_n \leq \frac{z}{\sqrt{n}}\right) = L(z)$$

where

$$L(z) = 1 - 2 \sum_{i=1}^{\infty} (-1)^{i-1} e^{-2i^2 z^2}$$

The function $L(z)$ has been tabulated in Smirnov (1948). Some of the results for the asymptotic approximation to $D_{n,\alpha} = z_\alpha/\sqrt{n}$ are:

$P(D_n > z_\alpha/\sqrt{n})$	0.20	0.15	0.10	0.05	0.01
z_α	1.07	1.14	1.22	1.36	1.63

The approximation has been found to be close enough for practical applications as long as n exceeds 35. A comparison of exact and asymptotic values of $D_{n,\alpha}$ for $\alpha = 0.01$ and 0.05 is given in Table 4.1.

Since the one-sided Kolmogorov-Smirnov statistics are also distribution-free, knowledge of their sampling distributions would make them useful in nonparametric statistical inference as well. Their exact sampling distributions are considerably easier to derive than that for D_n. Only the statistic D_n^+ is considered in the following theorem, but D_n^+ and D_n^- have identical distributions because of symmetry.

Theorem 4.4 *For $D_n^+ = \sup_x [S_n(x) - F_X(x)]$ where $F_X(x)$ is any continuous cdf, we have*

$$P(D_n^+ < c) = \begin{cases} 0 & \text{for } c \leq 0 \\ \int_{1-c}^{1} \int_{(n-1)/n-c}^{u_n} \cdots \int_{2/n-c}^{u_3} \int_{1/n-c}^{u_2} f(u_1, u_2, \ldots, u_n) \, du_1 \cdots du_n & \text{for } 0 < c < 1 \\ 1 & \text{for } c \geq 1 \end{cases}$$

Table 4.1 Exact and approximate values of $D_{n,\alpha}$ such that $P(D_n > D_{n,\alpha}) = \alpha$ for $\alpha = 0.01, 0.05$†

	Exact		Asymptotic		Ratio A/E	
n	0.05	0.01	0.05	0.01	0.05	0.01
2	0.8419	0.9293	0.9612	1.1509	1.142	1.238
3	0.7076	0.8290	0.7841	0.9397	1.108	1.134
4	0.6239	0.7341	0.6791	0.8138	1.088	1.109
5	0.5633	0.6685	0.6074	0.7279	1.078	1.089
10	0.4087	0.4864	0.4295	0.5147	1.051	1.058
20	0.2939	0.3524	0.3037	0.3639	1.033	1.033
30	0.2417	0.2898	0.2480	0.2972	1.026	1.025
40	0.2101	0.2521	0.2147	0.2574	1.022	1.021
50	0.1884	0.2260	0.1921	0.2302	1.019	1.018

† Z. W. Birnbaum (1952): Numerical Tabulation of the Distribution of Kolmogorov's Statistic for Finite Sample Size, *J. Am. Statist. Assoc.*, **47**: 431, table 2.

where

$$f(u_1, u_2, \ldots, u_n) = \begin{cases} n! & for\ 0 < u_1 < u_2 < \cdots < u_n < 1 \\ 0 & otherwise \end{cases}$$

Proof As before, we first assume without loss of generality that F_X is the uniform distribution on $(0,1)$. Then we can write

$$D_n^+ = \max\left[\max_{1 \leq i \leq n}\left(\frac{i}{n} - X_{(i)}\right), 0\right]$$

the form found in (4.3). For all $0 < c < 1$, we have

$$P(D_n^+ < c) = P\left[\max_{1 \leq i \leq n}\left(\frac{i}{n} - X_{(i)}\right) < c\right]$$

$$= P\left(\frac{i}{n} - X_{(i)} < c \text{ for all } i = 1, 2, \ldots, n\right)$$

$$= P\left(X_{(i)} > \frac{i}{n} - c \text{ for all } i = 1, 2, \ldots, n\right)$$

$$= \int_{1-c}^{\infty}\int_{(n-1)/n-c}^{\infty} \cdots \int_{2/n-c}^{\infty}\int_{1/n-c}^{\infty} f(x_1, x_2, \ldots, x_n)\, dx_1 \cdots dx_n$$

where

$$f(x_1, x_2, \ldots, x_n) = \begin{cases} n! & for\ 0 < x_1 < \cdots < x_n < 1 \\ 0 & otherwise \end{cases}$$

which is equivalent to the stated integral.

Another form of this result, due to Birnbaum and Tingey (1951), which is more computationally tractable, is

$$P(D_n^+ > c) = (1 - c)^n + c \sum_{j=1}^{[n(1-c)]} \binom{n}{j} \left(1 - c - \frac{j}{n}\right)^{n-j} \left(c + \frac{j}{n}\right)^{j-1}$$

(4.5)

The equivalence of the two forms can be shown by induction. Birnbaum and Tingey give a table of those values $D_{n,\alpha}^+$ which satisfy $P(D_n > D_{n,\alpha}^+) = \alpha$ for $\alpha = 0.01, 0.05, 0.10$ and selected values of n.

For large samples, we have the following theorem, which is given here without proof.

Theorem 4.5   *If F_X is any continuous distribution function, then for every $z \geq 0$,*

$$\lim_{n \to \infty} P\left(D_n^+ < \frac{z}{\sqrt{n}}\right) = 1 - e^{-2z^2}$$

As a result of this theorem, chi-square tables can be used for the distribution of a function of D_n^+ because of the following.

Corollary 4.5   *If F_X is any continuous distribution function, then for every $z \geq 0$, the limiting distribution of $V = 4nD_n^{+2}$, as $n \to \infty$, is the chi-square distribution with two degrees of freedom.*

Proof   We have $D_n^+ < z/\sqrt{n}$ if and only if $4nD_n^{+2} < 4z^2$ or $V < 4z^2$. Therefore

$$\lim_{n \to \infty} P(V < 4z^2) = 1 - e^{-2z^2}$$

$$\lim_{n \to \infty} P(V < c) \quad = 1 - e^{-c/2} \qquad \text{for all } c \geq 0$$

which is the chi-square distribution with two degrees of freedom.

As a numerical example of how this corollary enables us to approximate $D_{n,\alpha}^+$, consider $\alpha = 0.05$. A chi-square table gives 5.99 for the 0.05 critical point of chi square with two degrees of freedom. The procedure is to set $4nD_{n,0.05}^+{}^2 = 5.99$ and solve to obtain

$$D_{n,0.05}^+ = \sqrt{\frac{1.4975}{n}} = \frac{1.22}{\sqrt{n}}$$

4.5 APPLICATIONS OF THE KOLMOGOROV–SMIRNOV ONE-SAMPLE STATISTIC

The statistical use of the Kolmogorov-Smirnov statistic in a goodness-of-fit type of problem is obvious. Assume we have the random sample $X_1, X_2, \ldots, X_n$ and the hypothesis-testing situation

$$H_0: F_X(x) = F_0(x) \qquad \text{for all } x$$

where $F_0(x)$ is a completely specified continuous distribution function.

Since $S_n(x)$ is the statistical image of the population distribution $F_X(x)$, if the null hypothesis is true, the differences between $S_n(x)$ and $F_0(x)$ should be small for all x except for sampling variation. For the usual two-sided goodness-of-fit alternative

$$H_1: F_X(x) \neq F_0(x) \qquad \text{for some } x$$

large absolute values of these deviations tend to discredit the hypothesis. Therefore, the Kolmogorov-Smirnov goodness-of-fit test with significance level α is to reject H_0 when $D_n > D_{n,\alpha}$. From the Glivenko-Cantelli theorem (Theorem 3.2), we know that $S_n(x)$ converges to $F_X(x)$ with probability 1, which implies consistency.

With the statistics D_n^+ and D_n^-, it is possible to use Kolmogorov-Smirnov statistics for a one-sided goodness-of-fit test which would detect directional differences between $S_n(x)$ and $F_0(x)$. For the alternative

$$H_{1,+}: F_X(x) \geq F_0(x) \qquad \text{for all } x$$

the appropriate rejection region is $D_n^+ > D_{n,\alpha}^+$, and for the alternative

$$H_{1,-}: F_X(x) \leq F_0(x) \qquad \text{for all } x$$

H_0 is rejected when $D_n^- > D_{n,\alpha}^-$. Both these tests are consistent against their respective alternatives.

Two other useful applications of the Kolmogorov-Smirnov statistics relate to point and interval estimations of the unknown distribution F_X. In the case of a point estimate, the D_n statistic enables us to determine the minimum sample size required to state justifiably with a certain probability $1 - \alpha$ that the error in the estimate never exceeds a fixed value ϵ. In other words, we wish to find the minimum n such that

$$P(D_n < \epsilon) = 1 - \alpha$$

Since $\epsilon = D_{n,\alpha}$ is given, the value of n can be read directly from a table similar to Table 4.1 but more extensive. If it is found that no value of $n \leq 35$ will satisfy the desired accuracy, the asymptotic distribution of Theorem 4.3 can be used, and we solve $z_\alpha = \epsilon \sqrt{n}$ for n.

The other important statistical use of the D_n statistic is in finding confidence bands for $F_X(x)$ for all x. Using a table, we can find the number $D_{n,\alpha}$ such that

$$P(D_n > D_{n,\alpha}) = \alpha$$

This is equivalent to the statement

$$P[\sup_x |S_n(x) - F_X(x)| < D_{n,\alpha}] = 1 - \alpha$$

or

$$P[S_n(x) - D_{n,\alpha} < F_X(x) < S_n(x) + D_{n,\alpha} \text{ for all } x] = 1 - \alpha$$

We know that $0 \leq F_X(x) \leq 1$ for all x, whereas the inequality in this probability statement admits numbers outside this range. Thus we define

$$L_n(x) = \max[S_n(x) - D_{n,\alpha}, 0]$$

and

$$U_n(x) = \min[S_n(x) + D_{n,\alpha}, 1]$$

and call the region between $L_n(x)$ and $U_n(x)$ a confidence band for $F_X(x)$, with associated confidence coefficient $1 - \alpha$.

The simplest procedure in application is to graph the observed $S_n(x)$ as a step function and plot parallel lines at a distance $D_{n,\alpha}$ in either direction but always within the unit square. When $n > 35$, the value $D_{n,\alpha}$ can be determined from the asymptotic distribution. Of course this confidence-band procedure can be used to perform a test of the hypothesis $F_X(x) = F_0(x)$, since $F_0(x)$ lies wholly within the limits $L_n(x)$ and $U_n(x)$ if and only if the hypothesis cannot be rejected at a significance level α.

Similar applications of the D_n^- or D_n^+ statistics are obvious.

It should be noted that all the theoretical properties of the Kolmogorov-Smirnov statistics required the assumption that F_X be continuous, since this is necessary to guarantee their distribution-free nature. The properties of the empirical distribution function given in Sec. 3, including the Glivenko-Cantelli theorem, do not require this continuity assumption. Furthermore, it is certainly desirable to have a goodness-of-fit test which can be used when the hypothesized distribution is discrete. Noether [(1967), pp. 17–18] and others have shown that if the $D_{n,\alpha}$ values based on a continuous F_X are used in a discrete application, the true significance level is at most α. Therefore, we use the exact same procedure for discrete $F_0(x)$ as for continuous, remembering that the test is now conservative. The same conclusion applies to the construction of confidence bands for discrete distributions and for the corresponding statistical applications of the one-sided Kolmogorov-Smirnov statistics. For a more complete discussion of the discrete case, see David (1952).

4.6 COMPARISON OF CHI-SQUARE AND KOLMOGOROV-SMIRNOV TESTS FOR GOODNESS OF FIT

The chi-square test is specifically designed for use with categorical data, while the Kolmogorov-Smirnov (K-S) statistics are for random samples from continuous populations. However, when the data are not categorical as collected, these two goodness-of-fit tests can be used interchangeably. The reader is referred to Goodman (1954), Birnbaum (1952), and Massey (1951*b*) for discussions of their relative merits. Only a brief comparison will be made here, which is relevant whenever raw ungrouped measurement data are available.

The basic difference between the two tests is that chi square is sensitive to vertical deviations between the observed and expected histograms, whereas K-S procedures are based on vertical deviations between the observed and expected cumulative distribution functions. However, both types of deviations are useful in determining goodness of fit and probably are equally informative. Another obvious difference is that chi square requires grouped data whereas K-S does not. Therefore, when the hypothesized distribution is continuous, K-S allows us to examine the goodness of fit for each of the n observations, instead of only for k classes, where $k \leq n$. In this sense, K-S makes more complete use of the available data. Further, the chi-square statistic is affected by the number of classes and their widths, which are chosen arbitrarily by the experimenter.

One of the primary advantages of K-S is that the exact sampling distribution of D_n is known and tabulated, whereas the sampling distribution of Q is only approximately chi square for any finite n. K-S can be applied for any size sample, while the chi-square statistic should be used only for n large and each expected cell frequency not too small. When cells must be combined for chi-square application, the calculated value of Q is no longer unique, as it is affected by the scheme of combination. The K-S statistic is much more flexible than chi square, since it can be used in estimation to find minimum sample size and confidence bands. With the one-sided D_n^+ or D_n^- statistics, we can test for deviations in a particular direction, whereas chi square is always concerned equally with differences in either direction. In most cases, K-S is easier to apply.

Chi square also has some advantages over K-S. A hypothesized distribution which is discrete presents no problems for the chi-square test, while the exact properties of D_n are violated by the lack of continuity. As already stated, however, this is a minor problem with D_n which can generally be eliminated by replacing equalities by inequalities in the probabilities. Perhaps the main advantage of the chi square is that by simply reducing the number of degrees of freedom and replacing unknown parameters by consistent estimators, a goodness-of-fit test can be performed in the usual manner even when the hypothesized distribution is not completely specified. If the hypothesized $F_0(x)$ in D_n contains unspecified parameters which are estimated from the data, we obtain an estimate $\hat{D}_n$ whose sampling distribution is different from that of D_n. The test is conservative when the D_n critical values are used. Two papers by Lilliefors (1967, 1969) provide tables for the exact critical values of $\hat{D}_n$ (obtained by Monte Carlo calculations) when the hypothesized distributions are normal and exponential. His numerical investigations indicate that $\hat{D}_n$ provides a more powerful test than chi square for these populations.

As for relative performance, the power functions of the two statistics depend on different quantities. If $F_0(x)$ is the hypothesized cumulative distribution and $F_X(x)$ the true distribution, the power of K-S depends on

$$\sup_x |F_X(x) - F_0(x)|$$

while the power of chi square depends on

$$\sum_{i=0}^{k} \frac{\{[F_X(a_{i+1}) - F_X(a_i)] - [F_0(a_{i+1}) - F_0(a_i)]\}^2}{F_0(a_{i+1}) - F_0(a_i)}$$

where the a_i are the class limits in the numerical categories.

The power of the chi-square test can be improved by clever grouping in some situations. In particular, Cochran (1952) and others have shown that a choice of intervals which provide equal expected frequencies for all classes is a good procedure in this respect besides simplifying the computations. The number of classes k can be chosen such that the power is maximized in the vicinity of the point where power equals 0.5. This procedure also eliminates the arbitrariness of grouping. The expression for q in (2.1) reduces to $\left(k \sum_{i=1}^{k} f_i^2 - n^2\right)/n$ when $e_i = n/k$ for $i = 1, 2, \ldots, k$.

Some studies of power comparisons have been reported in the literature, but they do not seem to provide a definitive basis for choice in general. However, since the sampling distribution of Q is known only asymptotically, it must always be considered an approximate test. Thus at least when sample sizes are small, K-S would be the preferable statistic.

PROBLEMS

4.1. Two types of corn (golden and green-striped) carry recessive genes. When these were crossed, a first generation was obtained which was consistently normal (neither golden nor green-striped). When this generation was allowed to self-fertilize, four distinct types of plants were produced—normal, golden, green-striped, and golden-green-striped. In 1,200 plants this process produced the following distribution:

Normal: 670
Golden: 230
Green-striped: 238
Golden-green-striped: 62

A monk named Mendel has written an article theorizing that in a second generation of such hybrids, the distribution of plant types should be in a $9:3:3:1$ ratio. Are the above data consistent with the good monk's theory?

4.2. A group of four coins is tossed 160 times, and the following data are obtained:

Number of heads	0	1	2	3	4
Frequency	16	48	55	33	8

Do you think the four coins were balanced?

4.3. A certain genetic model suggests that the probabilities for a particular trinomial distribution are, respectively, $\theta_1 = p^2$, $\theta_2 = 2p(1-p)$, and $\theta_3 = (1-p)^2$, $0 < p < 1$. Assume that X_1, X_2, and X_3 represent the respective frequencies in a sample of n independent trials and that these numbers are known. Derive a chi-square goodness-of-fit test for this trinomial distribution if p is unknown.

4.4. According to a genetic model, the proportions of individuals having the four blood types should be related by

Type O: q^2
Type A: $p^2 + 2pq$
Type B: $r^2 + 2qr$
Type AB: $2pr$

where $p + q + r = 1$. Given the blood types of 1,000 individuals, how would you test the adequacy of the model?

4.5. If individuals are classified according to sex and color blindness, it is hypothesized that the distribution should be as follows:

	Male	*Female*
Normal	$\dfrac{p}{2}$	$\dfrac{p^2}{2} + pq$
Color blind	$\dfrac{q}{2}$	$\dfrac{q^2}{2}$

for some $p + q = 1$, where p denotes the proportion of defective genes in the relevant population, and therefore changes for each problem. How would the chi-square test be used to test the adequacy of the general model?

4.6. Show that in general, for Q defined as in (2.1),

$$E(Q) = E\left[\sum_{i=1}^{k} \frac{(F_i - e_i)^2}{e_i}\right] = \sum_{i=1}^{k}\left[\frac{n\theta_i(1 - \theta_i)}{e_i} + \frac{(n\theta_i - e_i)^2}{e_i}\right]$$

From this we see that if the null hypothesis is true, $n\theta_i = e_i$ and $E(Q) = k - 1$, the mean of the chi-square distribution.

4.7. Show algebraically that where $e_i = n\theta_i$

$$q = \sum_{i=1}^{2} \frac{(f_i - e_i)^2}{e_i} = \frac{(f_1 - n\theta_1)^2}{n\theta_1(1 - \theta_1)}$$

so that when $k = 2$, $\sqrt{q}$ is the statistic commonly used for testing a hypothesis concerning the parameter of the binomial distribution for large samples. By the central-limit theorem, $\sqrt{Q}$ approaches the standard normal distribution as $n \to \infty$, and the square of any standard normal variable is chi-square-distributed with one degree of freedom. Thus we have an entirely different argument for the distribution of Q when $k = 2$.

4.8. Define the random variable

$$\epsilon(x) = \begin{cases} 1 & \text{if } x \geq 0 \\ 0 & \text{if } x < 0 \end{cases}$$

Show that the random function defined by

$$F_n(x) = \sum_{i=1}^{n} \frac{\epsilon(x - X_i)}{n}$$

is the empirical distribution function of a sample $X_1, X_2, \ldots, X_n$, by showing that $F_n(x) = S_n(x)$ for all x.

4.9. Prove that $\mathrm{cov}[S_n(x), S_n(y)] = c[F_X(x), F_X(y)]/n$ where

$$c(s,t) = \min(s,t) - st = \begin{cases} s(1 - t) & \text{if } s \leq t \\ t(1 - s) & \text{if } s \geq t \end{cases}$$

and $S_n(\cdot)$ is the empirical distribution function of a random sample of size n from the population F_X.

4.10. Let $S_n(x)$ be the empirical distribution function for a random sample of size n from the uniform distribution on $(0,1)$. Define

$$X_n(t) = \sqrt{n}\,[S_n(t) - t]$$
$$Z_n(t) = (t + 1)X_n\left(\frac{t}{t + 1}\right) \qquad \text{for all } 0 \leq t \leq 1$$

Find $E[X_n(t)]$ and $E[Z_n(t)]$, $\mathrm{var}[X_n(t)]$ and $\mathrm{var}[Z_n(t)]$, and conclude that $\mathrm{var}[X_n(t)] \leq \mathrm{var}[Z_n(t)]$ for all $0 \leq t \leq 1$ and all n.

4.11. Give a simple proof that D_n, D_n^+, and D_n^- are completely distribution-free for any continuous F_X by appealing to the transformation $u = F_X(x)$ in the initial definitions of D_n, D_n^+, and D_n^-.

4.12. Prove that

$$D_n^- = \max \left\{ \max_{1 \le i \le n} \left[F_X(X_{(i)}) - \frac{i-1}{n} \right], 0 \right\}$$

4.13. Prove that the probability distribution of D_n^- is identical to the distribution of D_n^+:

(a) Using a derivation analogous to Theorem 4.4.

(b) Using a symmetry argument.

4.14. Using Theorem 4.3, verify that

$$\lim_{n \to \infty} P\left(D_n > \frac{1.07}{\sqrt{n}} \right) = 0.20$$

4.15. Find the minimum sample size n required such that $P(D_n < 0.05) \ge 0.99$.

4.16. Use Theorem 4.4 to verify directly that $P(D_5^+ > 0.447) = 0.10$. Calculate this same probability using the expression given in (4.5).

4.17. Related goodness-of-fit test. The Cramer–von Mises type of statistic is defined for continuous $F_X(x)$ by

$$\omega_n{}^2 = \int_{-\infty}^{\infty} [S_n(x) - F_X(x)]^2 f_X(x) \, dx$$

(a) Prove that $\omega_n{}^2$ is distribution-free.

(b) Explain how $\omega_n{}^2$ might be used for a goodness-of-fit test.

(c) Show that

$$n\omega_n{}^2 = \frac{1}{12n} + \sum_{i=1}^{n} \left[F_X(x_{(i)}) - \frac{2i-1}{2n} \right]^2$$

This statistic is discussed in Cramer (1928), von Mises (1931), Smirnov (1936), and Darling (1957).

5
Rank-order Statistics

5.1 INTRODUCTION

Many of the nonparametric procedures which will be covered in the following chapters are based on rank-order statistics. The rank-order statistics for a random sample are any set of constants which indicate the order of the observations. The actual magnitude of any observation is used only in the determination of its relative position in the sample array and thereafter ignored in any analysis based on rank-order statistics. Thus any statistical procedures based on rank-order statistics depend only on the relative magnitudes of the observations. If the jth element X_j is the ith smallest in the sample, the jth rank-order statistic must be the ith smallest rank-order statistic. Rank-order statistics might then be defined as the set of numbers which results when each original observation is replaced by the value of some order-preserving function. If the rank-order statistics for a random sample $X_1, X_2, \ldots, X_N$ are denoted by

$$r(X_1), r(X_2), \ldots, r(X_N)$$

r is any function such that $r(X_i) \leq r(X_j)$ whenever $X_i \leq X_j$. As with order statistics, rank-order statistics are invariant under monotone transformations, i.e., if $r(X_i) \leq r(X_j)$, then $r[F(X_i)] \leq r[F(X_j)]$, in addition to $F[r(X_i)] \leq F[r(X_j)]$, where F is any nondecreasing function.

For any set of N different sample observations, the simplest set of numbers to use to indicate relative positions is the first N positive integers. In order to eliminate the possibility of confusion and to simplify and unify the theory of rank-order statistics, we shall assume here that unless explicitly stated otherwise, the rank-order statistics are always a permutation of the first N integers. The ith rank-order statistic $r(X_i)$ then is called the rank of the ith observation in the unordered sample. The value it assumes, $r(x_i)$, is the number of observations $x_j, j = 1, 2, \ldots, N$, such that $x_j \leq x_i$. For example, the rank of the ith order statistic is equal to i, or $r(x_{(i)}) = i$. A functional definition of the rank of any x_i in a set of N different observations is provided by

$$r(x_i) = \sum_{j=1}^{N} S(x_i - x_j) = 1 + \sum_{j \neq i} S(x_i - x_j) \tag{1.1}$$

where

$$S(u) = \begin{cases} 0 & \text{if } u < 0 \\ 1 & \text{if } u \geq 0 \end{cases} \tag{1.2}$$

The random variable $r(X_i)$ is discrete and for a random sample from a continuous population it follows the discrete uniform distribution, or

$$P[r(X_i) = j] = \frac{1}{N} \quad \text{for } j = 1, 2, \ldots, N$$

Although admittedly the terminology may seem confusing at the outset, a function of the rank-order statistics will be called a *rank statistic*. Rank statistics are particularly useful in nonparametric inference since they are usually distribution-free. The methods are applicable to a wide variety of hypothesis-testing situations depending on the particular function used. The procedures are generally simple and quick to apply. These practical considerations often balance the loss of efficiency which is expected when actual magnitudes are available but ignored after ranks are determined. However, in most cases the efficiency loss is small. The first section of this chapter will define a possible measure of the strength of the relationship between the set of actual observations and their corresponding rank-order statistics. In populations where this relationship is strong, we may feel there is less loss of information when variate values are replaced by their ranks for the purpose of using some inference procedure based on rank statistics.

As long as the population from which the sample was drawn is assumed continuous, the probability of any two observations having identical magnitudes is zero. The set of ranks as defined in (1.1) then will be N different integers. The exact properties of most rank statistics depend upon this assumption. Two or more observations with the same magnitude are said to be *tied*. We may say only that *theoretically* no problem is presented by tied observations. However, in practice ties can certainly occur, either because the population is actually discrete or because of practical limitations on precision of measurement. Some of the conventional approaches to dealing with ties in assigning ranks will be discussed generally in this chapter, so that the problem can be ignored later in presenting the theory of some specific rank tests in the following chapters.

One additional point should be noted. Since rank statistics are functions only of the ranks of observations, only this information need be provided by the sample data. Actual measurements are often difficult, expensive, or even impossible to obtain, and they are superfluous here as long as a unique rank order can be designated for the observations. Therefore rank statistics may have wider applicability and more appeal than many other procedures which require more quantitative comparisons.

5.2 CORRELATION BETWEEN VARIATE VALUES AND RANKS

When actual measurements are impossible or not feasible to obtain but relative positions can be determined, rank-order statistics make full use of all the available information. However, if the fundamental data consist of variate values, how much information is lost by using the data only to determine relative magnitudes? One approach to a judgment concerning the potential loss of efficiency is to determine the correlation between the variate values and their assigned ranks. If the correlation is high, we would feel intuitively more justified in the replacement of actual values by ranks for the purpose of analysis. The hope is that inference procedures based on ranks alone will lead to conclusions which seldom differ from a corresponding inference based on actual variate values.

The ordinary product-moment correlation coefficient between two random variables X and Y is

$$\rho(X,Y) = \frac{E[(X - \mu_X)(Y - \mu_Y)]}{\sigma_X \sigma_Y} = \frac{E(XY) - E(X)E(Y)}{\sigma_X \sigma_Y}$$

Assume that for a continuous population F_X we would like to determine the correlation between the random variable X and its rank $r(X)$. Theoretically, a random variable from an infinite population cannot have a rank, since values on a continuous scale cannot be ordered. But an observation X_i, of a random sample of size N from this population, does have a rank $r(X_i)$ as defined in (1.1). The distribution of X_i is the same as the distribution of X and the $r(X_i)$ are identically distributed though not independent. Therefore, it is reasonable to define the population correlation coefficient between ranks and variate values as the correlation between X_i and $Y_i = r(X_i)$, or

$$\rho[X, r(X)] = \frac{E(X_i Y_i) - E(X)E(Y_i)}{\sigma_X \sigma_{Y_i}} \tag{2.1}$$

The marginal distribution of Y_i for any i is the discrete uniform, so that

$$f_{Y_i}(j) = \frac{1}{N} \qquad \text{for } j = 1, 2, \ldots, N \tag{2.2}$$

with moments

$$E(Y_i) = \sum_{j=1}^{N} \frac{j}{N} = \frac{N+1}{2} \tag{2.3}$$

$$E(Y_i^2) = \sum_{j=1}^{N} \frac{j^2}{N} = \frac{(N+1)(2N+1)}{6}$$

$$\text{var}(Y_i) = \frac{(N+1)(2N+1)}{6} - \frac{(N+1)^2}{4} = \frac{N^2 - 1}{12} \tag{2.4}$$

The joint distribution of X_i and its rank Y_i is

$$f_{X_i, Y_i}(x, j) = f_{X_i|Y_i=j}(x \mid j) f_{Y_i}(j) = \frac{f_{X_{(j)}}(x)}{N} \qquad \text{for } j = 1, 2, \ldots, N$$

where $X_{(j)}$ denotes the jth order statistic of a random sample of size N from the population F_X. From this expression we can write

$$E(X_i Y_i) = \frac{1}{N} \int_{-\infty}^{\infty} \sum_{j=1}^{N} jx f_{X_{(j)}}(x) \, dx = \sum_{j=1}^{N} \frac{jE(X_{(j)})}{N} \tag{2.5}$$

Substituting the results (2.3), (2.4), and (2.5) back into (2.1), we obtain

$$\rho[X, r(X)] = \left(\frac{12}{N^2 - 1}\right)^{1/2} \frac{\sum_{j=1}^{N} jE(X_{(j)}) - [N(N+1)/2]E(X)}{N\sigma_X} \tag{2.6}$$

Since the result here is independent of i, our definition in (2.1) may be considered a true correlation. The same result is obtained if the covariance between X and $r(X)$ is defined as the limit as M approaches infinity of the average of the M sample correlations which can be calculated when

M samples of size N are drawn from this population. This method will be left as an exercise for the reader.

The expression given in (2.6) can be written in another useful form. If the variate values X are drawn from a continuous population with distribution F_X, the following sum can be evaluated:

$$
\sum_{i=1}^{N} iE(X_{(i)}) = \sum_{i=1}^{N} \frac{iN!}{(i-1)!(N-i)!}
$$

$$
\int_{-\infty}^{\infty} x[F_X(x)]^{i-1}[1 - F_X(x)]^{N-i} f_X(x) \, dx
$$

$$
= \sum_{j=0}^{N-1} \frac{(j+1)N!}{j!(N-j-1)!}
$$

$$
\int_{-\infty}^{\infty} x[F_X(x)]^{j}[1 - F_X(x)]^{N-j-1} f_X(x) \, dx
$$

$$
= \sum_{j=1}^{N-1} \frac{N!}{(j-1)!(N-j-1)!}
$$

$$
\int_{-\infty}^{\infty} x[F_X(x)]^{j}[1 - F_X(x)]^{N-j-1} f_X(x) \, dx
$$

$$
+ \sum_{j=0}^{N-1} \frac{N!}{j!(N-j-1)!}
$$

$$
\int_{-\infty}^{\infty} x[F_X(x)]^{j}[1 - F_X(x)]^{N-j-1} f_X(x) \, dx
$$

$$
= N(N-1) \int_{-\infty}^{\infty} xF_X(x) \sum_{j=1}^{N-1} \binom{N-2}{j-1}
$$

$$
[F_X(x)]^{j-1}[1 - F_X(x)]^{N-j-1} f_X(x) \, dx
$$

$$
+ N \int_{-\infty}^{\infty} x \sum_{j=0}^{N-1} \binom{N-1}{j}
$$

$$
[F_X(x)]^{j}[1 - F_X(x)]^{N-j-1} f_X(x) \, dx
$$

$$
= N(N-1) \int_{-\infty}^{\infty} xF_X(x)f_X(x) \, dx + N \int_{-\infty}^{\infty} xf_X(x) \, dx
$$

$$
= N(N-1)E[XF_X(X)] + NE(X) \tag{2.7}
$$

If this quantity is now substituted in (2.6), the result is

$$
\rho[X, r(X)] = \left(\frac{12}{N^2 - 1}\right)^{1/2} \frac{1}{\sigma_X} \left\{ (N-1)E[XF_X(X)] \right.
$$

$$
\left. + E(X) - \frac{N+1}{2} E(X) \right\}
$$

$$
= \left(\frac{12}{N^2 - 1}\right)^{1/2} \frac{1}{\sigma_X} \left\{ (N-1)E[XF_X(X)] - \frac{N-1}{2} E(X) \right\}
$$

$$
= \left[\frac{12(N-1)}{N+1}\right]^{1/2} \frac{1}{\sigma_X} \left\{ E[XF_X(X)] - \tfrac{1}{2}E(X) \right\} \tag{2.8}
$$

and

$$\lim_{N \to \infty} \rho[X, r(X)] = \frac{2\sqrt{3}}{\sigma_X} \{E[XF_X(X)] - \tfrac{1}{2}E(X)\} \qquad (2.9)$$

Some particular evaluations of (2.9) are given in Stuart (1954).

5.3 TREATMENT OF TIES IN RANK TESTS

In a set of N observations which are *not* all different, arrangement in order of magnitude produces a set of t groups of different numbers, the ith different value occurring with frequency τ_i, where $\sum_{i=1}^{t} \tau_i = N$. Any group of numbers with $\tau_i \geq 2$ comprises a set of tied observations. The ranks are no longer well defined, and for any set of fixed ranks for N untied observations there are $\prod_{i=1}^{t} \tau_i!$ possible assignments of ranks to the entire sample with ties, each assignment leading to its own value for a rank test statistic. If a rank test is to be performed using a sample containing tied observations, we must have either a unique method of assigning ranks for ties so that the test statistic can be computed in the usual way or a method of combining the many possible values of the rank test statistic to reach one decision. Several acceptable methods will be discussed briefly.

RANDOMIZATION

In the method of randomization, one of the $\prod_{i=1}^{t} \tau_i!$ possible assignments of ranks is selected by some random procedure. For example, in the set of observations

3.0, 4.1, 4.1, 5.2, 6.3, 6.3, 6.3, 9

there are 2!(3!) or 12 possible assignments of the integer ranks 1 to 8 which this sample could represent. One of these 12 is selected by a supplementary random experiment and used as the unique assignment of ranks. Using this method, some theoretical properties of the rank statistic are preserved since each assignment occurs with equal probability. In particular the null probability distribution of the rank-order statistic, and therefore of the rank statistic, is unchanged, so that the test can be performed in the usual way. However, an additional element of chance is artificially imposed, affecting the probability distribution under alternatives.

MIDRANKS

The midrank method assigns to each member of a group of tied observations the simple average of the ranks they would have if distinguishable. Using this approach, tied observations are given tied ranks. The midrank method is perhaps the most frequently used, as it has much appeal experimentally. However, the null distribution of ranks is affected. Obviously, the mean rank is unchanged, but the variance of the ranks would be reduced. When the midrank method is used, for some tests a correction for ties can be incorporated into the test statistic.

AVERAGE STATISTIC

If one does not wish to choose a particular set of ranks as in the previous two methods, one may instead calculate the value of the test statistic for all the $\prod_{i=1}^{t} \tau_i!$ assignments and use their simple average as the single sample value. Again, the test statistic would have the same mean but smaller variance.

AVERAGE PROBABILITY

Instead of averaging the test statistic for each possible assignment of ranks, one could find the probability of each resulting value of the test statistic and use the simple average of these probabilities for the overall probability. This requires availability of tables of the exact null probability distribution of the test statistic rather than simply a table of critical values.

LEAST FAVORABLE STATISTIC

Having found all possible values of the test statistic, one might choose as a single value that one which minimizes the probability of rejection. This procedure leads to the most conservative test, i.e., the lowest probability of committing a type I error.

RANGE OF PROBABILITY

Alternatively, one could compute two values of the test statistic: the least favorable and the most favorable. However, unless both fall inside or both fall outside the rejection region, this method does not lead to a decision.

OMISSION OF TIED OBSERVATIONS

The final and most obvious possibility is to discard all tied observations and reduce the sample size accordingly. This method certainly leads to a loss of information, but if the number of observations to be omitted is small relative to the sample size, the loss may be minimal. The procedure generally introduces bias toward rejection of the null hypothesis.

The reader is referred to Savage's "Bibliography" (1962) for discussions of treatment of ties in relation to particular nonparametric rank test statistics.

PROBLEMS

5.1. Give a functional definition similar to (1.1) for the rank $r(X_i)$ of a random variable in any set of N independent observations where ties are dealt with by the midrank method. *Hint:* In place of $S(u)$ in (1.2), consider the function

$$c(u) = \begin{cases} 0 & \text{if } u < 0 \\ \frac{1}{2} & \text{if } u = 0 \\ 1 & \text{if } u > 0 \end{cases}$$

5.2. Find the correlation coefficient between variate values and ranks in a random sample of size N from:

    (a) The uniform distribution.

    (b) The standard normal distribution.

    (c) The exponential distribution.

6

Other One-sample and Paired-sample Techniques: The Sign Test and Signed-rank Test

6.1 INTRODUCTION

In the general one-sample problem, the data available consist of a single set of observations on which inferences can be based. The tests for randomness relate to inferences about a property of the joint probability distribution of a set of sample observations which are identically distributed but possibly dependent, i.e., the probability distribution of the data. The hypothesis in a goodness-of-fit study is concerned with the univariate population distribution from which a set of independent variables is drawn. Thus, although these two types of inference fall within the classification of the one-sample problem, the hypotheses are so general that no analogous counterparts exist within the realm of parametric statistics. The two tests are for nonparametric hypotheses. In a one-sample classical inference problem, the single-sample data are used to obtain information about some particular aspect of the population distribution, usually one or more of its parameters. Nonparametric techniques are useful here too, particularly when a location parameter is of interest.

In this chapter we shall be concerned with the nonparametric analog of the normal-theory test (variance known) or Student's t test (variance unknown) for the hypotheses H_0: $\mu = \mu_0$ and H_0: $\mu_X - \mu_Y = \mu_D = \mu_0$, for the one-sample and paired-sample problems, respectively. The classical tests are derived under the assumption that the single population or the population of differences is normal. For the nonparametric tests, only certain continuity assumptions about the populations need be postulated to determine sampling distributions. The hypotheses here are concerned with the median rather than the mean as the location parameter, but both are good indexes of central tendency and they do coincide for symmetric populations. The tests to be covered are the ordinary sign test and the Wilcoxon signed-rank test, including both hypothesis-testing and confidence-interval techniques. The complete discussion in each case will be given only for the single-sample case, since with paired-sample data once the differences of observations are formed, we have essentially only a single sample drawn from the population of differences and thus the methods of analysis are identical.

6.2 THE ORDINARY SIGN TEST

A random sample of N observations $X_1, X_2, \ldots , X_N$ is drawn from a population F_X with unknown median M, where F_X is assumed to be continuous in the vicinity of M. In other words, the assumptions are independent observations, and $P(X = M) = 0$. The hypothesis to be tested concerns the value of the population median

$$H_0: M = M_0$$

with a corresponding one- or two-sided alternative on the value of M.

For any distribution which satisfies $P(X = M) = 0$, by the definition of M we have $P(X > M) = P(X < M) = 0.50$. Since the hypothesis here states that M_0 is that value of X which divides the area under the frequency distribution into two equal parts, an equivalent symbolic representation of H_0 is

$$H_0: \theta = P(X > M_0) = P(X < M_0)$$

If the sample data are consistent with the hypothesized median value, on the average half of the sample observations will lie above the number M_0 and half below. Thus the number of observations above M_0, which will be denoted by K, can be used to test the validity of the null hypothesis. When the observations are dichotomized in this way, they constitute a set of N independent random variables from the Bernoulli population with parameter $\theta = P(X > M_0)$, regardless of the population F_X. The sampling distribution of the random variable K then is the binomial probability distribution with parameter θ, which equals 0.50 if the null hypothesis is true.   Since K is actually the number of plus signs among the N differences $X_i - M_0$, $i = 1, 2, \ldots , N$, the nonparametric test based on K is called the *ordinary sign test*.

The appropriate rejection region depends on the alternative.   A possible one-sided alternative of interest is that the true median exceeds the hypothesized value, which would be equivalent to the statement that the probability of a plus sign exceeds the probability of a minus sign. Symbolically, we write

$$H_1: M > M_0 \qquad \text{or} \qquad \theta = P(X > M_0) > P(X < M_0)$$

The sample will reflect this state if there is an excess of positive differences. Therefore, the rejection region for a test at significance level α is

$$K \in R \qquad \text{for } k \geq k_\alpha$$

where k_α is chosen to be the smallest integer which satisfies

$$\sum_{k=k_\alpha}^{N} \binom{N}{k} 0.5^N \leq \alpha$$

The desired significance level, called the nominal α, cannot usually be achieved exactly, because of the discreteness of the random sampling distribution.   The actual or exact α is equal to the sum

$$\sum_{k=k_\alpha}^{N} \binom{N}{k} 0.5^N$$

Ordinary tables of the cumulative binomial distribution [National Bureau of Standards (1949)] with $\theta = 0.5$ can be used to find the particular value of k_α for the given N and α.   Similarly, for a one-sided test with the alternative

$$H_1: M < M_0 \qquad \text{or} \qquad \theta = P(X > M_0) < P(X < M_0)$$

the rejection region for an α-level test is

$$K \in R \qquad \text{for } k \leq k'_\alpha$$

where k'_α is the largest integer satisfying

$$\sum_{k=0}^{k_\alpha'} \binom{N}{k} 0.5^N \leq \alpha$$

If the alternative is two-sided,

$$H_1: M \neq M_0 \qquad \text{or} \qquad \theta = P(X > M_0) \neq P(X < M_0)$$

the rejection region should consist of values of K which are either too large or too small. Since the binomial is symmetric when $\theta = 0.5$, the greatest power is achieved by choosing the two tails symmetrically. Thus we reject when $K \geq k_{\alpha/2}$ or $K \leq k'_{\alpha/2}$, where $k_{\alpha/2}$ and $k'_{\alpha/2}$ are, respectively, the smallest and largest integers satisfying

$$\sum_{k=k_{\alpha/2}}^{N} \binom{N}{k} 0.5^N \leq \frac{\alpha}{2} \quad \text{and} \quad \sum_{k=0}^{k'_{\alpha/2}} \binom{N}{k} 0.5^N \leq \frac{\alpha}{2} \tag{2.1}$$

Obviously, we have the relation $k'_{\alpha/2} = N - k_{\alpha/2}$.

The ordinary sign test statistics with these rejection regions are consistent against the respective one- and two-sided alternatives. Applying the criterion of consistency given in Chap. 1, since $E(K/N) = \theta$ and $\text{var}(K/N) = \theta(1 - \theta)/N \to 0$ as $N \to \infty$, K provides a consistent test statistic.

We know by the central-limit theorem that the binomial distribution approaches the normal distribution as $N \to \infty$. The normal approximation to the binomial is especially good when $\theta = 0.50$. Therefore, for moderate and large values of N (say at least 12), it is satisfactory to use the normal approximation to the binomial to determine rejection regions. Since this is a continuous approximation to a discrete distribution, a continuity correction of 0.5 may be incorporated in the calculations. For example, for the alternative $H_1: M > M_0$, if N is large H_0 is rejected for $K \geq k_\alpha$, where k_α satisfies

$$\frac{(k_\alpha - 0.5) - 0.5N}{0.5\sqrt{N}} = z_\alpha \tag{2.2}$$

and z_α is a standard normal deviate.

THE PROBLEM OF ZERO DIFFERENCES

Theoretically, no problem of zero differences exists because the population was assumed to be continuous in the vicinity of the median. In reality, of course, zero differences can and do occur. The usual procedure followed here is simply to ignore zero differences and reduce N accordingly. The inferences then are conditional upon the observed number of nonzero differences. Alternatively, half of the zeros may be treated as plus and half as minus, or plus and minus signs may be assigned at random to the zeros, or some strictly conservative approach may be followed like assigning to all zeros that sign which is least conducive to rejection of H_0.

POWER FUNCTION

In contrast to most nonparametric tests, the power function of the sign test is simple to determine. The random variable K follows the binomial probability distribution with parameter θ even when the null hypothesis is false.   The power is a function of θ, so that a general power curve can be plotted for, say, $H_1: \theta > 0.50$ by calculating for arbitrary $\theta > 0.50$

$$Pw(\theta) = \sum_{k=k_\alpha}^{N} \binom{N}{k} \theta^k (1 - \theta)^{N-k} \tag{2.3}$$

If the power function is desired for a more parametric type of situation where the population distribution is fully specified, the parameter $\theta = P(X > M_0)$ can be calculated.   This type of power function would be desirable for comparisons between the sign test and some parametric test for location.

As an example, let us calculate the power of the sign test of $H_0: M = 28$ versus $H_1: M > 28$ for $N = 16$, under the assumption that the population is normally distributed with mean 29.04 and known standard deviation 1.   Using binomial tables, we find the rejection region is $K \geq 12$.   Under the assumptions given, we can evaluate

$$\theta = P(X > 28) = P\left(\frac{X - \mu}{\sigma} > -1.04\right) = 1 - \Phi(-1.04)$$
$$= 0.8508$$

so that

$$Pw(0.85) = \sum_{k=12}^{16} \binom{16}{k} (0.85)^k (0.15)^{16-k} = 0.9211$$

This would be directly comparable with the normal theory test of $H_0: \mu = 28$ versus $H_1: \mu = 29.04$ with $\sigma = 1$, since the mean and median do coincide for the normal distributions.   The rejection region for this parametric test with $\alpha = 0.05$ is $\bar{X} \geq 28.41$, and the power is

$$Pw(29.04) = P(\bar{X} \geq 28.41 \mid \mu = 29.04)$$
$$= P\left(\frac{\bar{X} - \mu}{\sigma/\sqrt{n}} \geq -2.52\right) = 1 - \Phi(-2.52) = 0.9941$$

The only difficulty with this type of direct comparison is that the true probability of a type I error for the sign test is

$$Pw(0.50) = \sum_{k=12}^{16} \binom{16}{k} 0.5^{16} = 0.0383$$

If the exact α value were made to equal 0.05 by using a randomized test, the power of the sign test would increase.   The sign-test power of 0.9211 is more directly comparable with that of a normal-theory test where $\alpha = 0.0383$. Then the rejection region is $\bar{X} \geq 28.44$ and

$$Pw(29.04) = 0.9918$$

CONFIDENCE-INTERVAL PROCEDURE

The sign-test technique can be applied to obtain a confidence-interval estimate for the unknown population median. Let the order statistics for the random sample be $X_{(1)} < X_{(2)} < \cdots < X_{(N)}$. The order is preserved if each observation is reduced by the same fixed amount, but the number of plus signs in the resulting differences changes according to the magnitude of this fixed amount subtracted. We would accept the null hypothesis for a two-sided test with significance level α for all values M for which there are exactly K positive numbers among the N differences $X_{(i)} - M$, $i = 1, 2, \ldots, N$, for all K which satisfy the inequality

$$k'_{\alpha/2} + 1 \leq K \leq k_{\alpha/2} - 1 \tag{2.4}$$

where $k'_{\alpha/2}$ and $k_{\alpha/2}$ are integers chosen in accordance with (2.1).

In order to obtain a confidence-interval estimate of M, we need only translate the inequality in (2.4) to an equivalent statement involving order statistics and M. Applying the result given in (2.8.5) of Chap. 2 for confidence intervals for any population quantile to the specific case $p = 0.5$, we have

$$P(X_{(r)} < M < X_{(s)}) = P(r \leq K \leq s - 1)$$

Therefore our $(1 - \alpha)100$ percent confidence-interval estimate of M is

$$X_{(k'_{\alpha/2}+1)} < M < X_{(k_{\alpha/2})} \tag{2.5}$$

In an ordered array of the observations, the confidence-interval end points are then those numbers which are in the $(k'_{\alpha/2} + 1)$st positions from either end. For samples so large that N exceeds the range of available binomial tables, the normal approximation to the binomial can be used to set a $1 - \alpha$ probability statement on K, which in turn can be converted to a $(1 - \alpha)100$ percent confidence interval on M as above. We saw before that

$$P(k'_{\alpha/2} + 1 \leq K \leq k_{\alpha/2} - 1) = 1 - \alpha$$

if and only if

$$P(X_{(k'_{\alpha/2}+1)} < M < X_{(k_{\alpha/2})}) = 1 - \alpha$$

Using the normal approximation with continuity correction as in (2.2), we have

$$\frac{k_{\alpha/2} - 0.5 - 0.5N}{0.5 \sqrt{N}} = z_{\alpha/2}$$

$$k_{\alpha/2} = 0.5N + 0.5 + 0.5z_{\alpha/2} \sqrt{N}$$

$$\frac{k'_{\alpha/2} + 0.5 - 0.5N}{0.5 \sqrt{N}} = -z_{\alpha/2}$$

$$k'_{\alpha/2} = 0.5N - 0.5 - 0.5z_{\alpha/2} \sqrt{N}$$

PAIRED–SAMPLE PROCEDURES

The single-sample sign-test procedures for hypothesis testing and confidence-interval estimation of M are equally applicable to paired-sample data. From a random sample of N pairs

$$(X_1, Y_1), (X_2, Y_2), \ldots, (X_N, Y_N)$$

the N differences are formed

$$D_i = X_i - Y_i \qquad \text{for } i = 1, 2, \ldots, N$$

If the population of differences is assumed continuous at its median M so that $P(D = M) = 0$, and θ is defined as $\theta = P(D > M)$, the same procedures are clearly valid here with X_i replaced everywhere by D_i.

It should be emphasized that this is a test for the median difference, which is not necessarily the same as the difference of the two medians M_X and M_Y. The following simple example will serve to illustrate this often misunderstood fact. Let X and Y have the joint distribution

$$f_{X,Y}(x,y) = \tfrac{1}{2} \qquad \text{for } y - 1 \leq x \leq y, \ -1 \leq y \leq 1$$
$$\text{or } \ y + 1 \leq x \leq 1, \ -1 \leq y \leq 0$$

Then X and Y are uniformly distributed over the shaded region in Fig. 2.1. The marginal distributions of X and Y are identical, both being uniform

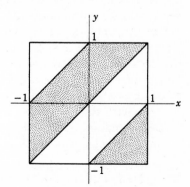

Fig. 2.1

on the interval $(-1,1)$, so that $M_X = M_Y = 0$. It is clear that where X and Y have opposite signs, in quadrants II and IV,

$$P(X < Y) = P(X > Y)$$

while in quadrants I and III, $X < Y$ always. For all pairs, then, we have $P(X < Y) = \frac{3}{4}$, which implies that the median of the population of differences is smaller than zero. It will be left as an exercise for the reader to show that the cumulative distribution function of $D = X - Y$ here is

$$F_D(d) = \begin{cases} 0 & d \le -1 \\ \dfrac{(d + 1)(d + 3)}{4} & -1 < d \le 0 \\ \frac{3}{4} & 0 < d \le 1 \\ \dfrac{d(4 - d)}{4} & 1 < d \le 2 \\ 1 & d > 2 \end{cases} \qquad (2.6)$$

The median difference is that value M such that $F_D(M) = \frac{1}{2}$, or $M = -2 + \sqrt{3}$.

In general, then, it is *not* true that $M = M_D = M_X - M_Y$. On the other hand, it is true that a mean of differences equals the difference of means. Since the mean and median coincide for symmetric distributions, if the X and Y populations are both symmetric and $M_X = M_Y$, and if the difference population is also symmetric,† then $M_D = M_X - M_Y$, and $M_X = M_Y$ is a necessary and sufficient condition for $M_D = 0$.

6.3 THE WILCOXON SIGNED-RANK TEST

Since the ordinary single-sample sign test utilizes only the signs of the differences between each observation and the hypothesized median M_0, the magnitudes of these observations relative to M_0 are ignored. Assuming that such information is available, a test statistic which takes into account these individual relative magnitudes might be expected to give better performance. If we are willing to make the assumption that the parent population is symmetric, the Wilcoxon signed-rank test statistic provides an alternative test of location which is affected by both the magnitudes and signs of these differences. The rationale and properties of this test will be treated in this section.

As with the single-sample situation of the previous section, we are given a random sample of N observations $X_1, X_2, \ldots, X_N$ from a continuous population with median M, but now we assume that this

† The difference population is symmetric if X and Y are symmetric and independent or if $f_{X,Y}(x,y) = f_{X,Y}(-x,-y)$.

population is symmetric.   Under the null hypothesis

$$H_0: M = M_0$$

the differences $D_i = X_i - M_0$ are symmetrically distributed about zero, so that positive and negative differences of equal absolute magnitude have the same probability of occurrence, i.e.,

$$P(D_i \leq -c) = P(D_i \geq c) = 1 - P(D_i \leq c)$$

With the assumption of a continuous population, we need not be concerned theoretically with zero or tied absolute differences $|D_i|$. Suppose we order these absolute differences $|D_1|, |D_2|, \ldots, |D_N|$ from smallest to largest and assign them ranks $1, 2, \ldots, N$, keeping track of the original signs of the differences D_i. If M_0 is the true median of the symmetrical population, the expectation of the sum of the ranks of the positive differences T^+ equals the expectation of the sum of the ranks of the negative differences T^-. Since the sum of all the ranks is a constant, $T^+ + T^- = \sum_{i=1}^{N} i = N(N + 1)/2$, test statistics based on T^+ only, T^- only, or $T^+ - T^-$ are linearly related and therefore equivalent criteria. In contrast to the ordinary one-sample sign test, the value of T^+, say, is influenced not only by the number of positive differences but also by their relative magnitudes.   When the symmetry assumption is justifiable, T^+ can therefore be expected to provide a more efficient test of location.

The derived sample data on which these test statistics are based consist of the set of N integer ranks $\{1, 2, \ldots, N\}$ and a corresponding set of N plus and minus signs.   The rank i is associated with a plus or minus sign according to the sign of $D_j = X_j - M_0$, where D_j occupies the ith position in the ordered array of absolute differences $|D_j|$. If we let $r(\cdot)$ denote the rank of a random variable, the Wilcoxon signed-rank statistics can be written symbolically as

$$T^+ = \sum_{i=1}^{N} Z_i r(|D_i|) \qquad T^- = \sum_{i=1}^{N} (1 - Z_i) r(|D_i|)$$

$$T^+ - T^- = 2 \sum_{i=1}^{N} Z_i r(|D_i|) - \frac{N(N + 1)}{2} \tag{3.1}$$

where

$$Z_i = \begin{cases} 1 & \text{if } D_i > 0 \\ 0 & \text{if } D_i < 0 \end{cases}$$

Since addition is commutative, it is clear that we can assume without loss of generality that the subscripts on the original sample data are such that $|D_1|, |D_2|, \ldots, |D_N|$ are order statistics and replace $r(|D_i|)$ by i

in (3.1) and Z_i by $Z_{(i)}$, where

$$Z_{(i)} = \begin{cases} 1 & \text{if the difference whose absolute value} \\ & \qquad \text{has rank } i \text{ is positive} \\ 0 & \text{if the difference whose absolute value} \\ & \qquad \text{has rank } i \text{ is negative} \end{cases}$$

Then, for example, we can write

$$T^+ = \sum_{i=1}^{N} i Z_{(i)} \qquad\qquad (3.2)$$

where the $Z_{(i)}$ are independent Bernoulli random variables but are not identically distributed. The moments of $Z_{(i)}$ then are

$$E(Z_{(i)}) = \theta_i \qquad \text{var}(Z_{(i)}) = \theta_i(1 - \theta_i) \qquad \text{cov}(Z_{(i)}, Z_{(j)}) = 0$$

where the Bernoulli parameter is given by

$$\theta_i = P(Z_{(i)} = 1) = P[r(|D_j|) = i \cap D_j > 0$$
$$\text{for some } j = 1, 2, \ldots, N]$$
$$= P(\text{the } i\text{th order statistic among } |D_1|, |D_2|, \ldots, |D_N|$$
$$\text{corresponds to a positive difference})$$
$$= \int_0^\infty \frac{N!}{(i-1)!(N-i)!} [F_{|D|}(u)]^{i-1}[1 - F_{|D|}(u)]^{N-i} f_D(u) \, du$$
$$= N \binom{N-1}{i-1} \int_0^\infty [F_D(u) - F_D(-u)]^{i-1}[1 - F_D(u)$$
$$+ F_D(-u)]^{N-i} f_D(u) \, du \quad (3.3)$$

Since T^+ is a linear combination of these independent random variables, the exact mean and variance of T^+ for any distribution are easily found if θ_i can be determined. In the null case, θ_i must be evaluated when X is symmetric about M_0, which implies that $D = X - M_0$ is symmetric about zero so that $F_D(0) = \frac{1}{2}$, $F_D(-u) = 1 - F_D(u)$. Substituting these results in (3.3), we have

$$\theta_i = N \binom{N-1}{i-1} \int_0^\infty [2F_D(u) - 1]^{i-1}[2 - 2F_D(u)]^{N-i} f_D(u) \, du$$
$$= \frac{N}{2} \binom{N-1}{i-1} \int_0^1 v^{i-1}(1 - v)^{N-i} \, dv$$
$$= \frac{N}{2} \binom{N-1}{i-1} B(i, N - i + 1) = \frac{1}{2}$$

The general moments of T^+ from (3.2) are

$$E(T^+) = \sum_{i=1}^{N} i\theta_i \qquad \text{var}(T^+) = \sum_{i=1}^{N} i^2\theta_i(1 - \theta_i) \qquad (3.4)$$

Therefore in the null case these reduce simply to

$$E(T^+ \mid H_0) = \frac{N(N + 1)}{4} \qquad \text{var}(T^+ \mid H_0) = \frac{N(N + 1)(2N + 1)}{24}$$

$$(3.5)$$

Since the θ_i are all functions of

$$\theta = F_D(0) = 1 - P(X - M_0 > 0) = P(X > M_0)$$

and $\theta = \frac{1}{2}$ under H_0, we can write

$$E\left[\frac{2T^+}{N(N + 1)}\right] = g(\theta) \qquad \text{where } g(\theta) = \frac{1}{2} \text{ under } H_0$$

Further we note that for any distribution, since $\theta_i(1 - \theta_i) \leq \frac{1}{4}$ for all $i = 1, 2, \ldots, N$, we have

$$\text{var}\left[\frac{2T^+}{N(N + 1)}\right] = 4 \sum_{i=1}^{N} \frac{i^2\theta_i(1 - \theta_i)}{N^2(N + 1)^2}$$

$$\leq \sum_{i=1}^{N} \frac{i^2}{N^2(N + 1)^2} = \frac{2N + 1}{6N(N + 1)} \to 0 \text{ as } N \to \infty$$

Therefore, using the method described in Chap. 1, the test with rejection region

$$T^+ \in R \qquad \text{for } \frac{2t^+}{N(N + 1)} - \frac{1}{2} \geq k$$

is consistent against alternatives of the form $\theta > \frac{1}{2}$. This result is reasonable since if the true population median exceeds M_0, the sample data would reflect this by having most of the larger ranks correspond to positive differences. A similar two-sided rejection region of T^+ centered on $N(N + 1)/4$ is consistent against alternatives $\theta \neq \frac{1}{2}$.

Actually the test statistic T^+ is consistent against a larger class of alternatives. This class can be defined by investigating the expected value of T^+ for any distribution. Since a two-sided rejection region centered on $\frac{1}{2}$ for the test statistic $2T^+/[N(N + 1)]$ is consistent against all alternatives for which $E\{2T^+/[N(N + 1)]\} \neq \frac{1}{2}$, or $E\{T^+/[N(N + 1)]\} \neq \frac{1}{4}$, the criterion for consistency is found as follows. Since

$$E(T^+) = \sum_{i=1}^{N} i\theta_i = N \int_0^\infty \sum_{i=1}^{N} [(i - 1) + 1]\binom{N - 1}{i - 1}$$

$$[F_{|D|}(u)]^{i-1}[1 - F_{|D|}(u)]^{N-i}f_D(u) \, du$$

$$= N \int_0^\infty [(N - 1)F_{|D|}(u) + 1]f_D(u) \, du$$

$$= N(N - 1) \int_0^\infty [F_D(u) - F_D(-u)]f_D(u) \, du$$

$$+ N[1 - F_D(0)] \quad (3.6)$$

the consistency condition is

$$E\left[\frac{2T^+}{N(N+1)}\right] = 2\frac{N-1}{N+1}\int_0^\infty [F_D(u) - F_D(-u)]f_D(u)\ du$$

$$+ \frac{2}{N+1}[1 - F_D(0)] \neq \tfrac{1}{2}$$

which for N large is approximately

$$\int_0^\infty [F_D(u) - F_D(-u)]f_D(u)\ du \neq \tfrac{1}{4}$$

In order to determine the rejection regions precisely for our consistent test, the probability distribution of T^+ must be determined under the null hypothesis

$$H_0: \theta = P(X > M_0) = 0.5$$

The extreme values of T^+ are zero and $N(N+1)/2$, occurring when all differences are of the same sign, negative and positive, respectively. The mean and variance were found in (3.5). Since T^+ is completely determined by the sign indicators $Z_{(i)}$ in (3.2), the sample space can be considered to be the set of all possible N-tuples $\{(z_1, z_2, \ldots, z_N)\}$ with components either 1 or 0, of which there are 2^N. Each of these distinguishable arrangements is equally likely under H_0. Therefore, the null probability distribution of T^+ is given by

$$P(T^+ = t) = \frac{u(t)}{2^N} \tag{3.7}$$

where $u(t)$ is the number of ways to assign plus and minus signs to the first N integers such that the sum of the positive integers equals t. Every assignment has a conjugate assignment with plus and minus signs interchanged, and T^+ for this conjugate is

$$\sum_{i=1}^N i(1 - Z_{(i)}) = \frac{N(N+1)}{2} - \sum_{i=1}^N iZ_{(i)}$$

Since every assignment occurs with equal probability, this implies that the null distribution of T^+ is symmetric about its mean $N(N+1)/4$.

Because of the symmetry property, only one-half of the null distribution need be determined. A systematic method of generating the complete distribution of T^+ for $N = 4$ is shown in Table 3.1.

$$f_{T^+}(t) = \begin{cases} \tfrac{1}{16} & t = 0, 1, 2, 8, 9, 10 \\ \tfrac{2}{16} & t = 3, 4, 5, 6, 7 \\ 0 & \text{otherwise} \end{cases}$$

Tables can be constructed in this way for all N.

Table 3.1

Value of T^+	Ranks associated with positive differences	Number of sample points $u(t)$
10	1, 2, 3, 4	1
9	2, 3, 4	1
8	1, 3, 4	1
7	1, 2, 4; 3, 4	2
6	1, 2, 3; 2, 4	2
5	1, 4; 2, 3	2

To use the rank sum statistics in hypothesis testing, the entire distribution is not necessary. In fact, one set of critical values is sufficient for even a two-sided test, because of the relationship $T^+ + T^- = N(N + 1)/2$ and the symmetry of T^+. Large values of T^+ correspond to small values of T^-, and T^+ and T^- are identically distributed since

$$\begin{aligned}
P(T^+ \geq c) &= P\left[T^+ - \frac{N(N + 1)}{4} \geq c - \frac{N(N + 1)}{4}\right] \\
&= P\left[T^+ - \frac{N(N + 1)}{4} \leq \frac{N(N + 1)}{4} - c\right] \\
&= P\left[T^+ - \frac{N(N + 1)}{2} \leq -c\right] \\
&= P(-T^- \leq -c) = P(T^- \geq c)
\end{aligned}$$

Since it is more convenient to work with smaller sums, tables of left-hand critical values are generally set up for the random variable T, which may denote either T^+ or T^-. If t_α is the number such that $P(T \leq t_\alpha) = \alpha$, the appropriate rejection regions for size α tests of $H_0: M = M_0$ are

$T^- \leq t_\alpha$ for $H_1: M > M_0$
$T^+ \leq t_\alpha$ for $H_1: M < M_0$
$T^+ \leq t_{\alpha/2}$ or $T^- \leq t_{\alpha/2}$ for $H_1: M \neq M_0$

Suppose that $N = 8$ and critical values are to be found for one- or two-sided tests at nominal $\alpha = 0.05$. Since $2^8 = 256$ and $256(0.05) = 12.80$, we need at least 13 cases of assignments of signs. We enumerate the small values of T^+ in Table 3.2. Since $P(T \leq 6) = 14/256 > 0.05$ and $P(T \leq 5) = 10/256 = 0.039$, $t_{0.05} = 5$ with exact probability of a type I error 0.039. Similarly we find $t_{0.025} = 3$ with exact $P(T \leq 3) = 0.0195$.

When the distribution is needed for several sample sizes, a simple recursive relation can be used to generate the probabilities. Let T_N^+

Table 3.2

Value of T^+	Ranks associated with positive differences	Number of sample points
0		1
1	1	1
2	2	1
3	3; 1, 2	2
4	4; 1, 3	2
5	5; 1, 4; 2, 3	3
6	6; 1, 5; 2, 4; 1, 2, 3	4

denote the sum of the ranks associated with positive differences D_i for a sample of N observations. Consider a set of $N - 1$ ordered $|D_i|$ with ranks 1, 2, . . . , $N - 1$ assigned, for which the null distribution of T_{N-1}^+ is known. To obtain the distribution of T_N^+ from this, an extra observation D_N is added, and we can assume without loss of generality that $|D_N| > |D_i|$ for all $i \leq N - 1$. The rank of $|D_N|$ is then N. If $D_N > 0$, the value of T_N^+ will exceed that of T_{N-1}^+ by the amount N for every arrangement of the $N - 1$ observations, but if $D_N < 0$, T_N^+ will be equal to T_{N-1}^+. Using the notation in (3.7), this can be stated as

$$P(T_N^+ = k) = \frac{u_N(k)}{2^N}$$

$$= \frac{u_{N-1}(k - N)P(D_N > 0) + u_{N-1}(k)P(D_N < 0)}{2^{N-1}}$$

$$= \frac{u_{N-1}(k - N) + u_{N-1}(k)}{2^N} \tag{3.8}$$

If N is moderate and systematic enumeration is desired, classification according to the number of positive differences D_i is often helpful. Define the random variable U as the number of positive differences. Then U follows the binomial distribution with parameter $\frac{1}{2}$, so that

$$P(T^+ = t) = \sum_{i=0}^{N} P(U = i \cap T^+ = t)$$

$$= \sum_{i=0}^{N} P(U = i)P(T^+ = t \mid U = i)$$

$$= \sum_{i=0}^{N} \binom{N}{i} \left(\frac{1}{2}\right)^N P(T^+ = t \mid U = i)$$

Tables of the left-hand critical values are available in many sources, e.g., Wilcoxon (1945, 1947, 1949) or Bradley (1968). For large samples, tables generally are not available, and the enumeration process is quite lengthy. With T^+ in the form of (3.2), we have a linear combination of N independent (though not identically distributed) Bernoulli random variables. From a generalization of the central-limit theorem, the asymptotic distribution of a standardized T^+ is the normal with mean zero and variance 1. In the null case, using the moments given in (3.5), the distribution of

$$Z = \frac{4T^+ - N(N + 1)}{\sqrt{2N(N + 1)(2N + 1)/3}} \tag{3.9}$$

approaches the standard normal as $N \to \infty$. The test for, say, $H_1: M > M_0$ can be performed for large N by computing (3.9) and rejecting H_0 for $Z \geq z_\alpha$. The approximation is generally adequate for N at least 15. A continuity correction of 0.5 may be used.

THE PROBLEM OF ZERO AND TIED DIFFERENCES

Since we assumed originally that the random sample was drawn from a continuous population, the problems of tied observations and zero differences could be ignored theoretically. In practice, generally any zero differences (observations equal to M_0) are ignored and N reduced accordingly, although the other procedures described for the ordinary sign test are equally applicable here. In the case where two or more absolute values of differences are equal, that is, $|d_i| = |dj|$ for at least one $i \neq j$, the observations are tied. The ties can be dealt with by any of the procedures described in Sec. 5.3. The midrank method is usually used, and the sign associated with the midrank of $|d_i|$ is determined by the original sign of d_i as before. The probability distribution of T is clearly not the same in the presence of tied ranks, but the effect is generally slight and thus no correction is made.

POWER FUNCTION

Since the indicator variables $Z_{(i)}$ of (3.2) are independent and follow the Bernoulli distribution with parameter θ_i regardless of whether the null hypothesis is true, the probability distribution of T^+ could be worked out under a specified alternative distribution, since then the θ_i of (3.3) can be evaluated. If exactly k specified ranks $r_1, r_2, \ldots, r_k$ have positive signs, the probability of this sample result is the product of their respective θ_j values times the respective $1 - \theta_j$ terms for the remaining ranks. The probability of any particular value of T^+ then is the sum

$$\sum \prod_{i=1}^{k} \theta_{r_i} \prod_{1 \leq j \neq r_i \leq N} (1 - \theta_j) \tag{3.10}$$

where the summation is extended over all sets of assignments of positive signs which lead to that value of T^+. From a listing of ranks associated with positive differences, like that given in Table 3.1, the exact power can then be found by summing the probabilities in (3.10) for those values of T^+ which are in the rejection region.

For large samples, an approximation to the power of the Wilcoxon signed-rank test can be found using a normal deviate with mean and variance calculated from (3.4).

It should be noted that the probability distribution of T^+ is no longer symmetric when the null hypothesis is not true, so that T^+ and T^- are not identically distributed under the alternative. For the probability of a particular value of T^-, we must interchange θ and $1 - \theta$ in (3.10) or use

$$P(T^- = k) = P\left[\frac{N(N + 1)}{2} - T^+ = k\right]$$

CONFIDENCE-INTERVAL PROCEDURES

As with the ordinary one-sample sign test, the Wilcoxon signed-rank procedure lends itself to confidence-interval estimation of the unknown population median M. In fact, there are two methods of interval estimation available here. Both will give the confidence limits as those values of M which do not lead to rejection of the null hypothesis, but one amounts to a trial-and-error procedure while the other is systematic and provides a unique interval. For any sample size N, we can find that number $t_{\alpha/2}$ such that if the true population median is M and T is calculated for the derived sample values $X_i - M$, then

$$P(T^+ \leq t_{\alpha/2}) = \frac{\alpha}{2} \quad \text{and} \quad P(T^- \leq t_{\alpha/2}) = \frac{\alpha}{2}$$

The null hypothesis will be accepted for all numbers M which make $T^+ > t_{\alpha/2}$ and $T^- > t_{\alpha/2}$. The confidence-interval technique is to find those two numbers, say M_1 and M_2, where $M_1 < M_2$, such that when T is calculated for the two sets of differences $X_i - M_1$ and $X_i - M_2$, at the significance level α, T^+, or T^-, whichever is smaller, is just short of significance, i.e., slightly larger than $t_{\alpha/2}$. Then the $100(1 - \alpha)$ percent confidence-interval estimate of M is $M_1 < M < M_2$.

In the trial-and-error procedure, we simply choose some suitable values of M and calculate the resulting values of T^+ or T^-, stopping whenever we get numbers slightly larger than $t_{\alpha/2}$. This generally does not lead to a unique interval, and the manipulations can be tedious even for moderate-sized samples. The technique is best illustrated by an example. The following eight observations are drawn from a continuous, symmetric population:

$$-1, 6, 13, 4, 2, 3, 5, 9$$

For $N = 8$ the two-sided rejection region of nominal size 0.05 was found on page 111 to be $t_{\alpha/2} = 3$ with exact significance level

$$\alpha = P(T^+ \leq 3) + P(T^- \leq 3) = {}^{10}\!/_{256} = 0.039$$

In Table 3.3 we try six different values for M and calculate T^+ or T^-, whichever is smaller, for the differences $X_i - M$. The example illustrates a number of difficulties which arise. In the first trial choice of M, the number 4 was subtracted and the resulting differences contained three sets of tied pairs and one zero even though the original sample contained neither ties nor zeros. If the zero difference is ignored, N must be reduced to 7 and then the $t_{\alpha/2} = 3$ is no longer accurate for $\alpha = 0.039$. The midrank method could be used to handle the ties, but this also disturbs the accuracy of $t_{\alpha/2}$. Since there seems to be no real solution to these problems, we try to avoid zeros and ties by judicious choices for our M values for subtraction. Since these data are all integers, a choice for M which is noninteger-valued obviously reduces the likelihood of ties and makes zero values impossible. Since T^- for the differences $X_i - 1.1$ was equal to $t_{\alpha/2}$, we know that M_1 will be larger than 1.1. Since $X_i - 1.5$ yields $T^- = 3.5$ using the midrank method, we shall choose $M_1 = 1.5$. The next three columns represent an attempt to find an M which makes T^+ around 4. They do illustrate the fact that M_1 and M_2 are far from unique. Clearly M_2 is in the vicinity of 9, but the differences $X_i - 9$ yield a zero. We conclude there is no need to go further. An approximate 96.1 percent confidence interval on M is given by $1.5 < M < 9$. The interpretation is that hypothesized values of M within this range will lead to acceptance of the null hypothesis for an exact significance level of 0.039.

Table 3.3

X_i	$X_i - 4$	$X_i - 1.1$	$X_i - 1.5$	$X_i - 9.1$	$X_i - 8.9$	$X_i - 8.95$
-1	-5	-2.1	-2.5	-10.1	-9.9	-9.95
6	2	4.9	4.5	-3.1	-2.9	-2.95
13	9	11.9	11.5	3.9	4.1	4.05
4	0	2.9	2.5	-5.1	-4.9	-4.95
2	-2	.9	.5	-7.1	-6.9	-6.95
3	-1	1.9	1.5	-6.1	-5.9	-5.95
5	1	3.9	3.5	-4.1	-3.9	-3.95
9	5	7.9	7.5	$-.1$	.1	.05
T^+ or T^-		3	3.5	3	5	5

The procedure is undoubtedly tedious, but the limits obtained are reasonably accurate. The numbers should be tried systematically to narrow down the range of possibilities. Thoughtful study of the intermediate results usually reduces the additional number of trials required.

A different method of construction which leads to a unique interval and is much easier to apply is described in Noether [(1967), pp. 57–58]. The procedure is to convert the interval $T^+ > t_{\alpha/2}$ and $T^- > t_{\alpha/2}$ to an equivalent statement on M whose end points are functions of the observations X_i. For this purpose we must analyze the comparisons involved in determining the ranks of the differences $r(|X_i - M_0|)$ and the signs of the differences $X_i - M_0$, since T^+ and T^- are functions of these comparisons. Recall from (5.1) that the rank of any random variable in a set $\{V_1, V_2, \ldots, V_N\}$ can be written symbolically as

$$r(V_i) = \sum_{k=1}^{N} S(v_i - v_k) = \sum_{k \neq i} S(v_i - v_k) + 1$$

where

$$S(u) = \begin{cases} 1 & \text{if } u \geq 0 \\ 0 & \text{if } u < 0 \end{cases}$$

To compute a rank, then, we make $\binom{N}{2}$ comparisons of pairs of different numbers and one comparison of a number with itself. To compute the set of all ranks, we make $\binom{N}{2}$ comparisons of pairs and N identity comparisons, a total of $\binom{N}{2} + N = N(N+1)/2$ comparisons. Substituting the rank function in (3.1), we obtain

$$T^+ = \sum_{i=1}^{N} Z_i r(|X_i - M_0|)$$

$$= \sum_{i=1}^{N} Z_i + \sum_{i=1}^{N} \sum_{k \neq i} Z_i S(|X_i - M_0| - |X_k - M_0|) \tag{3.11}$$

Therefore these comparisons affect T^+ as follows:

1. A comparison of $|X_i - M_0|$ with itself adds 1 to T^+ if $X_i - M_0 > 0$.
2. A comparison of $|X_i - M_0|$ with $|X_k - M_0|$ for any $i \neq k$ adds 1 to T^+ if $|X_i - M_0| > |X_k - M_0|$ and $X_i - M_0 > 0$, that is, $X_i - M_0 > |X_k - M_0|$. If $X_k - M_0 > 0$, this occurs when $X_i > X_k$, and if $X_k - M_0 < 0$, we have $X_i + X_k > 2M_0$ or $(X_i + X_k)/2 > M_0$. But when $X_i - M_0 > 0$ and $X_k - M_0 > 0$, we have $(X_i + X_k)/2 > M_0$ also.

Combining these two results, then, $(X_i + X_k)/2 > M_0$ is a necessary condition for adding 1 to T^+ for all i, k. Similarly if $(X_i + X_k)/2 < M_0$, then this comparison adds 1 to T^-. The relative magnitudes of the $N(N + 1)/2$ averages of pairs $(X_i + X_k)/2$ for all $i \leq k$ then determine the range of values for hypothesized numbers M_0 which will not lead to rejection of H_0. If these $N(N + 1)/2$ averages are arranged as order statistics, the two numbers which are in the $(t_{\alpha/2} + 1)$st position from either end are the end points of the $100(1 - \alpha)$ percent confidence interval on M. Note that this procedure is exactly analogous to the ordinary sign-test confidence interval given in (2.5) except that here the order statistics are for the averages of all pairs of observations instead of the original observations.

The data on page 115 for $N = 8$ arranged in order of magnitude are $-1, 2, 3, 4, 5, 6, 9, 13$, and the 36 averages are:

−1.0	0.5	1.0	1.5	2.0	2.5	4.0	6.0
2.0	2.5	3.0	3.5	4.0	5.5	7.5	
3.0	3.5	4.0	4.5	6.0	8.0		
4.0	4.5	5.0	6.5	8.5			
5.0	5.5	7.0	9.0				
6.0	7.5	9.5					
9.0	11.0						
13.0							

For exact $\alpha = 0.039$, we found before that $t_{\alpha/2} = 3$. Since the fourth largest numbers from either end are 1.5 and 9.0, the confidence interval is $1.5 < M < 9$. This result agrees exactly with that obtained by the previous method, but this will not always be the case since the trial-and-error procedure does not yield unique end points.

The process of determining a confidence interval on M by the above method is much facilitated by Tukey's graphical method of construction, which can be described as follows. Each of the N observations x_i is denoted by a dot on a horizontal scale. The closed interval $[x_{(1)}, x_{(N)}]$ then includes all dots. Form an isosceles triangle ABC by lines joining $x_{(1)}$ at A and $x_{(N)}$ at B each with a point C anywhere on the vertical line passing through the midrange value $(x_{(1)} + x_{(N)})/2$. Through each point x_i on the line segment AB draw lines parallel to AC and BC, marking each intersection with a dot. There will be $N(N + 1)/2$ intersections, the abscissas of which are all the $(x_i + x_k)/2$ values where $1 \leq i \leq k \leq N$. Vertical lines drawn through the $(t_{\alpha/2} + 1)$st intersection point from the left and right will allow us to read the respective confidence-interval end points on the horizontal scale. Figure 3.1 illustrates this method for the numerical data above.

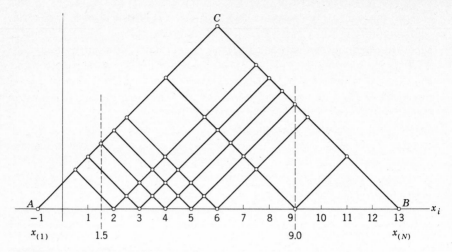

Fig. 3.1

PAIRED-SAMPLE PROCEDURES

The Wilcoxon signed-rank test was actually proposed for use with paired-sample data in making inferences concerning the value of the median of the population of differences. Given a random sample of N pairs

$$(X_1, Y_1), (X_2, Y_2), \ldots, (X_N, Y_N)$$

their differences are

$$X_1 - Y_1, X_2 - Y_2, \ldots, X_N - Y_N$$

We assume these are independent observations from a population of differences which is continuous and symmetric with median M. In order to test the hypothesis

$$H_0: M = M_0$$

form the N differences $D_i = X_i - Y_i - M_0$ and rank their absolute magnitudes from smallest to largest using integers $\{1, 2, \ldots, N\}$, keeping track of the original sign of each difference. Then the above procedures for hypothesis testing and confidence intervals are equally applicable here with the same notation, except that the parameter M must be interpreted now as the median of the population of differences.

USE OF WILCOXON STATISTICS TO TEST FOR SYMMETRY

The Wilcoxon signed-rank statistics can also be considered tests for symmetry if the only assumption made is that the random sample is drawn from a continuous distribution. If the null hypothesis states that the population is symmetric with median M_0, the null distributions of

T^+ and T^- are exactly the same as before. If the null hypothesis is rejected using a two-sided test, say, we must conclude here that either the population is symmetric with median not equal to M_0 or that the population is asymmetric.

6.4 GENERAL DISCUSSION OF THE TWO SIGN TESTS

The ordinary sign test and the signed-rank test are generally both useful in the same experimental situations regarding a single sample or paired samples. The assumptions required are minimal—independence of observations and a population which is continuous at M for the ordinary sign test and continuous everywhere and symmetric for the Wilcoxon test. Experimentally, both tests have the problem of zero differences, and the Wilcoxon test has the additional problem of ties. Both tests are applicable when quantitative measurements are impossible or not feasible, as when rating scales or preferences are used. For the Wilcoxon test, information concerning relative magnitudes as well as directions of differences is required. Only the ordinary sign test can be used for strictly dichotomous data, like yes-no observations. Both are very flexible and simple to use for hypothesis testing or constructing confidence intervals. The null distribution of the ordinary sign test is easier to work with since binomial tables are readily available. The normal approximation is quite accurate for even moderate N in both cases, and neither is particularly hampered by the presence of a moderate number of zeros or ties.

For hypothesis testing, in the paired-sample case the hypothesis need not state an actual median difference but only a relation between medians if both populations are assumed symmetric. For example, we might test the hypothesis that the X population values are on the average p percent larger than Y values. Assuming the medians are a reliable indication of size, we would write

$$H_0: M_X = (1 + 0.01p)M_Y$$

and take differences $D_i = X_i - (1 + 0.01p)Y_i$ and perform either test on these derived data as before.

Only the Wilcoxon signed-rank statistics are appropriate for tests of symmetry since the ordinary sign-test statistic is not at all related to the symmetry or asymmetry of the population. We have $P[(X_i - M) > 0] = \frac{1}{2}$ always, and the sole criterion in determining K is the number of positive signs, thus ignoring the magnitudes of the plus and minus differences. There are a number of other extensions and modifications of the sign-test type of criteria [see, for example, Walsh (1949a, b)].

If the population is symmetric, both sign tests can be considered to be tests for location of the population mean and therefore are direct nonparametric counterparts of Student's t test. Their asymptotic efficiencies relative to the t test under the assumption of a normal population are $2/\pi = 0.637$ for the ordinary sign test and $3/\pi = 0.955$ for the Wilcoxon signed-rank test. The term *asymptotic relative efficiency* was explained briefly in Chap. 1. A discussion of how these particular results were obtained will be given in Chap. 14. The fact that the ARE of the Wilcoxon test is higher is not surprising, as it uses information concerning relative magnitude of the differences in addition to direction of differences.

PROBLEMS

6.1. Verify the cumulative distribution function of differences given in (2.6) and the result $M = -2 + \sqrt{3}$. Find and graph the corresponding probability function of differences.

6.2. Answer parts (a) to (e) using (i) the ordinary sign-test procedure and (ii) the Wilcoxon signed-rank test procedure.

(a) Test at a significance level not exceeding 0.10 the null hypothesis H_0: $M = 2$ against the alternative H_1: $M > 2$, where M is the median of the continuous symmetric population from which is drawn the random sample:

$$-3, \ -6, \ 1, \ 9, \ 4, \ 10, \ 12$$

(b) Give the exact probability of a type I error in (a).

(c) On the basis of the following random sample of pairs:

X	126	131	153	125	119	102	116	163
Y	120	126	152	129	102	105	100	175

test at a significance level not exceeding 0.10 the null hypothesis H_0: $M = 2$ against the alternative H_1: $M \neq 2$, where M is the median of the continuous and symmetric population of differences $D = X - Y$.

(d) Give the exact probability of a type I error in (c).

(e) Give the confidence interval corresponding to the test in (c).

6.3. Generate the random-sampling distributions of T^+ and T^- under the null hypothesis for a random sample of six unequal and nonzero observations.

6.4. Show by calculations from tables that the normal distribution provides reasonably accurate approximations to the critical values of one-sided tests for $\alpha = 0.01$, 0.05, and 0.10 when:

(a) $N = 12$ for the ordinary sign test.

(b) $N = 16$ for the signed-rank test.

6.5. A random sample of 10 observations is drawn from a population which is normal with mean μ and variance 1. Instead of a normal-theory test, the ordinary sign test is used for $H_0: \mu = 0$, $H_1: \mu > 0$, with rejection region $K \in R$ for $k \geq 8$.

(a) Plot the power curve using the exact distribution of K.

(b) Plot the power curve using the normal approximation to the distribution of K.

(c) Discuss how the power function might help in the choice of an appropriate sample size for an experiment.

6.6. Prove that the Wilcoxon signed-rank statistic $T^+ - T^-$ based on a set of nonzero observations $X_1, X_2, \ldots, X_N$ can be written symbolically in the form

$$\sum\sum_{1 \leq i \leq j \leq N} \operatorname{sgn}(X_i + X_j)$$

where

$$\operatorname{sgn}(x) = \begin{cases} 1 & \text{if } x > 0 \\ -1 & \text{if } x < 0 \end{cases}$$

6.7. Let $D_1, D_2, \ldots, D_N$ be a random sample of N nonzero observations from some continuous population which is symmetric with median zero. Define

$$|D_i| = \begin{cases} X_i & \text{if } D_i > 0 \\ Y_i & \text{if } D_i < 0 \end{cases}$$

Assume there are m X values and n Y values, where $m + n = N$ and the X and Y values are independent. Show that the signed-rank test statistic T^+ calculated for these D_i is equal to the sum of the ranks of the X observations in the combined ordered sample of m X's and n Y's and also that $T^+ - T^-$ is the sum of the X ranks minus the sum of the Y ranks. This sum of the ranks of the X's is the test criterion for the Wilcoxon statistic in the two-sample problem to be discussed in Chap. 9. Show how T^+ might be used to test the hypothesis that the X and Y populations are identical.

7
The General Two-sample Problem

7.1 INTRODUCTION

For the matched-pairs sign and signed-rank tests of Chap. 6 the data consisted of two samples, but each element in one sample was linked with a particular element of the other sample by some unit of association. The sampling situation can be described as a case of two dependent samples or alternatively as a single sample of pairs from a bivariate population. When the inferences to be drawn are related only to the population of differences, the first step in the analysis usually consists of forming differences of the paired observations, which leaves only a single set of observations. Therefore, for most inference situations this type of data may be legitimately classified as a one-sample problem. In this chapter we shall be concerned with data which consist of two *independent* random samples; i.e., random samples are drawn independently from each of two populations. Not only are the elements within each sample independent, but also every element in the first sample is independent of every element in the second sample.

The universe consists of two populations, which we call the X and Y populations, with cumulative distribution functions denoted by F_X and F_Y, respectively. We have a random sample of size m drawn from the X population and another random sample of size n drawn independently from the Y population,

$$X_1, X_2, \ldots, X_m \qquad \text{and} \qquad Y_1, Y_2, \ldots, Y_n$$

Usually the hypothesis of interest in the two-sample problem is that the two samples are drawn from identical populations, i.e.,

$$H_0: F_Y(x) = F_X(x) \qquad \text{for all } x$$

If we are willing to make assumptions concerning the form of the common population, in particular that it is the normal distribution, and if, further, we assume that if a difference between the populations does exist, it is only between means or only between variances, the two-sample Student's t test for equality of means and the F test for equality of variances are respectively appropriate. The procedures and performance of these two tests are well known. However, these classical tests may be sensitive to violations of those assumptions which are inherent in the construction of the tests. Any conclusions reached using these tests are only as valid as the underlying assumptions made. If there is reason to suspect a violation of any of these postulates, or if sufficient information to judge their validity is not available, or if a completely general test of equality for unspecified distributions is desired, some nonparametric procedure is in order.

This two-sample situation is perhaps the most frequently discussed problem within the realm of nonparametric statistics. The null hypothesis is almost always formulated as identical populations with the common distribution completely unspecified except for the assumption that it is a continuous distribution function. In this null case, the two random samples can be considered a single random sample of size $N = m + n$ drawn from the common, continuous, but unspecified population. Then the combined ordered configuration of the m X and n Y random variables in the sample is one of the $\binom{m + n}{m}$ possible equally likely arrangements. The sample pattern of arrangement should provide information about the type of difference which may exist in the populations. Thus almost all the tests are based on some function of this combined arrangement. The type of function which is most appropriate depends on the class of difference one hopes to detect, which would be indicated by the alternative. There is an abundance of reasonable alternatives to H_0 which may be considered, but the type easiest to analyze using distribution-free techniques states some functional relationship between the distributions, as

$F_Y(x) = h[F_X(x)]$ for some function h. The most general two-sided alternative states simply

$$H_A: F_Y(x) \neq F_X(x) \qquad \text{for some } x$$

and a corresponding general one-sided function is

$$H_1: F_Y(x) \geq F_X(x) \qquad \text{for all } x$$
$$\quad F_Y(x) > F_X(x) \qquad \text{for some } x$$

In this latter case, we generally say that the random variable X is stochastically larger than the random variable Y.† An arbitrary representation of this relation is given in Fig. 1.1. Figure 1.2 is descriptive of this alternative H_1 where the populations F_X and F_Y are identical in every way except for their means and $\mu_X > \mu_Y$. In other words, the alternative that X is stochastically larger than Y is a family of alternatives which includes as a subclass the more specific alternative $\mu_X > \mu_Y$. (For the reverse inequality on F_X and F_Y, we say X is stochastically smaller than Y.)

If the particular alternative of interest is simply a difference in location, the function h can express this specifically as

$$H_L: F_Y(x) = F_X(x - \theta) \qquad \text{for all } x \text{ and some } \theta \neq 0$$

which is called the *location alternative*. Similarly a *scale alternative* which states a difference in scale would be

$$H_S: F_Y(x) = F_X(\theta x) \qquad \text{for all } x \text{ and some } \theta \neq 1$$

Although the three special alternatives H_1, H_L, and H_S are the most frequently encountered of all those included in the general class H_A, other types of relations may be considered. For example, the function $F_Y(x) = [F_X(x)]^k$ for some k and all x, called the *Lehmann type* of

† Some authors define the term *stochastically larger* somewhat differently, by saying that X is stochastically larger than Y if $P(X > Y) > P(X < Y)$.

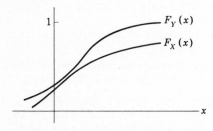

Fig. 1.1

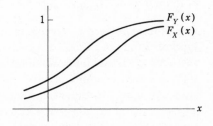

Fig. 1.2

alternative, states that the Y random variables are distributed as the largest of k of the X variables.

The available statistical literature on the two-sample problem is quite extensive. A multitude of tests have been proposed for a wide variety of functional alternatives, but only a few of the best-known tests have been selected for inclusion here. The Wald-Wolfowitz runs test, the Kolmogorov-Smirnov two-sample test, the median test, and the Mann-Whitney U test will be treated in this chapter. Chapters 9 and 10 are concerned with a specific class of tests particularly useful for the location and scale alternatives, respectively.

7.2 THE WALD–WOLFOWITZ RUNS TEST

Let the two sets of independent random variables $X_1, X_2, \ldots, X_m$ and $Y_1, Y_2, \ldots, Y_n$ be combined into a single ordered sequence from smallest to largest, keeping track of which observations correspond to the X sample and which to the Y. Assuming that their probability distributions are continuous, a unique ordering is always possible, since theoretically ties do not exist. For example, with $m = 4$ and $n = 5$, the arrangement might be

$$X \; Y \; Y \; X \; X \; Y \; X \; Y \; Y$$

which indicates that in the pooled sample the smallest element was an X, the second smallest a Y, etc., and the largest a Y. Under the null hypothesis of identical distributions

$$H_0: F_Y(x) = F_X(x) \qquad \text{for all } x$$

we expect the X and Y random variables to be well mixed in the ordered configuration, since the $m + n = N$ random variables constitute a single random sample of size N from the common population. With a run defined as in Chap. 3 as a sequence of identical letters preceded and followed by a different letter or no letter, the total number of runs in the ordered pooled sample is indicative of the degree of mixing. A pattern of arrangement with too few runs would suggest that this group of N is not a single random sample but instead is composed of two samples from two distinguishable populations. For example, if all the elements in the X sample are less than all the elements in the Y sample, there would be only two runs. This particular configuration might indicate not only that the populations are not equivalent but also that the X's are stochastically smaller than the Y's. However, the reverse ordering also contains only two runs, and therefore a test criterion based solely on the total number of runs cannot distinguish these cases. The runs test is appropriate primarily then when the alternative is completely general and two-sided, as in

$$H_A: F_Y(x) \neq F_X(x) \qquad \text{for some } x$$

We define the random variable R as the total number of runs in the combined ordered arrangement of m X and n Y random variables. Since too few runs tend to discredit the null hypothesis when the alternative is H_A, the Wald-Wolfowitz (1940) runs test for significance level α generally has the rejection region

$$R \leq c_\alpha$$

where c_α is chosen to be the largest integer satisfying

$$P(R \leq c_\alpha) \leq \alpha \qquad \text{when } H_0 \text{ is true}$$

Since the X and Y observations are two types of objects arranged in a completely random sequence if H_0 is true, the null probability distribution of R is exactly the same as was found in Chap. 3 for the runs test for randomness. The distribution is given in Theorem 3.2.2 with n_1 and n_2 replaced by m and n, respectively, assuming the X's are called type 1 objects and Y's are the type 2 objects. The other properties of R discussed in that chapter, including the moments and asymptotic null distribution, are also unchanged. The only difference here is that a one-sided test is being used.

THE PROBLEM OF TIES

Ideally, no ties should occur because of the assumption of continuous populations. In practice, ties do not present a problem in counting the number of runs unless the tie is across samples; i.e., two or more observations from different samples have exactly the same magnitude. For a conservative test, we can break all ties in all possible ways and compute the total number of runs for each resolution of all ties. Then the actual r used is the largest computed value, since that is the one least likely to lead to rejection of H_0. For each group of ties across samples, where there are s x's and t y's of equal magnitude for some $s \geq 1, t \geq 1$, there are $\binom{s+t}{s}$ ways to break the ties. Thus if there are k groups of ties, the total number of values of r to be computed is the product

$$\prod_{i=1}^{k} \binom{s_i + t_i}{s_i}$$

DISCUSSION

The Wald-Wolfowitz runs test is extremely general, being consistent against all types of differences in populations [Wald and Wolfowitz

(1940)]. The very generality of the test weakens its performance against specific alternatives. Asymptotic power can be evaluated using the normal distribution with appropriate moments under the alternative, which are given in Wolfowitz (1949). Since power, whether exact or asymptotic, can be calculated only for completely specified alternatives, numerical power comparisons are not really indicative of this test's performance. Its primary usefulness is in preliminary analyses of data when no particular form of alternative is yet formulated. Then, if the hypothesis is rejected, further studies can be made with other tests in an attempt to classify the type of difference between populations.

7.3 THE KOLMOGOROV–SMIRNOV TWO–SAMPLE TEST

The Kolmogorov-Smirnov statistic is another one-sample test which can be adapted to the two-sample problem. Recall that as a goodness-of-fit criterion, this test compared the empirical distribution function of a random sample with a hypothesized cumulative distribution. In the two-sample case, the comparison is made between the empirical distribution functions of the two samples.

For two random samples of sizes m and n from continuous populations F_X and F_Y, their order statistics are

$$X_{(1)}, X_{(2)}, \ldots, X_{(m)} \quad \text{and} \quad Y_{(1)}, Y_{(2)}, \ldots, Y_{(n)}$$

Their respective empirical distribution functions, denoted by $S_m(x)$ and $T_n(x)$, are defined as before:

$$S_m(x) = \begin{cases} 0 & \text{if } x < X_{(1)} \\ \dfrac{k}{m} & \text{if } X_{(k)} \leq x < X_{(k+1)} \quad \text{for } k = 1, 2, \ldots, m-1 \\ 1 & \text{if } x \geq X_{(m)} \end{cases}$$

$$T_n(x) = \begin{cases} 0 & \text{if } x < Y_{(1)} \\ \dfrac{k}{n} & \text{if } Y_{(k)} \leq x < Y_{(k+1)} \quad \text{for } k = 1, 2, \ldots, n-1 \\ 1 & \text{if } x \geq Y_{(n)} \end{cases}$$

In a combined ordered arrangement of the $m + n$ random variables, $S_m(x)$ and $T_n(x)$ are the respective proportions of X and Y random variables which do not exceed the number x.

The empirical distribution functions for the X and Y samples should be reasonable estimates of their respective population distributions. If the null hypothesis

$$H_0: F_Y(x) = F_X(x) \quad \text{for all } x$$

is true, the population distributions are identical and we have two samples from the same population. Therefore, allowing for sampling variation, under H_0 there should be reasonable agreement between the two empirical distributions. The two-sided Kolmogorov-Smirnov two-sample test criterion, denoted by $D_{m,n}$, is the maximum absolute difference between the two empirical distributions

$$D_{m,n} = \max_x |S_m(x) - T_n(x)|$$

Since here only the magnitudes, and not the directions, of the deviations are considered, $D_{m,n}$ is appropriate for a general two-sided alternative

$$H_A: F_Y(x) \neq F_X(x) \qquad \text{for some } x$$

Because of the Glivenko-Cantelli theorem (Theorem 4.3.2), the test statistic is consistent here with the rejection region defined by

$$D_{m,n} \geq c_\alpha$$

As with the one-sample Kolmogorov-Smirnov statistic, $D_{m,n}$ is completely distribution-free for any continuous common population distribution since order is preserved under a monotone transformation. That is, if we let $z = F(x)$ for the common distribution F, we have $S_m(z) = S_m(x)$ and $T_n(z) = T_n(x)$, where the random variable Z has the uniform distribution on the unit interval.

The exact null probability distribution of $D_{m,n}$ will not be derived here. It is usually attributed to Gnedenko and the Russian School [see, for example, Gnedenko (1954)]. Tables for both equal and unequal sample sizes are available in Massey (1951a, 1952). For calculations, several methods are possible, generally involving recursive formulas. Drion (1952) derived a closed expression for exact probabilities in the case $m = n$ by applying random-walk techniques.

Several approaches to calculating the probability of a $D_{m,n}$ value greater than or equal to that observed are summarized in Hodges (1958). Only one of these methods, which is particularly useful for small sample sizes, will be presented here as an aid to understanding. To compute $P(D_{m,n} \geq d)$, where d is the observed value of $\max_x |S_m(x) - T_n(x)|$, we first arrange the combined sample of $m + n$ observations in increasing order of magnitude. The arrangement can be depicted graphically on a Cartesian coordinate system by a path which starts at the origin and moves one step to the right for an x observation and one step up for a y observation, ending at (m,n). For example, the sample arrangement $x\ y\ y\ x\ x\ y\ y$ is represented in Fig. 3.1. The observed values of $mS_m(x)$ and $nT_n(x)$ are, respectively, the coordinates of all points (u,v) on the path

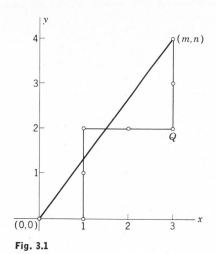

Fig. 3.1

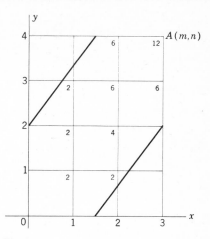

Fig. 3.2

where u and v are integers.   The number d is the largest of the differences $|u/m - v/n| = |nu - mv|/mn$.   If a line is drawn connecting the points $(0,0)$ and (m,n) on this graph, the equation of the line is $nx - my = 0$ and the vertical distance from any point (u,v) on the path to this line is $|v - nu/m|$.   Therefore, nd for the observed sample is the distance from the diagonal to that point on the path which is farthest from the diagonal line.   In Fig. 3.1 the farthest point is labeled Q, and the value of d is $\frac{2}{4}$.

The total number of arrangements of m X and n Y random variables is $\binom{m+n}{m}$, and under H_0 each of the corresponding paths is equally likely.   The probability of an observed value of $D_{m,n}$ not less than d then is the number of paths which have points at a distance from the diagonal not less than nd, divided by $\binom{m+n}{m}$.

In order to count this number, we draw another figure of the same dimension as before and mark off two lines at vertical distance nd from the diagonal, as in Fig. 3.2.   Denote by $A(m,n)$ the number of paths from $(0,0)$ to (m,n) which lie entirely *within* (not on) these boundary lines.   Then the desired probability is

$$P(D_{m,n} \geq d) = 1 - P(D_{m,n} < d) = 1 - \frac{A(m,n)}{\binom{m+n}{m}}$$

$A(m,n)$ can easily be counted in the manner indicated in Fig. 3.2.   The number $A(u,v)$ at any intersection (u,v) clearly satisfies the recursion

For explicit formula for the case $m=n$, see Pratt & Gibbons, Concepts of Nonparametric Theory, §4.3 (pp. 325-328). Springer-Verlag, 1981.

relation

$$A(u,v) = A(u-1, v) + A(u, v-1)$$

with boundary conditions

$$A(0,v) = A(u,0) = 1$$

Thus $A(u,v)$ is the sum of the numbers at the intersections where the previous point on the path could have been while still within the boundaries. This procedure is shown in Fig. 3.2 for the arrangement $x\ y\ y\ x\ x\ y\ y$, where $nd = 2$. Since here $A(3,4) = 12$, we have

$$P(D_{3,4} \geq 0.5) = 1 - \frac{12}{\binom{7}{4}} = \frac{23}{35} = 0.34286$$

For the asymptotic null distribution, that is, m and n approach infinity in such a way that m/n remains constant, Smirnov (1939) proved the result

$$\lim_{m,n \to \infty} P\left(\sqrt{\frac{mn}{m+n}} D_{m,n} \leq z\right) = L(z)$$

where

$$L(z) = 1 - 2 \sum_{i=1}^{\infty} (-1)^{i-1} e^{-2i^2 z^2}$$

Note that the asymptotic distribution of $\sqrt{mn/(m+n)}\ D_{m,n}$ is exactly the same as the asymptotic distribution of $\sqrt{N}\ D_N$ in Theorem 4.4.3. This is not surprising, since we know from the Glivenko-Cantelli theorem that as $n \to \infty$, $T_n(x)$ converges to $F_Y(x)$, which can be relabeled $F_X(x)$ as in Theorem 4.4.3. Then the only difference here is in the normalizing factor $\sqrt{mn/(m+n)}$, which replaces $\sqrt{N}$.

A one-sided two-sample maximum-unidirectional-deviation test can also be defined, based on the statistic

$$D_{m,n}^+ = \max_x [S_m(x) - T_n(x)]$$

For an alternative that the X random variables are stochastically smaller than the Y's,

$$H_1: F_Y(x) \leq F_X(x) \qquad \text{for all } x$$
$$F_Y(x) < F_X(x) \qquad \text{for some } x$$

the rejection region should be

$$D_{m,n}^+ \geq c_\alpha$$

$D_{m,n}^{+}$ is also distribution-free and consistent against this alternative. Since either sample may be labeled the X sample, it is not necessary to define another one-sided statistic for the alternative that X is stochastically larger than Y.

The graphic method described for $D_{m,n}$ can be applied here to calculate $P(D_{m,n}^{+} \geq d)$. The point Q^{+}, corresponding to Q, would be the point farthest *below* the diagonal line, and $A(m,n)$ is the number of paths lying entirely *above* the lower boundary line (see Prob. 7.1). Tables of the null distribution of $D_{m,n}^{+}$ are available in Goodman (1954) for $m = n$.

As with the two-sided statistic, the asymptotic distribution of $\sqrt{mn/(m+n)}\, D_{m,n}^{+}$ is equivalent to the asymptotic distribution of $\sqrt{N}\, D_{N}^{+}$, which was given in Theorem 4.4.5 as

$$\lim_{m,n \to \infty} P\left(\sqrt{\frac{mn}{m+n}}\, D_{m,n}^{+} \leq z\right) = 1 - e^{-2z^2}$$

DISCUSSION

The Kolmogorov-Smirnov tests are very easy to apply, using the exact distribution for any m and n within the range of the available tables and using the asymptotic distribution for larger samples. They are useful mainly for the general alternatives H_A and H_1, since the test statistic is sensitive to all types of differences between the cumulative distribution functions. Their primary application then should be for preliminary studies of data, as was the runs test. The Kolmogorov-Smirnov tests are more powerful than the runs tests when compared for large samples against the Lehmann (1953) type of nonparametric alternatives. The large-sample performance of the Kolmogorov-Smirnov tests against specific location or scale alternatives varies considerably according to the population sampled. Capon (1965) has made a study of these properties. Goodman (1954) has shown that when applied to data from discrete distributions, these tests are conservative.

7.4 THE MEDIAN TEST

In order to test the hypothesis of identical populations with two independent samples, the Kolmogorov-Smirnov two-sample test compares the proportions of observations from each sample which do not exceed some number x for *all* real numbers x. The test criterion was the maximum difference (absolute or unidirectional) between the two empirical distributions, which are defined for all x. Suppose that instead of using all possible differences, we choose some arbitrary but specific number δ and compare only the proportions of observations from each sample which are strictly less than δ. As before, the two independent samples are denoted by

$$X_1, X_2, \ldots, X_m \qquad \text{and} \qquad Y_1, Y_2, \ldots, Y_n$$

Each of the $m + n = N$ observations is to be classified according to whether it is less than δ or not. Let U and V denote the respective numbers of X and Y observations less than δ. Since the random variables in each sample have been dichotomized, U and V both follow the binomial probability distribution with parameters

$$p_X = P(X < \delta) \qquad \text{and} \qquad p_Y = P(Y < \delta)$$

and numbers of trials m and n, respectively. For two independent samples, the joint distribution of U and V then is

$$f_{U,V}(u,v) = \binom{m}{u}\binom{n}{v} p_X{}^u p_Y{}^v (1 - p_X)^{m-u}(1 - p_Y)^{n-v}$$

$$u = 0, 1, \ldots, m$$
$$v = 0, 1, \ldots, n \qquad (4.1)$$

The random variables U/m and V/n are unbiased point estimates of the parameters p_X and p_Y. The difference $U/m - V/n$ then is appropriate for testing the null hypothesis

$$H_0: p_X - p_Y = 0$$

The exact null probability distribution of $U/m - V/n$ can easily be found from (4.1), and for m and n large its distribution can be approximated by the normal. The test statistic in either case depends on the common value $p = p_X = p_Y$, but the test can be performed by replacing p by its unbiased estimate $(u + v)/(m + n)$. Otherwise there is no difficulty in constructing a test (although approximate) based on the criterion of difference of proportions of observations less than δ. This is essentially a modified sign test for two independent samples. The hypothesis really is that δ is the pth quantile point in both populations, where p is unspecified but estimated from the data.

The above test will not be pursued here since it is approximate and is not always appropriate to the general two-sample problem, where we are primarily interested in the hypothesis of identical populations. If the two populations are the same, the pth quantile points are equal for every value of p. However, two populations may be quite disparate even though some particular quantile points are equal. The value of δ, which is supposedly chosen without knowledge of the observations, then affects the sensitivity of the test criterion. If δ is chosen too small or too large, both U and V will have too small a range to be reliable. We cannot hope to have reasonable power for the general test without a judicious choice of δ. A test where the experimenter chooses a particular value of p (rather than δ), preferably a central value, would be more appropriate for our general hypothesis, especially if the type of difference one hopes to detect is primarily in location. In other words, we would rather control the *position* of δ, regardless of its actual value, but p and δ are hopelessly interrelated in the common population.

When the populations are assumed identical but unspecified, we cannot choose p and then determine the corresponding δ. Yet δ must be known at least positionally to classify each sample observation as less than δ or not. Therefore, suppose we decide to control the position of δ relative to the magnitudes of the *sample* observations. If the quantity $U + V$ is fixed by the experimenter prior to sampling, p is to some extent controlled since $(u + v)/(m + n)$ is an estimate of the common p. If p denotes the probability that any observation is less than δ, the probability distribution of $T = U + V$ is

$$f_T(t) = \binom{m + n}{t} p^t(1 - p)^{m+n-t} \qquad t = 0, 1, \ldots, m + n \qquad (4.2)$$

The conditional distribution of U given $T = t$ is (4.1) divided by (4.2). In the null case where $p_X = p_Y = p$, the result is simply

$$f_{U|T}(u \mid t) = \frac{\binom{m}{u}\binom{n}{t - u}}{\binom{m + n}{t}} \qquad u = 0, 1, \ldots, t \qquad (4.3)$$

which is the hypergeometric probability distribution. This result could also have been argued directly as follows. Each of the $m + n$ observations is dichotomized according to whether it is less than δ or not. Among all the observations, if $p_X = p_Y = p$, every one of the $\binom{m + n}{t}$ sets of t numbers is equally likely to comprise the less-than-δ group. The number of sets which have exactly u from the X sample is $\binom{m}{u}\binom{n}{t - u}$.

Since U/m is an estimate of p_X, if the hypothesis $p_X = p_Y = p$ is true, u/m should be close to $t/(m + n)$. A test criterion can then be found using the conditional distribution of U in (4.3) for any chosen t.

So far nothing has been said about the value of δ, since once t is chosen, δ really need not be specified to perform the test. Any number greater than the tth and not greater than the $(t + 1)$st order statistic in the combined ordered sample will yield the same value of u. In practice, the experimenter would probably rather choose the fraction $t/(m + n)$ since he is interested in controlling the value of p. Suppose we decide that if the populations differ at all, it is only in location or that we are most interested in detecting a difference in location. Then a reasonable choice of $t/(m + n)$ is $\frac{1}{2}$. But $N = m + n$ may be odd or even, while t must be an integer. To eliminate inconsistencies in application, δ can be defined as the $[(N + 1)/2]$nd order statistic if N is odd, and any number between the $(N/2)$nd and $[(N + 2)/2]$nd order statistics for N even. Then a unique value of u is obtained for any set of N observations, and δ is actually defined to be the median of the combined samples. The probability distribution of U is given in (4.3), where $t = N/2$ for N even and $t = (N - 1)/2$ for N odd. The test based on U, the number of observations from the X sample which are less than the combined sample median, is called the *median test*. It is attributed mainly to Mood (1950) and Westenberg (1948).

The fact that δ cannot be determined before the samples are taken may be disturbing, since it implies that δ should be treated as a random variable. In deriving (4.3) we treated δ as a constant, but the same result is obtained for δ defined as the sample median value. Denote the combined sample median by the random variable Z and the continuous distributions of the X and Y populations by F_X and F_Y, and assume that N is odd. The median Z can be either an X or a Y random variable, and these possibilities are mutually exclusive. The joint density of U and Z for t observations less than the sample median where $t = (N - 1)/2$ is the limit as Δz approaches zero of the sum of the probabilities that (1) the X's are divided into three classifications, u less than z, one between z and $z + \Delta z$, and the remainder greater than $z + \Delta z$, and the Y's are divided such that $t - u$ are less than z, and (2) exactly u X's are less than z, and the Y's are divided such that $t - u$ are less than z, one is between z and $z + \Delta z$, and the remainder are greater than $z + \Delta z$. The result then is

$$f_{U,Z}(u,z) = \binom{m}{u,\, 1,\, m-1-u} [F_X(z)]^u f_X(z)[1 - F_X(z)]^{m-1-u}$$
$$\binom{n}{t-u} [F_Y(z)]^{t-u}[1 - F_Y(z)]^{n-t+u}$$
$$+ \binom{m}{u} [F_X(z)]^u[1 - F_X(z)]^{m-u} \binom{n}{t-u,\, 1,\, n-t+u-1}$$
$$[F_Y(z)]^{t-u}f_Y(z)[1 - F_Y(z)]^{n-t+u-1}$$

The marginal density of U is obtained by integrating the above expression over all z, and if $F_X(z) = F_Y(z)$ for all z, the result is

$$f_U(u) = \left[m \binom{m-1}{u} \binom{n}{t-u} + n \binom{m}{u} \binom{n-1}{t-u} \right]$$
$$\int_{-\infty}^{\infty} [F(z)]^t[1 - F(z)]^{m+n-t-1} f(z)\, dz$$

$$= \binom{m}{u}\binom{n}{t-u}[(m-u)+(n-t+u)]B(t+1, m+n-t)$$

$$= \binom{m}{u}\binom{n}{t-u}\frac{t!(m+n-t)!}{(m+n)!}$$

which agrees with the expression in (4.3).

Because of this result, we might say then that before sampling, i.e., before the value of δ is determined, the median test statistic is appropriate for the general hypothesis of identical populations and after the samples are obtained, the hypothesis tested is that δ is the pth quantile value in both populations, where p is a number close to $\frac{1}{2}$. The null distributions of the test statistic are the same for both hypotheses, however.

Even though the foregoing discussion may imply that the median test has some statistical and philosophical limitations in conception, it is well known and accepted within the context of the general two-sample problem. The procedure for two independent samples of measurements is to arrange the combined samples in increasing order of magnitude and determine the sample median δ, the observation with rank $(N+1)/2$ if N is odd and any number between the observations with rank $N/2$ and $(N+2)/2$ if N is even. A total of t observations is then less than δ, where $t = (N-1)/2$ or $N/2$ according as N is odd or even. Let U denote the number of X observations less than δ. If the two samples are drawn from identical continuous populations, the probability distribution of U for fixed t is

$$f_U(u) = \frac{\binom{m}{u}\binom{n}{t-u}}{\binom{m+n}{t}} \qquad \begin{array}{l} u = 0, 1, \ldots, t \\ t = [N/2] \end{array} \tag{4.4}$$

where $[x]$ denotes the largest integer not exceeding the value x. If the null hypothesis is true, then $P(X < \delta) = P(Y < \delta)$ for all δ, and in particular the two populations have a common median, which is estimated by δ.

Since U/m is an estimate of $P(X < \delta)$, which is approximately one-half under H_0, a test based on the value of U will be most sensitive to differences in location. The general location alternative is

$$H_L: F_Y(x) = F_X(x - \theta) \qquad \text{for all } x \text{ and some } \theta \neq 0$$

If U is much larger than $m/2$, most of the X values are less than most of the Y values. This lends credence to the relation $P(X < \delta) > P(Y < \delta)$, or that the median of the X population is smaller than the median of the Y population, or that $\theta > 0$. If U is too small relative to $m/2$, the opposite conclusion is implied. The appropriate rejection regions for nominal significance level α then are:

Alternative	Rejection Region
$\theta > 0$ or $M_X < M_Y$	$u \geq c'_\alpha$
$\theta < 0$ or $M_X > M_Y$	$u \leq c_\alpha$
$\theta \neq 0$ or $M_X \neq M_Y$	$u \leq c$ or $u \geq c'$

where c_α and c'_α are, respectively, the largest and smallest integers such that $P(U \leq c_\alpha) \leq \alpha$ and $P(U \geq c'_\alpha) \leq \alpha$ and c and c' are any two integers such that

$$P(U \leq c) + P(U \geq c') \leq \alpha$$

The critical points can easily be found from (4.4) or from tables of the hypergeometric distribution [Lieberman and Owen (1961)] or using tables of the binomial coefficients. If N is even, we choose $c'_\alpha = m - c_\alpha$. Since the distribution in (4.4) is not symmetric for $m \neq n$ if N is odd, the choice of an optimum rejection region for a two-sided test is not clear for this case. It could be chosen such that α is divided equally or that the range of u is symmetric, or neither.

If m and n are so large that calculation or use of tables to find critical values is not feasible, an approximation to the hypergeometric distribution can be used. Using the technique of factorial moments for the distribution in (4.4), the mean and variance of U are easily found to be

$$E(U \mid t) = \frac{mt}{N} \qquad \mathrm{var}(U \mid t) = \frac{mnt(N - t)}{N^2(N - 1)} \tag{4.5}$$

If m and n approach infinity in such a way that m/n remains constant, this hypergeometric distribution approaches the binomial distribution for t trials with parameter m/N, which in turn approaches the normal distribution. For N large, the variance of U in (4.5) is approximately

$$\mathrm{var}(U \mid t) = \frac{mnt(N - t)}{N^3}$$

and thus the asymptotic distribution of

$$Z = \frac{U - mt/N}{[mnt(N - t)/N^3]^{1/2}} \tag{4.6}$$

is approximately standard normal. A continuity correction of 0.5 may be used.

A test based on this statistic z is equivalent to the normal-theory test for the difference of two independent proportions in classical statistics.

This can be shown by algebraic manipulation of (4.6) with $t = u + v$ as follows:

$$z = \frac{Nu - mt}{\sqrt{mnt(N - t)/N}} = \frac{nu - m(t - u)}{\sqrt{mnN(t/N)(1 - t/N)}}$$

$$= \frac{u/m - v/n}{\sqrt{[(u + v)/N][1 - (u + v)/N]N/mn}}$$

$$= \frac{u/m - v/n}{\sqrt{\hat{p}(1 - \hat{p})(1/m + 1/n)}}$$

If a success is defined as an observation being less than δ, u/m and v/n are the observed sample proportions of successes, and $\hat{p} = (u + v)/N$ is the best sample estimate of the common proportion. This then is the same approximate test statistic that was described at the beginning of this section for large samples, except that here $u + v = t$, a constant which is fixed by the choice of δ as the sample median.

The presence of ties either within or across samples presents no problem for the median test except in two particular cases. If N is odd and more than one observation is equal to the sample median, or if N is even and the $(N/2)$nd and $[(N + 2)/2]$nd order statistics are equal, t cannot be defined as before unless the ties are broken. The conservative approach is recommended, where the ties are broken in all possible ways and the value of u chosen for decision is the one which is least likely to lead to rejection of H_0.

CONFIDENCE–INTERVAL PROCEDURE

The median-test procedure can be adapted to a confidence-interval estimate for a parameter which corresponds to the shift in location in the following manner. Suppose that the two populations are identical in every way except for their medians. Denote these unknown parameters by M_X and M_Y, respectively, and the difference $M_Y - M_X$ by θ. From the original samples, if θ were known, we could form the derived random variables

$$X_1, X_2, \ldots, X_m \quad \text{and} \quad Y_1 - \theta, Y_2 - \theta, \ldots, Y_n - \theta$$

and these would constitute samples from identical populations or, equivalently, a single sample of size $m + n$ from the common population. According to the median-test criterion with significance level α, the null hypothesis of identical distributions would be accepted for these derived samples if U, the number of X observations less than the median of the combined sample of derived observations, lies in the interval $c < u < c'$.

The integers c and c' are chosen such that

$$\sum_{u=0}^{c} \frac{\binom{m}{u}\binom{n}{t-u}}{\binom{m+n}{t}} + \sum_{u=c'}^{t} \frac{\binom{m}{u}\binom{n}{t-u}}{\binom{m+n}{t}} = \alpha$$

where $t = N/2$ or $(N - 1)/2$ according as N is even or odd. Since H_0 is accepted for all $c + 1 \leq u \leq c' - 1$, a confidence-interval estimate for θ would consist of all values of θ for which the derived sample observations will give values of U which lie in this interval.

To find this range of θ, we order the two derived samples separately from smallest to largest,

$$X_{(1)}, X_{(2)}, \ldots, X_{(m)} \quad \text{and} \quad Y_{(1)} - \theta, Y_{(2)} - \theta, \ldots, Y_{(n)} - \theta$$

The t smallest observations of the $m + n$ total number are made up of exactly i X and $t - i$ Y variables if each observation of the set

$$x_{(1)}, \ldots, x_{(i)}, y_{(1)} - \theta, \ldots, y_{(t-i)} - \theta$$

is less than each observation of the set

$$x_{(i+1)}, \ldots, x_{(m)}, y_{(t-i+1)} - \theta, \ldots, y_{(n)} - \theta$$

The value of i is at least $c + 1$ if and only if for $i = c + 1$, the largest x in the first set is less than the smallest y in the second set, that is, $x_{(c+1)} < y_{(t-c)} - \theta$. Similarly, $x_{(c')} > y_{(t-c'+1)} - \theta$ is a necessary and sufficient condition for having at most $c' - 1$ X observations among the t smallest of the $m + n$ total. Therefore, the null hypothesis of no difference would be accepted at significance level α if and only if

$$x_{(c+1)} < y_{(t-c)} - \theta \quad \text{and} \quad x_{(c')} > y_{(t-c'+1)} - \theta$$

or

$$y_{(t-c)} - x_{(c+1)} > \theta \quad \text{and} \quad y_{(t-c'+1)} - x_{(c')} < \theta$$

This inequality provides our confidence-interval estimate, and we can make the probability statement

$$P(Y_{(t-c'+1)} - X_{(c')} < \theta < Y_{(t-c)} - X_{(c+1)}) = 1 - \alpha$$

The confidence-interval estimate of θ is found simply from the order statistics of the respective random samples.

POWER OF THE MEDIAN TEST

Under the assumption that

$$F_Y(x) = F_X(x - \theta)$$

for some continuous distribution, the null hypothesis is that θ equals zero. From the confidence-interval procedure developed above we know that for the median test a necessary and sufficient condition for accepting this hypothesis is that the number zero be included in the random interval

$$[(Y_{(t-c'+1)} - X_{(c')}),(Y_{(t-c)} - X_{(c+1)})]$$

The power function of the median test, the probability of rejecting the hypothesis when it is false, is then the probability that this interval does not cover zero when $\theta \neq 0$, that is,

$$Pw(\theta) = P(Y_{(t-c'+1)} - X_{(c')} > 0 \text{ or } Y_{(t-c)} - X_{(c+1)} < 0 \text{ when } \theta \neq 0)$$

The two events in the union here are mutually exclusive since if $Y_{(t-c'+1)} > X_{(c')}$ and $c' > c$, then since $X_{(c')} \geq X_{(c+1)}$ and $Y_{(t-c'+1)} = Y_{[t-(c'-1)]} \leq Y_{(t-c)}$, we have $Y_{(t-c)} > X_{(c+1)}$. As a result, the power function can be expressed as the sum of two probabilities involving order statistics

$$Pw(\theta) = P(Y_{(t-c'+1)} > X_{(c')}) + P(Y_{(t-c)} < X_{(c+1)})$$

Since every X random variable is independent of every Y random variable, the joint distribution of, say, $X_{(r)}$ and $Y_{(s)}$ is the product of their marginal distributions, which can be easily found using the methods of Chap. 2 for completely specified populations F_X and F_Y or, equivalently, F_X and θ since $F_Y(x) = F_X(x - \theta)$. In order to calculate the power function then, we need only evaluate two double integrals of the following type

$$P(Y_{(s)} < X_{(r)}) = \int_{-\infty}^{\infty} \int_{-\infty}^{x} f_{Y_{(s)}}(y)f_{X_{(r)}}(x) \, dy \, dx$$

The power function for a one-sided test is simply one integral of this type. For large samples, since the marginal distribution of any order statistic approaches the normal distribution and the order statistics $X_{(r)}$ and $Y_{(s)}$ are independent here, the distribution of their difference $Y_{(s)} - X_{(r)}$ approaches the normal distribution with mean and variance

$$E(Y_{(s)}) - E(X_{(r)}) \qquad \text{and} \qquad \text{var}(Y_{(s)}) + \text{var}(X_{(r)})$$

Given the specified distribution functions, from the results in (2.6.5) and (2.6.6), we can approximate these quantities by

$$E(X_{(r)}) = F_X^{-1}\left(\frac{r}{m+1}\right) \qquad E(Y_{(s)}) = F_Y^{-1}\left(\frac{s}{n+1}\right)$$

$$\text{var}(X_{(r)}) = \frac{r(m-r+1)}{(m+1)^2(m+2)} \left\{ f_X\left[F_X^{-1}\left(\frac{r}{m+1}\right)\right]\right\}^{-2}$$

$$\text{var}(Y_{(s)}) = \frac{s(n-s+1)}{(n+1)^2(n+2)} \left\{ f_Y\left[F_Y^{-1}\left(\frac{s}{n+1}\right)\right]\right\}^{-2}$$

and an approximation to the power function can be found using normal probability tables.

DISCUSSION

The median test is one of the simplest to use of all the two-sample nonparametric tests. Numerical measurements are not really required so long as the t smallest elements in the combined sample can be identified. The test is primarily sensitive to differences in location, while the two tests previously considered were sensitive to any type of difference. Its asymptotic efficiency relative to Student's t test for normal populations is $2/\pi = 0.637$ (see Chap. 14). As a test for location, this is relatively poor performance. The Mann-Whitney test, discussed in the next section, generally has greater power.

7.5 THE MANN-WHITNEY U TEST

Like the Wald-Wolfowitz runs test, the Mann-Whitney U test [Mann and Whitney (1947)] is based on the idea that the particular pattern exhibited when m X random variables and n Y random variables are arranged together in increasing order of magnitude provides information about the relationship between their populations. However, instead of measuring the tendency to cluster by the total number of runs, the Mann-Whitney criterion is based on the magnitudes of the Y's in relation to the X's, that is, the positions of the Y's in the combined ordered sequence. A sample pattern of arrangement where most of the Y's are greater than most of the X's, or vice versa, or both, would be evidence against a random mixing and thus tend to discredit the null hypothesis of identical distributions.

The Mann-Whitney U statistic is defined as the number of times a Y precedes an X in the combined ordered arrangement of the two independent random samples

$$X_1, X_2, \ldots, X_m \qquad \text{and} \qquad Y_1, Y_2, \ldots, Y_n$$

into a single sequence of $m + n = N$ variables increasing in magnitude. We assume that the two samples are drawn from continuous distributions, so that the possibility $X_i = Y_j$ for some (i,j) need not be considered. If the mn indicator random variables are defined as

$$D_{ij} = \begin{cases} 1 & \text{if } Y_j < X_i \quad \text{for all } i = 1, 2, \ldots, m \\ 0 & \text{if } Y_j > X_i \quad\qquad j = 1, 2, \ldots, n \end{cases} \tag{5.1}$$

a symbolic representation of the Mann-Whitney U statistic is

$$U = \sum_{i=1}^{m} \sum_{j=1}^{n} D_{ij} \tag{5.2}$$

The logical rejection region for the one-sided alternative that the X's are stochastically smaller than the Y's,

$$H_1: F_Y(x) \leq F_X(x) \qquad \text{strict inequality for some } x$$

would clearly be small values of U. The fact that this is a consistent test criterion can be shown by investigating the convergence of U/mn to a certain parameter where H_0 can be written as a statement concerning the value of that parameter.

For this purpose, we define

$$\pi = P(Y < X) = \int_{-\infty}^{\infty} \int_{-\infty}^{x} f_Y(y)f_X(x) \, dy \, dx = \int_{-\infty}^{\infty} F_Y(x)f_X(x) \, dx \tag{5.3}$$

If $H_0: F_Y(x) = F_X(x)$ for all x is true, then

$$\pi = \int_{-\infty}^{\infty} F_X(x)f_X(x) \, dx = \frac{1}{2} \tag{5.4}$$

and if H_1 is true, $\pi \leq \frac{1}{2}$ for all x and $\pi < \frac{1}{2}$ for some x. Therefore the hypothesis of identical distributions has been parameterized to $H_0: \pi = \frac{1}{2}$. Then the mn random variables defined in (5.1) are Bernoulli variables, with moments

$$E(D_{ij}) = E(D_{ij}^2) = \pi \qquad \text{var}(D_{ij}) = \pi(1 - \pi) \tag{5.5}$$

For the joint moments we note that these random variables are not independent whenever the X subscripts or the Y subscripts are common, so that

$$\begin{aligned}
&\text{cov}(D_{ij},D_{hk}) = 0 \qquad \text{for } i \neq h \text{ and } j \neq k \\
&\text{cov}(D_{ij},D_{ik}) = \pi_1 - \pi^2 \qquad \text{cov}(D_{ij},D_{hj}) = \pi_2 - \pi^2 \\
&\phantom{\text{cov}(D_{ij},D_{ik})} {\scriptstyle j \neq k} \qquad\qquad \phantom{\text{cov}(D_{ij},D_{hj})} {\scriptstyle i \neq h}
\end{aligned} \tag{5.6}$$

where the additional parameters introduced are

$$\begin{aligned}
\pi_1 &= P(Y_j < X_i \cap Y_k < X_i) \\
&= \int_{-\infty}^{\infty} \int_{-\infty}^{x_i} \int_{-\infty}^{y_k} f_Y(y_j)f_Y(y_k)f_X(x_i) \, dy_j \, dy_k \, dx_i \\
&\qquad + \int_{-\infty}^{\infty} \int_{-\infty}^{x_i} \int_{-\infty}^{y_j} f_Y(y_k)f_Y(y_j)f_X(x_i) \, dy_k \, dy_j \, dx_i \\
&= \int_{-\infty}^{\infty} [F_Y(x)]^2 f_X(x) \, dx \tag{5.7} \\
\pi_2 &= P(X_i > Y_j \cap X_h > Y_j) = \int_{-\infty}^{\infty} [1 - F_X(y)]^2 f_Y(y) \, dy \tag{5.8}
\end{aligned}$$

Since in (5.2) U is defined as a linear combination of these mn random variables, the mean and variance of U are

$$E(U) = \sum_{i=1}^{m} \sum_{j=1}^{n} E(D_{ij}) = mn\pi \tag{5.9}$$

$$\mathrm{var}(U) = \sum_{i=1}^{m} \sum_{j=1}^{n} \mathrm{var}(D_{ij}) + \sum_{i=1}^{m} \sum_{1 \leq j \neq k \leq n} \mathrm{cov}(D_{ij}, D_{ik})$$

$$+ \sum_{j=1}^{n} \sum_{1 \leq i \neq h \leq m} \mathrm{cov}(D_{ij}, D_{hj}) + \sum_{1 \leq i \neq h \leq m} \sum_{1 \leq j \neq k \leq n} \mathrm{cov}(D_{ij}, D_{hk}) \tag{5.10}$$

Now substituting (5.5) and (5.6) in (5.10), this variance is

$$\mathrm{var}(U) = mn\pi(1 - \pi) + mn(n - 1)(\pi_1 - \pi^2)$$
$$+ nm(m - 1)(\pi_2 - \pi^2)$$
$$= mn[\pi - \pi^2(N - 1) + (n - 1)\pi_1 + (m - 1)\pi_2] \tag{5.11}$$

Since $E(U/mn) = \pi$ and $\mathrm{var}(U/mn) \to 0$ as $m,\, n \to \infty$, U/mn is a consistent estimator for π. Using the method described in Chap. 1, the Mann-Whitney test is consistent in the following cases:

Subclass of Alternative		Rejection Region	
$\pi < \frac{1}{2}$	$F_Y(x) \leq F_X(x)$	$U - \dfrac{mn}{2} < k_1$	
$\pi > \frac{1}{2}$	$F_Y(x) \geq F_X(x)$	$U - \dfrac{mn}{2} > k_2$	(5.12)
$\pi \neq \frac{1}{2}$	$F_Y(x) \neq F_X(x)$	$\left\| U - \dfrac{mn}{2} \right\| > k_3$	

In order to determine the size α critical regions of the Mann-Whitney test, we must now find the null probability distribution of U. Under H_0, each of the $\binom{m+n}{m}$ arrangements of the random variables into a combined sequence occurs with equal probability, so that

$$f_U(u) = P(U = u) = \frac{r_{m,n}(u)}{\dbinom{m+n}{m}} \tag{5.13}$$

where $r_{m,n}(u)$ is the number of distinguishable arrangements of the m X and n Y random variables such that in each sequence the number of times a Y precedes an X is exactly u. The values of u for which $f_U(u)$ is nonzero range between zero and mn, for the two most extreme orderings in which every x precedes every y and every y precedes every x, respectively. We first note that the probability distribution of U is symmetric about the mean $mn/2$ in the null case. This property may be argued as follows. For every particular arrangement z of the m x and n y letters, define the conjugate arrangement z' as the sequence z written backward. In other words, if z denotes a set of numbers written from smallest to largest, z' denotes the same numbers written from largest to smallest. Every y that precedes an x in z then follows that x in z', so that if u is the value of the Mann-Whitney statistic for z, $mn - u$ is the value for z'. Therefore $r_{m,n}(u) = r_{m,n}(mn - u)$ or, equivalently,

$$P\left(U - \frac{mn}{2} = u\right) = P\left(U = \frac{mn}{2} + u\right)$$

$$= P\left[U = mn - \left(\frac{mn}{2} + u\right)\right] = P\left(U - \frac{mn}{2} = -u\right)$$

Because of this symmetry property, only lower tail critical values need be found for either a one- or two-sided test. We define the random variables U' as the number of times an X precedes a Y or, in the notation of (5.1),

$$U' = \sum_{i=1}^{m} \sum_{j=1}^{n} (1 - D_{ij})$$

The rejection regions for size α tests corresponding to (5.12) are

Alternative	Rejection Region
$F_Y(x) \leq F_X(x)$	$U \leq c_\alpha$
$F_Y(x) \geq F_X(x)$	$U' \leq c_\alpha$
$F_Y(x) \neq F_X(x)$	$U \leq c_{\alpha/2}$ or $U' \leq c_{\alpha/2}$

To determine the number c_α for any m and n, we can enumerate the cases starting with $u = 0$ and work up until at least $\alpha \binom{m+n}{m}$ cases are counted. For example, for $m = 4$, $n = 5$, the arrangements with the smallest values of u, that is, where most of the X's are smaller than most of the Y's, are as shown in Table 5.1. The rejection regions for this one-sided test for nominal significance levels of 0.01 and 0.05 would then be $U \leq 0$ and $U \leq 2$, respectively.

Table 5.1

Ordering	u	
X X X X Y Y Y Y Y	0	
X X X Y X Y Y Y Y	1	
X X Y X X Y Y Y Y	2	$P(U \leq 0) = \frac{1}{126} = 0.008$
X X X Y Y X Y Y Y	2	$P(U \leq 1) = \frac{2}{126} = 0.016$
X Y X X X Y Y Y Y	3	$P(U \leq 2) = \frac{4}{126} = 0.032$
X X Y X Y X Y Y Y	3	$P(U \leq 3) = \frac{7}{126} = 0.056$
X X X Y Y Y X Y Y	3	

Even though it is relatively easy to guess which orderings will lead to the smallest values of u, $\binom{m+n}{m}$ increases rapidly as m, n increase. Some more systematic method of generating critical values is needed to eliminate the possibility of overlooking some arrangements with u small and to increase the feasible range of sample sizes and significance levels for constructing tables. A particularly simple and useful recurrence relation can be employed for the Mann-Whitney statistic. Consider a sequence of $m + n$ letters being built up by adding a letter to the right of a sequence of $m + n - 1$ letters. If the $m + n - 1$ letters consist of m x and $n - 1$ y letters, the extra letter must be a y. But if a y is added to the right, the number of times a y precedes an x is unchanged. If the additional letter is an x, which would be the case for $m - 1$ x and n y letters in the original sequence, all the y's will precede this new x and there are n of them, so that u is increased by n. These two possibilities are mutually exclusive. Using the notation of (5.13) again, this recurrence relation can be expressed as

$$r_{m,n}(u) = r_{m,n-1}(u) + r_{m-1,n}(u - n)$$

and

$$f_U(u) = p_{m,n}(u) = \frac{r_{m,n-1}(u) + r_{m-1,n}(u - n)}{\binom{m+n}{m}}$$

$$= \frac{n}{m+n}\frac{r_{m,n-1}(u)}{\binom{m+n-1}{n-1}} + \frac{m}{m+n}\frac{r_{m-1,n}(u-n)}{\binom{m+n-1}{m-1}}$$

or

$$(m + n)p_{m,n}(u) = np_{m,n-1}(u) + mp_{m-1,n}(u - n) \tag{5.14}$$

This recursive relation holds for all $u = 0, 1, 2, \ldots, mn$ and all integer-valued m and n if the following initial and boundary conditions are defined for all $i = 1, 2, \ldots, m$ and $j = 1, 2, \ldots, n$:

$$r_{i,j}(u) = 0 \quad \text{for all } u < 0$$
$$r_{i,0}(0) = 1 \quad r_{0,i}(0) = 1$$
$$r_{i,0}(u) = 0 \quad \text{for all } u \neq 0$$
$$r_{0,i}(u) = 0 \quad \text{for all } u \neq 0$$

If the smaller-sized sample is always labeled the X sample, tables are needed only for $m \leq n$ and left tail critical points. Such tables are widely available, for example, in Auble (1953) or Mann and Whitney (1947).

When m and n are too large for the existing tables, the asymptotic probability distribution can be used. Since U is the sum of identically distributed (though dependent) random variables, a generalization of the central-limit theorem allows us to conclude that the null distribution of the standardized U approaches the standard normal as m, $n \to \infty$ in such a way that m/n remains constant [Mann and Whitney (1947)]. To make use of this approximation, the mean and variance of U under the null hypothesis must be determined. When $F_Y(x) = F_X(x)$, the integrals in (5.7) and (5.8) are evaluated as $\pi_1 = \pi_2 = \frac{1}{3}$. Substituting these results in (5.9) and (5.11) along with the value $\pi = \frac{1}{2}$ from (5.4) gives

$$E(U \mid H_0) = \frac{mn}{2} \qquad \operatorname{var}(U \mid H_0) = \frac{mn(N+1)}{12} \tag{5.15}$$

The large-sample test statistic then is

$$Z = \frac{U - mn/2}{\sqrt{mn(N+1)/12}}$$

whose distribution is approximately standard normal. This approximation has been found reasonably accurate for equal sample sizes as small as 6. Since U can assume only integer values, a continuity correction of 0.5 can be used.

THE PROBLEM OF TIES

The definition of U in (5.2) was adopted for presentation here because most tables of critical values are designed for use in the way described above. Since D_{ij} is not defined for $X_i = Y_j$, this expression does not allow for the possibility of ties across samples. If ties occur within one or both of the samples, a unique value of U is obtained. However, if one or more X is tied with one or more Y, our definition requires that the ties be broken in some way. The conservative approach may be adopted, which means that all ties are broken in all possible ways and the largest resulting value of u (or u') is used in reaching the decision. When there are many ties (as might be the case when each random variable can assume only a few ordered values such as very strong, strong, weak, very weak), an alternative approach may be preferable.

A common definition of the Mann-Whitney statistic which does allow for ties is

$$U_T = \sum_{i=1}^{m} \sum_{j=1}^{n} D_{ij}$$

where

$$D_{ij} = \begin{cases} 1 & \text{if } X_i > Y_j \\ 0 & \text{if } X_i = Y_j \\ -1 & \text{if } X_i < Y_j \end{cases} \tag{5.16}$$

If the two parameters π^+ and π^- are defined as

$$\pi^+ = P(X > Y) \qquad \text{and} \qquad \pi^- = P(X < Y)$$

U_T may be considered an estimate of its mean

$$E(U_T) = mn(\pi^+ - \pi^-)$$

A standardized U_T is asymptotically normally distributed. Since under the null hypothesis $\pi^+ = \pi^-$, we have $E(U_T \mid H_0) = 0$ whether ties occur or not. The presence of ties does affect the variance, however. The variance of U_T conditional upon the observed ties can be calculated in a manner similar to the steps leading to (5.11) by introducing some additional parameters. Then a correction for ties can be incorporated in the standardized variable used for the test statistic. The result is

$$\text{var}(U_T \mid H_0) = \frac{mn(N + 1)}{3}\left[1 - \frac{\Sigma(t^3 - t)}{N(N^2 - 1)}\right]$$

where t denotes the multiplicity of a tie and the sum is extended over all sets of t ties. The details will be left to the reader as an exercise [or see Noether (1967), pp. 32–35].

CONFIDENCE–INTERVAL PROCEDURE

If the populations from which the X and Y samples are drawn are identical in every respect except location, say $F_Y(x) = F_X(x - \theta)$ for all x and some θ, we say that the Y population is the same as the X population but shifted by an amount θ, which may be either positive or negative, the sign indicating the direction of the shift. We wish to use the Mann-Whitney test procedure to find a confidence interval for θ, the amount of shift. Under the assumption that $F_Y(x) = F_X(x - \theta)$ for all x and some θ, the sample observations $X_1, X_2, \ldots, X_m$ and $Y_1 - \theta, Y_2 - \theta, \ldots, Y_n - \theta$ come from identical populations. By a confidence interval for θ with confidence coefficient $1 - \alpha$ we mean the range of values of θ for which the null hypothesis of identical populations will be accepted at significance level α.

To apply the Mann-Whitney procedure to this problem, the random variable U now denotes the number of times a $y - \theta$ precedes an x, that is, the number of pairs $(x_i, y_j - \theta)$, $i = 1, 2, \ldots, m$ and $j = 1, 2, \ldots, n$, for which $x_i > y_j - \theta$, or equivalently, $y_j - x_i < \theta$. If a table of critical values for a two-sided U test at level α gives a rejection region of $u \leq k$, say, we reject H_0 when there are no more than k differences $y_j - x_i$ which are less than the value θ, and accept H_0 for more than k differences less than θ. There are a total of mn differences $y_j - x_i$. If these differences are ordered from smallest to largest according to actual (not absolute) magnitude, denoted by $d_{(1)}, d_{(2)}, \ldots, d_{(mn)}$, there are exactly k differences less than θ if θ is the $(k + 1)$st ordered difference, $d_{(k+1)}$. Any number exceeding this $(k + 1)$st difference will produce more than k differences less than θ. Therefore the lower limit of the confidence interval for θ is $d_{(k+1)}$. Similarly, since the probability distribution of U is symmetric, an upper confidence limit is given by that difference which is $(k + 1)$st from the largest, that is, $d_{(mn-k)}$. The confidence interval with coefficient $1 - \alpha$ then is

$$d_{(k+1)} < \theta < d_{(mn-k)} \qquad (5.17)$$

The procedure is simply illustrated by the following numerical example. Suppose that $m = 3$, $n = 5$, $\alpha = 0.10$. By simple enumeration, we find $P(U \leq 1) = \frac{2}{56} = 0.036$ and $P(U \leq 2) = \frac{3}{56} = 0.054$, and so the critical value for $\alpha/2 = 0.05$ is 1, with the exact probability of a type I error 0.072. The confidence interval will then be $d_{(2)} < \theta < d_{(14)}$. Suppose the sample results are X: 1, 6, 7; Y: 2, 4, 9, 10, 12. In order to find $d_{(2)}$ and $d_{(14)}$ systematically, we first order the x and y data separately, then subtract from each y, starting with the smallest y, the successive values of x as shown in Table 5.2, and order the differences. The interval here is

$$-4 < \theta < 9$$

with an exact confidence coefficient of 0.928.

The so-called *Moses' graphical approach* can be used to simplify the procedure of constructing intervals here. Each of the $m + n$ sample observations is plotted on a graph, the X observations on the abscissa and Y on the ordinate. Then all mn pairings of observations can be easily indicated by dots at all possible intersections. The line $y - x = \theta$ with slope 1 for any number θ divides the pairings into two groups: those on the left and above have $y - x < \theta$, and those on the right and below have $y - x > \theta$. Thus if the rejection region for a size α test is $U \leq k$, two lines with slope 1 such that k dots lie on each side of the included band will determine the appropriate values of θ. If the two lines are drawn through the $(k + 1)$st dots from the upper left and lower right, respectively, the values on the vertical axis where these lines have x intercept

Table 5.2

$y_j - 1$	$y_j - 6$	$y_j - 7$
1	-4	-5
3	-2	-3
8	3	2
9	4	3
11	6	5

zero determine the confidence-interval end points. In practice, it is often convenient to add or subtract an arbitrary constant from each observation before the pairs are plotted, the number chosen so that all observations are positive and the smallest is close to zero. This does not change the resulting interval for θ, since the parameter θ is invariant under a change in location in both the X and Y populations. This method is illustrated in Fig. 5.1 for the above example where $k = 1$ for $\alpha = 0.072$.

DISCUSSION

The Mann-Whitney U test is a frequently used nonparametric test that is equivalent to another well-known test, the Wilcoxon sum-of-ranks test, which will be presented independently in Sec. 9.2. The discussion here applies equally to both tests.

Only independence and continuous distributions need be assumed to test the null hypothesis of identical populations. The test is simple to use for any size samples, and tables of the exact null distribution are widely available. The large-sample approximation is quite adequate for

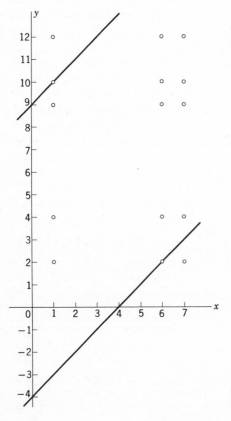

Fig. 5.1

most practical purposes, and corrections for ties can be incorporated in the test statistic. The test has been found to perform particularly well as a test for equal means (or medians), since it is especially sensitive to differences in location. In order to reduce the generality of the null hypothesis in this way, however, we must feel that we can legitimately assume that the populations are identical in form, at least if their locations are the same. A particular advantage of the test procedure in this case is that it can be adapted to confidence-interval estimation of the difference in location.

When the populations are assumed to differ only in location, the Mann-Whitney test is directly comparable with Student's t test for means. The asymptotic relative efficiency of U relative to t is *never* less than 0.864, and if the populations are normal, the ARE is quite high at $3/\pi = 0.9550$ (see Chap. 14). Many statisticians consider the Mann-Whitney (or equivalently the Wilcoxon) test the best nonparametric test for location. Therefore power functions for smaller sample sizes and/or other distributions are of interest. To calculate exact power, we sum the probabilities under the alternative for those arrangements of m X and n Y random variables which are in the rejection region. For any combined arrangement Z where the X random variables occur in the positions $r_1, r_2, \ldots, r_m$ and the Y's in positions $s_1, s_2, \ldots, s_n$, this probability is

$$P(Z) = m!n! \int_{-\infty}^{\infty} \int_{-\infty}^{u_N} \cdots \int_{-\infty}^{u_3} \int_{-\infty}^{u_2} \prod_{i=1}^{m} f_X(u_{r_i})$$

$$\prod_{j=1}^{n} f_Y(u_{s_j}) \, du_1 \cdots du_N \quad (5.18)$$

which is generally extremely tedious to evaluate. The asymptotic normality of U holds even in the nonnull case, and the mean and variance of U in (5.9) and (5.11) depend only on the parameters π, π_1, and π_2 if the distributions are continuous. Thus, approximations to power can be found if the integrals in (5.3), (5.7), and (5.8) are evaluated. Unfortunately, even under the more specific alternative $F_Y(x) = F_X(x - \theta)$ for some θ, these integrals depend on both θ and F_X, so that calculating even the approximation to power requires that the basic parent population be specified.

PROBLEMS

7.1. Use the graphic method of Hodges to find $P(D_{m,n}^+ \geq d)$, where d is the observed value of $D_{m,n}^+ = \max_{x} [S_m(x) - T_n(x)]$ in the arrangement $x \, y \, y \, x \, y \, x$.

7.2. Using (4.4), derive the complete null distribution of U for $m = 6$, $n = 7$, and set up one- and two-sided critical regions for the median-test statistic when $\alpha = 0.01, 0.05$, and 0.10.

7.3. Find the large-sample approximation to the power function of a two-sided median test for $m = 6$, $n = 7$, $\alpha = 0.10$ when F_X is the standard normal distribution.

7.4. Use the recursion relation for the Mann-Whitney test statistic given in (5.14) to generate the complete null probability distribution of U for all $m + n \leq 4$.

7.5. Verify the expressions given in (5.15) for the moments of U under H_0.

7.6. Answer parts (a) to (c) using (i) the median-test procedure and (ii) the Mann-Whitney test procedure (use tables).

(a) Given the following two independent random samples drawn from continuous populations which have the same form but possibly a difference of θ in their locations:

X	79	13	138	129	59	76	75	53
Y	96	141	133	107	102	129	110	104

Using the significance level 0.10, test

$$H_0: \theta = 0 \qquad \text{versus} \qquad H_1: \theta \neq 0$$

(b) Give the exact level of the test in (a).

(c) Give a confidence interval on θ with an exact confidence coefficient corresponding to the exact level noted in (b).

7.7. Represent a sample of m X and n Y random variables by a path of $m + n$ steps, the ith step being one unit up or to the right according as the ith from the smallest observation in the combined sample is an X or a Y, respectively. What is the algebraic relation between the area under the path and the Mann-Whitney statistic?

7.8. Can you think of other functions of the difference $S_m(x) - T_n(x)$ (besides the maximum) which could also be used for distribution-free tests of the equality of two population distributions?

8

Linear Rank Statistics and the General Two-sample Problem

8.1 INTRODUCTION

In Chap. 7 the general two-sample problem was described and some tests presented which were all based on various criteria related to the combined ordered arrangement of the two sets of sample observations. Many statistical procedures applicable to the two-sample problem are based on the rank-order statistics for the combined samples, since various functions of these rank-order statistics can provide information about the possible differences between populations. For example, if the X population has a larger mean than the Y population, the sample values will reflect this difference if most of the ranks of the X values exceed the ranks of the Y values.

Many commonly used two-sample rank tests can be classified together as linear combinations of certain indicator variables for the combined ordered samples. Such functions are often called *linear rank statistics*. This unifying concept will be defined in the next section, and then some of the general theory of these linear rank statistics will be presented.   Particular linear rank tests will then be treated in Chaps. 9 and 10 for the location and scale problems.

8.2 DEFINITION OF LINEAR RANK STATISTICS

Assume we have two independent random samples, X_1, X_2, . . . , X_m and Y_1, Y_2, . . . , Y_n, drawn from populations with continuous cumulative distribution functions F_X and F_Y, respectively.   Under the null hypothesis

$$H_0: F_X(x) = F_Y(x) \qquad \text{for all } x, F \text{ unspecified}$$

we then have a single set of $m + n = N$ random observations from the common but unknown population, to which the integer ranks $1, 2, . . . , N$ can be assigned.

In accordance with the definition given in (5.1.1) for the rank of an observation in a single sample, a functional definition of the rank of an observation in the combined sample with no ties can be given as

$$r(x_i) = \sum_{k=1}^{m} S(x_i - x_k) + \sum_{k=1}^{n} S(x_i - y_k)$$
$$r(y_i) = \sum_{k=1}^{m} S(y_i - x_k) + \sum_{k=1}^{n} S(y_i - y_k)$$

$$(2.1)$$

where

$$S(u) = \begin{cases} 0 & \text{if } u < 0 \\ 1 & \text{if } u \geq 0 \end{cases}$$

However, it is easier to denote the combined ordered sample by a vector of indicator random variables as follows.   Let

$$Z = (Z_1, Z_2, . . . , Z_N)$$

where $Z_i = 1$ if the ith random variable in the combined ordered sample is an X and $Z_i = 0$ if it is a Y, for $i = 1, 2, . . . , N$, with $N = m + n$. The rank of the observation for which Z_i is an indicator is i, and therefore the vector Z indicates the rank-order statistics of the combined samples and in addition identifies the sample to which each observation belongs.

For example, given observations

$$(X_1, X_2, X_3, X_4) = (2,9,3,4) \qquad (Y_1, Y_2, Y_3) = (1,6,10)$$

the combined ordered sample is (1,2,3,4,6,9,10) or $(Y_1, X_1, X_3, X_4, Y_2, X_2, Y_3)$, and the corresponding Z vector is (0,1,1,1,0,1,0).   Since $Z_6 = 1$, for example, an X observation (in particular X_2) had rank 6 in the combined ordered array.

Many of the statistics based on rank-order statistics which are useful in the two-sample problem can be easily expressed in terms of this notation. An important class of statistics of this type is called the *linear rank statistic,* which is defined by

$$T_N(Z) = \sum_{i=1}^{N} a_i Z_i \tag{2.2}$$

where the a_i are given numbers. It should be noted that T_N is linear in the indicator variables and no similar restriction is implied for the constants which are called weights or scores.

8.3 DISTRIBUTION PROPERTIES OF THE LINEAR RANK STATISTIC

T_N has been introduced here for convenience and as a unifying concept. Therefore, we shall now prove some general properties of T_N in order to facilitate the study of particular linear-rank-statistic tests later.

Theorem 3.1 *Under the null hypothesis $H_0\colon F_X(x) = F_Y(x)$ for all x, we have for all $i = 1, 2, \ldots, N$,*

$$E(Z_i) = \frac{m}{N} \quad \text{var}(Z_i) = \frac{mn}{N^2} \quad \text{cov}(Z_i, Z_j) = \frac{-mn}{N^2(N-1)} \tag{3.1}$$

Proof Since

$$f_{Z_i}(z_i) = \begin{cases} \dfrac{m}{N} & \text{if } z_i = 1 \\[2mm] \dfrac{n}{N} & \text{if } z_i = 0 \qquad \text{for } i = 1, 2, \ldots, N \\[2mm] 0 & \text{otherwise} \end{cases}$$

is the Bernoulli distribution, the mean and variance are

$$E(Z_i) = \frac{m}{N} \quad \text{var}(Z_i) = \frac{mn}{N^2}$$

For the joint moments, we have

$$\underset{i \neq j}{E}(Z_i Z_j) = \underset{i \neq j}{P}(Z_i = 1 \cap Z_j = 1)$$

$$= \frac{\dbinom{m}{2}}{\dbinom{N}{2}} = \frac{m(m-1)}{N(N-1)}$$

so that

$$\operatorname*{cov}_{i \neq j}(Z_i, Z_j) = \frac{m(m-1)}{N(N-1)} - \left(\frac{m}{N}\right)^2$$

$$= \frac{-mn}{N^2(N-1)}$$

Theorem 3.2 *Under the null hypothesis H_0: $F_X(x) = F_Y(x)$ for all x,*

$$E(T_N) = m \sum_{i=1}^{N} \frac{a_i}{N}$$

$$\operatorname{var}(T_N) = \frac{mn}{N^2(N-1)} \left[N \sum_{i=1}^{N} a_i^2 - \left(\sum_{i=1}^{N} a_i\right)^2 \right] \tag{3.2}$$

Proof $E(T_N) = \sum_{i=1}^{N} a_i E(Z_i) = m \sum_{i=1}^{N} \frac{a_i}{N}$

$$\operatorname{var}(T_N) = \sum_{i=1}^{N} a_i^2 \operatorname{var}(Z_i) + \sum\sum_{i \neq j} a_i a_j \operatorname{cov}(Z_i, Z_j)$$

$$= \frac{mn \sum_{i=1}^{N} a_i^2}{N^2} - \frac{mn \sum\sum_{i \neq j} a_i a_j}{N^2(N-1)}$$

$$= \frac{mn}{N^2(N-1)} \left(N \sum_{i=1}^{N} a_i^2 - \sum_{i=1}^{N} a_i^2 - \sum\sum_{i \neq j} a_i a_j \right)$$

$$= \frac{mn}{N^2(N-1)} \left[N \sum_{i=1}^{N} a_i^2 - \left(\sum_{i=1}^{N} a_i\right)^2 \right]$$

Theorem 3.3 *If $B_N = \sum_{i=1}^{N} b_i Z_i$ and $T_N = \sum_{i=1}^{N} a_i Z_i$ are two linear rank statistics, under the null hypothesis H_0: $F_X(x) = F_Y(x)$ for all x,*

$$\operatorname{cov}(B_N, T_N) = \frac{mn}{N^2(N-1)} \left(N \sum_{i=1}^{N} a_i b_i - \sum_{i=1}^{N} a_i \sum_{i=1}^{N} b_i \right)$$

Proof

$$\operatorname{cov}(B_N, T_N) = \sum_{i=1}^{N} a_i b_i \operatorname{var}(Z_i) + \sum\sum_{i \neq j} a_i b_j \operatorname{cov}(Z_i, Z_j)$$

$$= \frac{mn}{N^2} \sum_{i=1}^{N} a_i b_i - \frac{mn}{N^2(N-1)} \sum\sum_{i \neq j} a_i b_j$$

$$= \frac{mn}{N^2(N-1)} \left(N \sum_{i=1}^{N} a_i b_i - \sum_{i=1}^{N} a_i b_i - \sum_{i \neq j}\sum a_i b_j \right)$$

$$= \frac{mn}{N^2(N-1)} \left(N \sum_{i=1}^{N} a_i b_i - \sum_{i=1}^{N} a_i \sum_{i=1}^{N} b_i \right)$$

Using these theorems, the exact moments under the null hypothesis can be found for any linear rank statistics. The exact null probability distribution of T_N depends upon the probability distribution of the vector Z, which indicates the ranks of the X and Y random variables. This distribution was given in Eq. (7.5.18) for any distributions F_X and F_Y. In the null case, $F_X = F_Y = F$, say, and (7.5.18) reduces to

$$P(Z) = m!n! \int_{-\infty}^{\infty} \int_{-\infty}^{u_N} \cdots \int_{-\infty}^{u_2} \prod_{i=1}^{m} f(u_{r_i}) \prod_{j=1}^{n} f(u_{s_j}) \, du_1 \cdots du_N$$

where $r_1, r_2, \ldots, r_m$ and $s_1, s_2, \ldots, s_n$ are the ranks of the X and Y random variables, respectively, in the arrangement Z. Since the distributions are identical, the product in the integrand is the same for all subscripts, or

$$P(Z) = m!n! \int_{-\infty}^{\infty} \int_{-\infty}^{u_N} \cdots \int_{-\infty}^{u_2} \prod_{i=1}^{N} f(u_i) \, du_1 \cdots du_N = \frac{m!n!}{N!}$$

$$(3.3)$$

The final result follows from the fact that except for the terms $m!n!/N!$, $P(Z)$ is the integral over the entire region of the density function of the N order statistics for a random sample from the population F. Since $\binom{m+n}{m} = \binom{N}{m}$ is the total number of distinguishable Z vectors, i.e., distinguishable arrangements of m 1s and n 0s, the result in (3.3) implies that all vectors Z are equally likely under H_0.

Since each Z occurs with probability $1 / \binom{N}{m}$, the exact null probability distribution of any linear rank statistic can always be found by direct enumeration. The values of $T_N(Z)$ are calculated for each Z, and the probability of a particular value k is the number of Z vectors which lead to that number k divided by $\binom{N}{m}$. In other words, we have

$$P[T_N(Z) = k] = \frac{t(k)}{\binom{N}{m}}$$

$$(3.4)$$

where $t(k)$ is the number of arrangements of m X and n Y random variables such that $T_N(Z) = k$. Naturally, the tediousness of enumeration increases rapidly as m and n increase. For some statistics, recursive methods are possible.

When the null distribution of a linear rank statistic is known to be symmetric, only one-half of the distribution need be generated. The statistic is symmetric about its mean μ if for every $k \neq 0$

$$P[T_N(Z) - \mu = k] = P[T_N(Z) - \mu = -k]$$

or

$$t(\mu + k) = t(\mu - k)$$

Suppose that for every vector Z of m 1s and n 0s, there exists a conjugate vector Z', $Z' \neq Z$, of m 1s and n 0s such that whenever $T_N(Z) = \mu + k$, we have $T_N(Z') = \mu - k$. Then the frequency of the number $\mu + k$ is the same as that of $\mu - k$, and the distribution is symmetric. The condition for symmetry of a linear rank statistic then is that

$$T_N(Z) + T_N(Z') = 2\mu$$

The following theorem establishes a simple relation between the scores which will ensure the symmetry of $T_N(Z)$.

Theorem 3.4 *The null distribution of $T_N(Z)$ is symmetric about its mean*

$$\mu = m \sum_{i=1}^{N} \frac{a_i}{N} \text{ whenever the weights satisfy the relation}$$

$$a_i + a_{N-i+1} = c \qquad c = const, \text{ for } i = 1, 2, \ldots, N$$

Proof For any vector $Z = (Z_1, Z_2, \ldots, Z_N)$ of m 1s and n 0s, define the conjugate vector $Z' = (Z_1', Z_2', \ldots, Z_N')$, where $Z_i' = Z_{N-i+1}$.

Then

$$T_N(Z) + T_N(Z') = \sum_{i=1}^{N} a_i Z_i + \sum_{i=1}^{N} a_i Z_{N-i+1} = \sum_{i=1}^{N} a_i Z_i + \sum_{j=1}^{N} a_{N-j+1} Z_j$$

$$= \sum_{i=1}^{N} (a_i + a_{N-i+1}) Z_i = c \sum_{i=1}^{N} Z_i = cm$$

Since $E[T_N(Z)] = E[T_N(Z')]$, we must have $cm = 2\mu$, or

$$c = \frac{2\mu}{m} = 2 \sum_{i=1}^{N} \frac{a_i}{N}$$

The next theorem establishes the symmetry of *any* linear rank statistic when $m = n$.

Theorem 3.5 *The null distribution of $T_N(Z)$ is symmetric for any set of weights if $m = n = N/2$.*

Proof Since $m = n$, we can define our conjugate Z' with components $Z'_i = 1 - Z_i$. Then

$$T_N(Z) + T_N(Z') = \sum_{i=1}^{N} a_i Z_i + \sum_{i=1}^{N} a_i(1 - Z_i) = \sum_{i=1}^{N} a_i = 2\mu$$

A rather special but useful case of symmetry is given as follows.

Theorem 3.6 *The null distribution of $T_N(Z)$ is symmetric about its mean μ if N is even and the weights are $a_i = i$ for $i \le N/2$ and $a_i = N - i + 1$ for $i > N/2$.*

Proof The appropriate conjugate Z' has components $Z'_i = Z_{N/2+i}$ for $i \le N/2$ and $Z'_i = Z_{i-N/2}$ for $i > N/2$. Then

$$T_N(Z) + T_N(Z') = \sum_{i=1}^{N/2} iZ_i + \sum_{i=N/2+1}^{N} (N - i + 1)Z_i + \sum_{i=1}^{N/2} iZ_{N/2+i}$$

$$+ \sum_{i=N/2+1}^{N} (N - i + 1)Z_{i-N/2}$$

$$= \sum_{i=1}^{N/2} iZ_i + \sum_{i=N/2+1}^{N} (N - i + 1)Z_i$$

$$+ \sum_{j=N/2+1}^{N} \left(j - \frac{N}{2}\right)Z_j + \sum_{i=1}^{N/2} \left(\frac{N}{2} - j + 1\right)Z_j$$

$$= \sum_{i=1}^{N/2} \left(\frac{N}{2} + 1\right)Z_i + \sum_{i=N/2+1}^{N} \left(\frac{N}{2} + 1\right)Z_i$$

$$= m\left(\frac{N}{2} + 1\right) = 2\mu$$

In determining the frequency $t(k)$ for any value k which is assumed by the linear-rank-test statistic, the number of calculations required may be reduced considerably by the following properties of $T_N(Z)$, which are easily verified.

Theorem 3.7

Property 1: Let

$$T = \sum_{i=1}^{N} a_i Z_i \quad and \quad T' = \sum_{i=1}^{N} a_i Z_{N-i+1}$$

See also Hájek, A Course in Nonparametric Statistics (1969), p. 10, Theorem 3D.

158   NONPARAMETRIC STATISTICAL INFERENCE

Then $T = T'$ *if* $a_i = a_{N-i+1}$ *for* $i = 1, 2, \ldots, N$

Property 2: Let

$$T = \sum_{i=1}^{N} a_i Z_i \quad and \quad T' = \sum_{i=1}^{N} a_i (1 - Z_i)$$

Then $T + T' = \sum_{i=1}^{N} a_i$

Property 3: Let

$$T = \sum_{i=1}^{N} a_i Z_i \quad and \quad T' = \sum_{i=1}^{N} a_i (1 - Z_{N-i+1})$$

Then $T + T' = \sum_{i=1}^{N} a_i$ *if* $a_i = a_{N-i+1}$ *for* $i = 1, 2, \ldots, N$

For large samples, that is, $m \to \infty$ and $n \to \infty$ in such a way that m/n remains constant, an approximation exists which is applicable to the distribution of almost all linear rank statistics. Since T_N is a linear combination of the Z_i, which are identically distributed (though dependent) random variables, a generalization of the central-limit theorem allows us to conclude that the probability distribution of the standardized linear rank statistic

$$\frac{T_N - E(T_N)}{\sigma(T_N)}$$

approaches the standard normal probability distribution subject to certain regularity conditions.

These foregoing properties of the linear rank statistic hold only in the hypothesized case of identical populations. Chernoff and Savage (1958) have proved that the asymptotic normality property is valid also in the nonnull case, subject to certain regularity conditions relating mainly to the smoothness and size of the weights. The expressions for the mean and variance will be given here, since they are also useful in investigating consistency and efficiency properties of most two-sample linear rank statistics.

If the weights for the linear rank statistic are functions of the ranks, an equivalent representation of $T_N = \sum_{i=1}^{N} a_i Z_i$ in terms of a Stieltjes integral is

$$T_N = m \int_{-\infty}^{\infty} J_N[H_N(x)] \, dS_m(x)$$

where the notation is defined as follows:

1. $S_m(x)$ and $T_n(x)$ are the empirical distribution functions of the X and Y samples, respectively.
2. $\lambda_N = m/N$, $0 < \lambda_N < 1$.
3. $H_N(x) = \lambda_N S_m(x) + (1 - \lambda_N) T_n(x)$, so that $H_N(x)$ is the proportion of observations from either sample which do not exceed the value x, or the empirical distribution function of the combined sample.
4. $J_N(i/N) = a_i$.

This Stieltjes integral form is given here because it appears frequently in the journal literature and is useful for proving theoretical properties. Since the theorems following are given here without proof anyway, the student not familiar with Stieltjes integrals can consider the following equivalent representation:

$$T_N = m \sum_{\substack{\text{over all } x \text{ such} \\ \text{that } p(x)>0}} J_N[H_N(x)]p(x)$$

where

$$p(x) = \begin{cases} \dfrac{1}{m} & \text{if } x \text{ is the observed value of an } X \text{ random variable} \\ 0 & \text{otherwise} \end{cases}$$

For example, in the simplest case where $a_i = i/N$, $J_N[H_N(x)] = H_N(x)$ and

$$T_N = m \int_{-\infty}^{\infty} H_N(x)\, dS_m(x) = \frac{m}{N} \int_{-\infty}^{\infty} [mS_m(x) + nT_n(x)]\, dS_m(x)$$

$$= \frac{m}{N} \int_{-\infty}^{\infty} (\text{number of observations} \le x)\ (1/m \text{ if } x \text{ is the value}$$

$$\text{of an } X \text{ random variable and } 0 \text{ otherwise})$$

$$= \frac{1}{N} \sum_{i=1}^{N} iZ_i$$

Now when the X and Y samples are drawn from the continuous populations F_X and F_Y, respectively, we define the combined population cdf as

$$H(x) = \lambda_N F_X(x) + (1 - \lambda_N) F_Y(x)$$

The Chernoff and Savage theorem follows.

Theorem 3.8 *Subject to certain regularity conditions, the most important of which is that for $J(H) = \lim\limits_{N \to \infty} J_N(H)$,*

$$|J^{(r)}(H)| = |d^r J(H)/dH^r| \le K|H(1 - H)|^{-r-\frac{1}{2}+\delta}$$

for $r = 0, 1, 2$ *and some* $\delta > 0$ *and* K *any constant which does not depend on* m, n, N, F_X, *or* F_Y, *then for* λ_N *fixed*

$$\lim_{N \to \infty} P\left(\frac{T_N/m - \mu_N}{\sigma_N} \leq t\right) = \Phi(t)$$

where

$$\mu_N = \int_{-\infty}^{\infty} J[H(x)]f_X(x)\, dx$$

$$N\sigma_N{}^2 = 2\frac{1 - \lambda_N}{\lambda_N}\left\{\lambda_N \iint_{-\infty < x < y < \infty} F_Y(x)[1 - F_Y(y)]J'[H(x)]\right.$$
$$J'[H(y)]f_X(x)f_X(y)\, dx\, dy + (1 - \lambda_N)\iint_{-\infty < x < y < \infty}$$
$$\left.F_X(x)[1 - F_X(y)]J'[H(x)]J'[H(y)]f_Y(x)f_Y(y)\, dx\, dy\right\}$$

Corollary 3.8   *If* X *and* Y *are identically distributed with common distribution* $F(x) = F_X(x) = F_Y(x)$, *we have*

$$\mu_N = \int_0^1 J(u)\, du$$

$$N\lambda_N\sigma_N{}^2 = 2(1 - \lambda_N)\iint_{0 < x < y < 1} x(1 - y)J'(x)J'(y)\, dx\, dy$$

$$= 2(1 - \lambda_N)\iiiint_{0 < u < x < y < v < 1} J'(x)J'(y)\, dx\, dy\, du\, dv$$

$$= 2(1 - \lambda_N)\iint_{0 < u < v < 1} \int_u^v \int_x^v J'(x)J'(y)\, dy\, dx\, du\, dv$$

$$= 2(1 - \lambda_N)\iint_{0 < u < v < 1} \int_u^v [J(v) - J(x)]J'(x)\, dx\, du\, dv$$

$$= 2(1 - \lambda_N)\iint_{0 < u < v < 1} \left[J(v)J(x) - \frac{J^2(x)}{2}\right]\Big|_u^v\, du\, dv$$

$$= (1 - \lambda_N)\iint_{0 < u < v < 1} [J^2(v) - 2J(v)J(u) + J^2(u)]\, du\, dv$$

$$= (1 - \lambda_N)\left[\int_0^1 vJ^2(v)\, dv\right.$$
$$\left. + \int_0^1 (1 - u)J^2(u)\, du - \int_0^1 J(u)\, du\int_0^1 J(v)\, dv\right]$$

$$= (1 - \lambda_N)\left\{\int_0^1 J^2(u)\, du - \left[\int_0^1 J(u)\, du\right]^2\right\}$$

These expressions are equivalent to those given in Theorem 3.2 for $a_i = J_N(i/N)$.

8.4 USEFULNESS IN INFERENCE PROBLEMS

The general alternative to the null hypothesis in the two-sample problem is simply that the populations are not identical, i.e.,

$$F_X(x) \neq F_Y(x) \qquad \text{for some } x$$

or the analogous one-sided general alternative, which states a directional inequality as

$$F_X(x) \leq F_Y(x) \qquad \text{for all } x$$

The two-sample tests considered in Chap. 7, namely, the Kolmogorov-Smirnov, Wald-Wolfowitz runs, Mann-Whitney and median tests, are all appropriate for these alternatives. In most parametric two-sample situations, the alternatives are much more specific, as in the t and F tests for comparison of means and variances, respectively. Although all the two-sample rank tests are for the same null hypothesis, particular test statistics may be especially sensitive to a particular form of alternative, thus increasing their power against that type of alternative.

Since any set of scores $a_1, a_2, \ldots, a_N$ may be employed for the coefficients in a linear rank statistic, this form of test statistic lends itself particularly well to more specific types of alternatives. The appropriateness of choice depends on the type of difference between populations one hopes to detect. The simplest type of situation to deal with is where the statistician has enough information about the populations to feel that if a difference exists, it is only in location or only in scale. These will be called, respectively, the *two-sample location problem* and the *two-sample scale problem*. In the following two chapters we shall discuss briefly some of the better-known and more widely accepted linear rank statistics useful in these problems. No attempt will be made to provide recommendations regarding which to use. The very generality of linear rank tests makes it difficult to make direct comparisons of power functions, since calculation of power requires more specification of the alternative probability distributions and moments. A particular test might have high power against normal alternatives but perform poorly for the gamma distribution. Furthermore, calculation of the power of rank tests is usually extremely tiresome. We must be able to determine the probability distribution of the statistic $T_N(Z)$ or the arrangement Z as in (7.5.18) under the specified alternative and sum these probabilities over those arrangements Z in the rejection region specified by the test. Isolated and specific comparisons of power between nonparametric tests receive much attention in the literature, and the reader is referred to Savage's "Bibliography" (1962) for references. However, calculation of asymptotic relative efficiency of linear rank tests versus the t and F tests for normal alternatives is not particularly difficult. Therefore, information regarding the AREs of the tests presented here for the location and scale problems will be provided.

PROBLEMS

8.1. One of the simplest linear rank statistics is defined as

$$W_N = \sum_{i=1}^{N} iZ_i$$

This is the Wilcoxon statistic to be discussed in the next chapter. Use Theorem 3.2 to evaluate the mean and variance of W_N.

8.2. Express the two-sample median-test statistic U defined in Sec. 7.4 in the form of a linear rank statistic and use Theorem 3.2 to find its mean and variance. *Hint:* For the appropriate argument k, use the function $S(k)$ defined as for (2.1).

8.3. Prove the three properties stated in Theorem 3.7.

9

Linear Rank Tests for the Location Problem

9.1 INTRODUCTION

Suppose that independent samples of sizes m and n are drawn from two continuous populations. We wish to test the null hypothesis of identical distributions. The location alternative is that the populations are of the same form but with a different measure of central tendency. This can be expressed symbolically as follows:

$$H_0: F_Y(x) = F_X(x) \qquad \text{for all } x$$
$$H_L: F_Y(x) = F_X(x - \theta) \qquad \text{for all } x \text{ and some } \theta \neq 0$$

The cumulative distribution function of the Y population under H_L is the same as that of the X population but shifted to the left if $\theta < 0$ and shifted to the right if $\theta > 0$, as in Fig. 1.1.   For $\theta < 0$, the median of the X population is larger than the median of the Y population.

If it is reasonable to assume that F_X is the cumulative normal distribution, then the mean and median coincide and a one-sided normal-

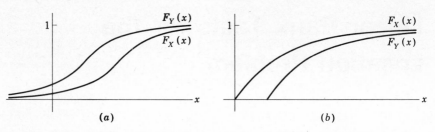

Fig. 1.1 $F_Y(x) = F_X(x - \theta)$.   (a) F_X normal, $\theta < 0$; (b) F_X exponential, $\theta > 0$.

theory test with equal but unknown variances of the hypothesis

$$\mu_Y - \mu_X = 0 \qquad \text{versus} \qquad \mu_Y - \mu_X < 0$$

is equivalent to the general location alternative with $\theta = \mu_Y - \mu_X < 0$. The appropriate parametric test alternative is the t statistic with $m + n - 2$ degrees of freedom:

$$t_{m+n-2} = \frac{\bar{X} - \bar{Y}}{\sqrt{\dfrac{(m-1)S_X{}^2 + (n-1)S_Y{}^2}{m+n-2}} \sqrt{\dfrac{m+n}{mn}}}$$

The t test statistic has been shown to be robust for the assumptions of normality and equal variances. However, there are many good and simple nonparametric tests for the general location problem which do not require any assumptions other than independent samples from continuous populations. Many of these are rank statistics since the ranks of the X's relative to the ranks of the Y's provide information about the relative size of the population medians. In the form of a linear rank statistic, any set of scores which are nondecreasing or nonincreasing in magnitude would allow the statistic to reflect a combined ordered sample in which most of the X's are larger than the Y's, or vice versa. The Wilcoxon test is one of the best known and easiest to use, since it employs scores which are positive integers. The other tests which will be covered here are the Terry test, inverse-normal-scores test, and percentile modified rank tests. There are many others to be found in the literature.

9.2 THE WILCOXON TEST

The ranks of the X's in the combined ordered arrangement of the two samples would generally be larger than the ranks of the Y's if the median of the X population exceeds the median of the Y population. Therefore, Wilcoxon (1945) proposed a test where we accept the one-sided location alternative $H_L: \theta < 0$ if the sum of the ranks of the X's is too large or H_L: $\theta > 0$ if the sum of the ranks of the X's is too small and the two-sided location alternative H_L: $\theta \neq 0$ if the sum of the ranks of the X's is either too large or too small. This function of the ranks expressed as a linear rank statistic has the simple weights $a_i = i$, $i = 1, 2, \ldots, N$, $N = m + n$. In other words, the Wilcoxon test statistic is

$$W_N = \sum_{i=1}^{N} i Z_i \tag{2.1}$$

where the Z_i are the indicator random variables as defined for (8.2.2). See pp. 152-153

If there are no ties, the exact mean and variance of W_N under the null hypothesis of equal distributions are easily found from Theorem 8.3.2 to be

$$E(W_N) = \frac{m(N+1)}{2} \qquad \mathrm{var}(W_N) = \frac{mn(N+1)}{12}$$

The value of W_N has a minimum of $\displaystyle\sum_{i=1}^{m} i = \frac{m(m+1)}{2}$ and a maximum

of $\displaystyle\sum_{i=N-m+1}^{N} i = \frac{m(2N-m+1)}{2}$. Furthermore, from Theorem 8.3.4, since

$$a_i + a_{N-i+1} = N + 1 \qquad \text{for } i = 1, 2, \ldots, N$$

the statistic is symmetric about its mean. The exact null probability distribution here can be obtained systematically by enumeration using these properties. For example, suppose $m = 3$, $n = 4$. There are $\binom{7}{4} = 35$ possible distinguishable configurations of 1s and 0s in the vector Z, but these need not be enumerated individually. W_N will range between 6 and 18, symmetric about 12, the values occurring in conjunction with the ranks in Table 2.1, from which the complete probability distribution is easily found.

Table 2.1

Value of W_N	Ranks of X's	Frequency
18	5, 6, 7	1
17	4, 6, 7	1
16	3, 6, 7; 4, 5, 7	2
15	2, 6, 7; 3, 5, 7; 4, 5, 6	3
14	1, 6, 7; 2, 5, 7; 3, 4, 7; 3, 5, 6	4
13	1, 5, 7; 2, 4, 7; 2, 5, 6; 3, 4, 6	4
12	1, 4, 7; 2, 3, 7; 1, 5, 6; 2, 4, 6; 3, 4, 5	5

Several recursive schemes are also available for generation of the distribution. The simplest to understand is analogous to the recursion relations given in (6.3.8) for the Wilcoxon signed-rank statistic and (7.5.14) for the Mann-Whitney statistic. If $r_{m,n}(k)$ denotes the number of arrangements of m X and n Y random variables such that the sum of the X ranks is equal to k, it is evident that

$$r_{m,n}(k) = r_{m-1,n}(k - N) + r_{m,n-1}(k)$$

and

$$f_{W_N}(k) = p_{m,n}(k) = \frac{r_{m-1,n}(k - N) + r_{m,n-1}(k)}{\binom{m + n}{m}}$$

or

$$(m + n)p_{m,n}(k) = mp_{m-1,n}(k - N) + np_{m,n-1}(k) \qquad (2.2)$$

Tables of critical values for $N \leq 20$ are given in Wilcoxon (1947) and several other sources. Either one- or two-sided tests may be performed.

For larger samples, generation of the exact probability distribution is rather irksome. However, the normal approximation to the distribution or rejection regions can be used because of the asymptotic normality of the general linear rank statistic (Theorem 8.3.8). The normal approximation for W_N has been shown to be accurate enough for most practical purposes for combined sample sizes as small as 12.

The midrank method is easily applied to handle the problem of ties. The presence of a moderate number of tied observations seems to have little effect on the probability distribution. Corrections for ties have been thoroughly investigated [see, for example, Noether (1967), pp. 32–35].

The Wilcoxon test is actually the same as the Mann-Whitney U test discussed previously, since a linear relationship exists between the two test statistics. With U defined as the number of times a Y precedes an X, as in (7.5.2), we have

$$U = \sum_{i=1}^{m} \sum_{j=1}^{n} D_{ij} = \sum_{i=1}^{m} (D_{i1} + D_{i2} + \cdots + D_{in})$$

where

$$D_{ij} = \begin{cases} 1 & \text{if } Y_j < X_i \\ 0 & \text{if } Y_j > X_i \end{cases}$$

Then $\sum_{j=1}^{n} D_{ij}$ is the number of values of j for which $Y_j < X_i$, or the rank of X_i reduced by n_i, the number of X's which are less than or equal to X_i. Thus we can write

$$
\begin{aligned}
U = \sum_{i=1}^{m} [r(X_i) - n_i] &= \sum_{i=1}^{m} r(X_i) - (n_1 + n_2 + \cdots + n_m) \\
&= \sum_{i=1}^{N} iZ_i - (1 + 2 + \cdots + m) \\
&= \sum_{i=1}^{N} iZ_i - \frac{m}{2}(m + 1) \\
&= W_N - \frac{m}{2}(m + 1)
\end{aligned}
\tag{2.3}
$$

Therefore, all the properties of the tests are the same, including consistency and the ARE of 0.955. A confidence-interval procedure based on the Wilcoxon statistic need not be discussed, since it leads to the same results as the Mann-Whitney procedure.

The Wilcoxon statistic is also equivalent to an ordinary analysis of variance of ranks (see Prob. 11.5), a procedure which is easily extended to the case of more than two samples. This problem will be discussed in Chap. 11.

9.3 OTHER LOCATION TESTS

Generally, almost any set of monotone-increasing weights a_i which are adopted for the linear rank statistic will provide a consistent test for shift in location. Only a few of the better-known ones will be mentioned here.

TERRY–HOEFFDING TEST

The Terry (1952) c_1 and Hoeffding (1951) test uses the weights $a_i = E(\xi_{(i)})$, and so the test statistic is

$$
c_1 = \sum_{i=1}^{N} E(\xi_{(i)})Z_i
\tag{3.1}
$$

where $\xi_{(i)}$ is the ith order statistic from a standard normal population. These expected values of standard normal order statistics are tabulated for $N \leq 100$ and some larger sizes in Harter (1961), so that the exact null distribution can be found by enumeration. Tables of the distribution of the test statistic are given in Terry (1952) and Klotz (1964). The

Terry test statistic is symmetric about the origin, and its variance is

$$\sigma^2 = mn \frac{\sum_{i=1}^{N} [E(\xi_{(i)})]^2}{N(N-1)}$$

The normal distribution provides a good approximation to the null distribution for larger samples. An approximation based on the t distribution is even closer. This statistic is $r(N-2)^{1/2}/(1-r^2)^{1/2}$, where $r = c_1/[\sigma^2(N-1)]^{1/2}$, and the distribution is approximately Student's t with $N-2$ degrees of freedom.

The Terry test is asymptotically optimum against the alternative that the populations are both normal distributions, with the same variance but different means. Under the classical assumptions for a test of location then, its ARE is 1 relative to Student's t test. For certain other families of continuous distributions, the Terry test is more efficient than Student's t test (ARE > 1) [Chernoff and Savage (1958)].

The weights employed for the Terry test $E(\xi_{(i)})$ are often called *expected normal scores*, since the order statistics of a sample from the standard normal population are commonly referred to as normal scores. The idea of using expected normal scores instead of integer ranks as rank-order statistics is appealing generally, since for many populations the expected normal scores may be more "representative" of the raw data or variate values. This could be investigated by comparing the correlation coefficients between (1) variate values and expected normal scores and (2) variate values and integer ranks for particular families of distributions. The limiting value of the correlation between variate values from a normal population and the expected normal scores is equal to 1, for example. Since inferences based on rank-order statistics are really conclusions about transformed variate values, that transformation which most closely approximates the actual data should be most efficient when these inferences are extended to the actual data.

Since the Terry test statistic is the sum of the expected normal scores of the variables in the X sample, it may be interpreted as identical to the Wilcoxon two-sample test of the last section when the normal-scores transformation is used instead of the integer-rank transformation. Other linear rank statistics for location can be formed in the same way by using different sets of rank-order statistics for the combined samples. An obvious possibility suggested by the Terry test is to use the rank-order statistics $\Phi^{-1}[i/(N+1)]$, where $\Phi(x)$ is the cumulative standard normal distribution, since we showed in (2.6.5) that $\Phi^{-1}[i/(N+1)]$ is a first approximation to $E(\xi_{(i)})$. If κ_p is the pth quantile point of the standard normal distribution, $\Phi(\kappa_p) = p$ and $\kappa_p = \Phi^{-1}(p)$. Therefore here the ith order statistic in the combined ordered sample is replaced by the $[i/(N+1)]$st quantile point of the standard normal. This is usually called the *inverse-normal-scores transformation*.

VAN DER WAERDEN TEST

When the Wilcoxon criterion is applied to these rank-order statistics, we obtain the van der Waerden (1952, 1953) X_N test, where

$$X_N = \sum_{i=1}^{N} \Phi^{-1}\left(\frac{i}{N+1}\right) Z_i \tag{3.2}$$

In other words, the constant a_i in the general linear rank statistic is the value on the abscissa of the graph of a standard normal density function such that the area to the left of a_i is equal to $i/(N+1)$. These weights a_i are easily found for any N from tables of the cumulative normal distribution. Tables of critical values are given in van der Waerden and Nievergelt (1956) for $N \leq 50$. The X_N statistic is symmetric about the origin and has variance

$$mn \frac{\sum_{i=1}^{N} \left[\Phi^{-1}\left(\frac{i}{N+1}\right) \right]^2}{N(N-1)}$$

For larger samples, the null distribution of the standardized X_N is well approximated by the normal.

The X_N test is perhaps easier to use than the Terry test because of the ready access to tables which will supply the values of the weights for any N. Otherwise, there is little basis for choice between them. In fact, the van der Waerden test is asymptotically equivalent to the Terry test. Since $\text{var}(\xi_{(i)}) \to 0$ as $N \to \infty$, $\xi_{(i)}$ converges in probability to $E(\xi_{(i)})$, the weights for the Terry statistic. However, by the probability-integral transformation, $\Phi(\xi_{(i)})$ is the ith order statistic of a sample of size N from the uniform distribution. Therefore from (2.5.2) and (2.5.3), $E[\Phi(\xi_{(i)})] = i/(N+1)$ and

$$\text{var}[\Phi(\xi_{(i)})] = \frac{i(N-i+1)}{(N+1)^2(N+2)} \to 0$$

as $N \to \infty$. This implies that $\Phi(\xi_{(i)})$ converges in probability to $i/(N+1)$ or $\xi_{(i)}$ converges to $\Phi^{-1}[i/(N+1)]$. We may conclude that the expected normal scores and corresponding inverse normal scores are identical for N infinite. The large-sample properties, including the ARE, are thus identical for the Terry and van der Waerden tests.

It should be noted that the expected normal scores and inverse normal scores may be useful in any procedures based on rank-order statistics. For example, in the one-sample and paired-sample Wilcoxon signed-rank test discussed in Sec. 6.3, the rank of the absolute value of the difference $|D_i|$ can be replaced by the corresponding expected value of the absolute value of that normal score $E(|\xi_{(i)}|)$ (which is not equal to the absolute value of the expected normal score). The sum of those "ranks" which correspond to positive differences D_i is then employed as a test statistic. This statistic provides the asymptotically optimum test of location when the population of differences is normal and thus has an ARE of 1 relative to Student's t test in this case. Expected normal scores are also useful in rank-correlation methods, which will be covered in Chap. 12.

PERCENTILE MODIFIED RANK TESTS

Another interesting linear rank statistic for the two-sample location problem is a member of the class of so-called percentile modified linear rank tests [Gastwirth (1965)]. The idea is as follows. We select two numbers p and r, both between 0 and 1, and then score only the data in the upper pth and lower rth percentiles of the combined sample. In other words, a linear rank statistic is formed in the usual way except that a score of zero is assigned to a group of observations in the middle. Symbolically, we let $P = [Np] + 1$ and $R = [Nr] + 1$, where $[x]$ denotes the largest integer not exceeding the number x. Define B_r and T_p as

N odd:

$$B_r = \sum_{i=1}^{R} (R - i + 1)Z_i \quad \text{and} \quad T_p = \sum_{i=N-P+1}^{N} [i - (N - P)]Z_i$$

N even: (3.3)

$$B_r = \sum_{i=1}^{R} (R - i + \tfrac{1}{2})Z_i$$

and

$$T_p = \sum_{i=N-P+1}^{N} [i - (N - P) - \tfrac{1}{2}]Z_i$$

The combination $T_p - B_r$ provides a test for location, and $T_p + B_r$ is a test for scale, which will be discussed in the next chapter. It is easily seen that if N is even and $P = R = N/2$, so that no observations are assigned a score of zero, $T - B$ is equivalent to the Wilcoxon test. When N is odd and all the sample data are used, the tests differ slightly because of the different way of handling the middle observation $z_{(N+1)/2}$.

The mean and variance of the $T_p \pm B_r$ statistics can be calculated using Theorem 8.3.2 alone if $P + R \leq N$, remembering that $a_i = 0$ for $R + 1 \leq i \leq N - P$. Alternatively, Theorems 8.3.2 and 8.3.3 can be used on the pieces T_p and B_r along with the fact that

$$\text{var}(T_p \pm B_r) = \text{var}(T_p) + \text{var}(B_r) \pm 2 \text{cov}(T_p, B_r)$$

The results for N even and $P = R$ are

$$E(T_p - B_r) = 0 \qquad \text{var}(T_p - B_r) = \frac{P(4P^2 - 1)}{6N(N - 1)} \tag{3.4}$$

By Theorem 8.3.4, the null distribution is symmetric about the origin for any m and n when $P = R$. Tables of the null distribution for $m = n \leq 6$ are given in Gibbons and Gastwirth (1966). It is also shown there empirically that for significance levels not too small, say at least 0.025, the normal distribution can be used to define critical regions with sufficient accuracy for most practical purposes when $m = n \geq 6$.

One of the main advantages of this test is that a judicious choice of p and r may lead to a test which attains higher power than the Wilcoxon test without the necessity of introducing complicated scoring systems. For example, any knowledge of asymmetry in the populations might be incorporated into the test statistic. The asymptotic relative efficiency of this test against normal alternatives reaches its maximum value of 0.968 when $p = r = 0.42$; when $p = r = 0.5$, the ARE is 0.955, as for the Wilcoxon statistic.

PROBLEMS

9.1. Given independent samples of m X and n Y variables, define the following random variables for $i = 1, 2, \ldots, m$:

$K_i = $ rank of X_i among $X_1, X_2, \ldots, X_m$
$R_i = $ rank of X_i among $X_1, X_2, \ldots, X_m, Y_1, Y_2, \ldots, Y_n$

Use K_i and R_i to prove the linear relation between the Mann-Whitney and Wilcoxon statistics given in (2.3).

9.2. A single random sample $D_1, D_2, \ldots, D_N$ of size N is drawn from a population which is continuous and symmetric. Assume there are m positive values, n negative values, and no zero values. Define the $m + n = N$ random variables

$X_i = D_i$ if $D_i > 0$
$Y_i = |D_i|$ if $D_i < 0$

Then the $X_1, X_2, \ldots, X_m$ and $Y_1, Y_2, \ldots, Y_n$ constitute two independent random samples of sizes m and n.

   (a) Show that the two-sample Wilcoxon statistic W_N of (2.1) for these two samples equals the Wilcoxon signed-rank statistic T^+ defined in (6.3.1).

(b) If these two samples are from identical populations, the median of the symmetric D population must be zero. Therefore the null distribution of W_N is identical to the null distribution of T^+ conditional upon the observed number of plus and minus signs. Explain fully how tables of the null distribution of W_N could be used to find the null distribution of T^+. Since for N large, m and n will both converge to the constant value $N/2$ in the null case, these two test statistics have equivalent properties asymptotically.

9.3. Generate by enumeration the exact null probability distribution of $T_p - B_r$ as defined in (3.3) for $m = n = 3$, all $P = R < 3$, and compare the rejection regions for $\alpha \leq 0.10$ with those for the Wilcoxon test W_N when $m = n = 3$.

9.4. (a) Verify the results given in (3.4) for the mean and variance of $T_p - B_r$ when $P = R$ and N is even.

(b) Derive a similar result for $P = R$ when N is odd.

10

Linear Rank Tests for the Scale Problem

10.1 INTRODUCTION

Consider again the situation of Chap. 9, where the null hypothesis is that two independent samples are drawn from identical populations; however, now suppose that we are interested in detecting differences in variability or dispersion instead of location. Some of the tests presented in Chaps. 7 and 9, namely, the median, Mann-Whitney, Wilcoxon, Terry, van der Waerden, and $T_p - B_r$ tests, were noted to be particularly sensitive to differences in location when the populations are identical otherwise, a situation described by the relation $F_Y(x) = F_X(x - \theta)$. These tests cannot be expected to perform especially well against other alternatives. The general two-sample tests, like the Wald-Wolfowitz runs test or Kolmogorov-Smirnov tests, are affected by any type of difference in the populations and therefore cannot be relied upon as efficient for detecting differences in variability. Some other nonparametric tests are needed for the dispersion problem.

The classical test for which we are seeking an analog is the test for equality of variances, $H_0: \sigma_X = \sigma_Y$ against one- or two-sided alternatives.

If it is reasonable to assume that the two populations are both normal distributions, the parametric test statistic is

$$F_{m-1,n-1} = \frac{\sum\limits_{i=1}^{m} \dfrac{(X_i - \bar{X})^2}{m-1}}{\sum\limits_{i=1}^{n} \dfrac{(Y_i - \bar{Y})^2}{n-1}}$$

which has Snedecor's F distribution with $m-1$ and $n-1$ degrees of freedom. The F test is not particularly robust with respect to the normality assumption. If there is reason to question the postulates inherent in the construction of the test, a nonparametric test of dispersion is appropriate.

The F test does not require any assumption regarding the locations of the two normal populations. The magnitudes of the two sample variances are directly comparable since they are each computed as measures of deviations around the respective sample means. The traditional concept of dispersion is a measure of spread around some population central value. The model for the relationship between the two normal populations assumed for the F test might be written

$$F_{Y-\mu_Y}(x) = F_{X-\mu_X}\left(\frac{\sigma_X}{\sigma_Y}\,x\right) = F_{X-\mu_X}(\theta x) \qquad \text{for all } x \text{ and some } \theta > 0$$

$$(1.1)$$

where $\theta = \sigma_X/\sigma_Y$ and $F_{(X-\mu_X)/\sigma_X}(x) = \Phi(x)$, and the null hypothesis to be tested is $H_0\colon \theta = 1$. We could say then that we assume that the distributions of $X - \mu_X$ and $Y - \mu_Y$ differ only by the scale factor θ for any μ_X and μ_Y, which need not be specified. The relationship between the respective moments is

$$E(X - \mu_X) = \theta E(Y - \mu_Y) \qquad \text{and} \qquad \text{var}(X) = \theta^2 \,\text{var}(Y)$$

Since medians are the customary location parameters in distribution-free procedures, if nonparametric dispersion is defined as spread around the respective medians, the nonparametric model corresponding to (1.1) is

$$F_{Y-M_Y}(x) = F_{X-M_X}(\theta x) \qquad \text{for all } x \text{ and some } \theta > 0 \qquad (1.2)$$

Suppose that the test criterion we wish to formulate for this model is to be based on the configuration of the X and Y random variables in the combined ordered sample, as in a linear rank test. The characteristics of respective locations and dispersions are inextricably mixed in the combined sample ordering, and possible location differences may mask dispersion differences. If the population medians M_X and M_Y were known, the model (1.2) suggests that the sample observations should be adjusted by

$$X_i' = X_i - M_X \quad \text{and} \quad Y_j' = Y_j - M_Y \quad \text{for } i = 1, 2, \ldots, m$$
$$\text{and } j = 1, 2, \ldots, n$$

Then the X_i' and Y_j' populations both have zero medians, and the arrangement of X' and Y' random variables in the combined ordered sample should indicate dispersion differences as unaffected by location differences. The model is $F_{Y'}(x) = F_{X'}(\theta x)$. In practice, M_X and M_Y would probably not be known, so that this is not a workable approach. If we simply assume that $M_X = M_Y = M$ unspecified, the combined sample arrangement of the unadjusted X and Y should still reflect dispersion differences. Since the X and Y populations differ only in scale, the logical model for this situation would seem to be the alternative

$$H_S: F_Y(x) = F_X(\theta x) \qquad \text{for all } x \text{ and some } \theta > 0, \theta \neq 1 \qquad (1.3)$$

This is appropriately called the *scale alternative* because the cumulative distribution function of the Y population is the same as that of the X population but with a compressed or enlarged scale according as $\theta > 1$ or $\theta < 1$, respectively.

In Fig. 1.1a, the relation $F_Y(x) = F_X(\theta x)$ is shown for $F_X(x) = \Phi(x)$, the standard normal, and $\theta > 1$. Since $\mu_X = M_X = \mu_Y = M_Y = 0$ and $\theta = \sigma_X/\sigma_Y$, this model is a special case of (1.1) and (1.2). Figure 1.1b illustrates the difficulty in thinking any arbitrary distribution may be taken for the scale alternative in (1.3) to be interpreted exclusively as a dispersion alternative. Here we have a representation of the exponential distribution in H_S for $\theta < 1$, for example, $f_X(x) = e^{-x}$, $x > 0$, so that $f_Y(x) = \theta e^{-\theta x}$ for some $\theta < 1$. Since $\text{var}(X) = 1$ and $\text{var}(Y) = 1/\theta^2$, it

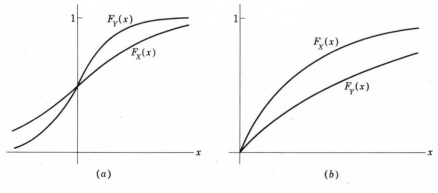

Fig. 1.1 $F_Y(x) = F_X(\theta x)$. (a) F_X normal, $\theta > 1$; (b) F_X exponential, $\theta < 1$.

is true that $\sigma_Y > \sigma_X$. However, $E(X) = 1$ and $E(Y) = 1/\theta > E(X)$, and further $M_X = \ln 2$ while $M_Y = \ln(2/\theta) > M_X$ for all $\theta < 1$. The combined ordered arrangement of samples from these exponential populations will be reflective of both the location and dispersion differences. The scale alternative in (1.3) should be interpreted as a dispersion alternative only if the population locations are identical or very nearly the same.

Actually, the scale model $F_Y(x) = F_X(\theta x)$ is not general enough even when the locations are the same. The relationship implies that $E(X) = \theta E(Y)$ and $M_X = \theta M_Y$, so that the locations are identical for all θ only if $\mu_X = \mu_Y = 0$ or $M_X = M_Y = 0$. A more general scale alternative can be written in the form

$$H_S : F_{Y-M}(x) = F_{X-M}(\theta x) \qquad \text{for all } x \text{ and some } \theta > 0, \theta \neq 1 \quad (1.4)$$

where M is interpreted to be the common median. Both (1.3) and (1.4) are called the scale alternatives applicable to the two-sample scale problem, but in (1.3) we essentially assume without loss of generality that $M = 0$.

Many tests based on the ranks of the observations in a combined ordering of the two samples have been proposed for the scale problem. If they are to be useful for detecting differences in dispersion, we must assume that the medians (or means) of the two populations are equal but unknown or that the sample observations can be adjusted to have equal locations, by subtracting the respective location parameters from each set or by subtracting the difference between the parameters from one set. Under these assumptions, an appropriate set of weights for a linear rank-test statistic will provide information about the relative spread of the observations about their common central value. If the X population has a larger dispersion, the X values should be positioned approximately symmetrically at both extremes of the Y values. Therefore the weights a_i should be symmetric; for example, small weights in the middle and large at the two extremes, or vice versa. We shall consider several choices for simple sets of weights of this type which provide linear rank tests particularly sensitive to scale differences only. These are basically the best-known tests—the Mood test, the Freund-Ansari-Bradley-David-Barton tests, the Siegel-Tukey test, the Klotz normal-scores test, the percentile modified rank tests, and the Sukhatme test. Many other tests have also been proposed in the literature, a few of which will be covered in the last section.

10.2 THE MOOD TEST

In the combined ordered sample of N variables with no ties, the average rank is the mean of the first N integers, $(N + 1)/2$. The deviation of the rank of the ith ordered variable about its mean rank is $i - (N + 1)/2$, and the amount of deviation is an indication of relative spread. However, as in the case of defining a measure of sample dispersion in classical descriptive statistics, the fact that the deviations are equally divided between positive and negative numbers presents a problem in using these actual deviations as weights in constructing a linear rank statistic. For example, if $m = n = 3$, the ordered arrangements

$$X \; Y \; X \; Y \; X \; Y \quad \text{and} \quad X \; X \; Y \; Y \; Y \; X$$

both have $\sum_{i=1}^{6} \left(i - \dfrac{N + 1}{2} \right) z_i = -1.5$, whereas the first arrangement suggests the variances are equal and the second suggests the X's are more dispersed than the Y's. The natural solution is to use as weights either the absolute values or the squared values of the deviations to give equal weight to deviations on either side of the central value.

The Mood (1954) test is based on the sum of squares of the deviations of the X ranks from the average combined rank, or

$$M_N = \sum_{i=1}^{N} \left(i - \frac{N + 1}{2} \right)^2 Z_i \tag{2.1}$$

A large value of M_N would imply that the X's are more widely dispersed, and M_N small implies the opposite conclusion. Specifically, the set of weights is as shown in Tables 2.1 and 2.2. The larger weights are in the tails of the arrangement. When N is odd, the median of the combined sample is assigned a weight of zero. In that case, therefore, the

median rank $= \dfrac{N+1}{2}$

Table 2.1 N **even**

i	1	2	3	$\cdots$	$\dfrac{N}{2} - 1$	$\dfrac{N}{2}$
a_i	$\left(\dfrac{N-1}{2}\right)^2$	$\left(\dfrac{N-3}{2}\right)^2$	$\left(\dfrac{N-5}{2}\right)^2$	$\cdots$	$\left(\dfrac{3}{2}\right)^2$	$\left(\dfrac{1}{2}\right)^2$

i	$\dfrac{N}{2} + 1$	$\dfrac{N}{2} + 2$	$\cdots$	$N - 2$	$N - 1$	N
a_i	$\left(\dfrac{1}{2}\right)^2$	$\left(\dfrac{3}{2}\right)^2$	$\cdots$	$\left(\dfrac{N-5}{2}\right)^2$	$\left(\dfrac{N-3}{2}\right)^2$	$\left(\dfrac{N-1}{2}\right)^2$

Table 2.2 N **odd**   { median rank

i	1	2	3	$\cdots$	$\dfrac{N-1}{2}$	$\dfrac{N+1}{2}$
a_i	$\left(\dfrac{N-1}{2}\right)^2$	$\left(\dfrac{N-3}{2}\right)^2$	$\left(\dfrac{N-5}{2}\right)^2$	$\cdots$	1^2	0

i	$\dfrac{N+3}{2}$	$\cdots$	$N-2$	$N-1$	N
a_i	1^2	$\cdots$	$\left(\dfrac{N-5}{2}\right)^2$	$\left(\dfrac{N-3}{2}\right)^2$	$\left(\dfrac{N-1}{2}\right)^2$

middle observation is essentially ignored, but this is necessary to achieve perfectly symmetric weights.

     The moments of M_N under the null hypothesis are easily found from Theorem 8.3.2 as follows:

p.154

$$NE(M_N) = m \sum_{i=1}^{N} \left(i - \frac{N+1}{2}\right)^2$$

$$= m \left[\sum i^2 - (N+1) \sum i + \frac{N(N+1)^2}{4}\right]$$

$$= m \left[\frac{N(N+1)(2N+1)}{6} - \frac{N(N+1)^2}{2} + \frac{N(N+1)^2}{4}\right]$$

$$12NE(M_N) = mN(N+1)(N-1)$$

$$E(M_N) = \frac{m(N^2-1)}{12} \tag{2.2}$$

$$N^2(N-1)\,\mathrm{var}(M_N) = mn \left\{N \sum_{i=1}^{N} \left(i - \frac{N+1}{2}\right)^4\right.$$

$$\left. - \left[\sum_{i=1}^{N}\left(i - \frac{N+1}{2}\right)^2\right]^2\right\}$$

$$= mn \left\{N\left[\sum i^4 - 4\frac{N+1}{2}\sum i^3\right.\right.$$

$$\left. + 6\frac{(N+1)^2}{4}\sum i^2 - 4\frac{(N+1)^3}{8}\sum i\right.$$

$$\left.\left. + \frac{N(N+1)^4}{16}\right] - \left[\frac{N(N^2-1)}{12}\right]^2\right\}$$

Using the following relations easily proved by induction

$$\sum_{i=1}^{N} i^3 = \left[\frac{N(N+1)}{2}\right]^2$$

$$\sum_{i=1}^{N} i^4 = \frac{N(N+1)(2N+1)(3N^2+3N-1)}{30}$$

and simplifying, the desired result is

$$\operatorname{var}(M_N) = \frac{mn(N+1)(N^2-4)}{180} \tag{2.3}$$

The exact null probability distribution of M_N can be derived by enumeration in small samples. The labor is somewhat reduced by noting that since $a_i = a_{N-i+1}$, the properties of Theorem 8.3.7 apply. From Theorem 8.3.5 the distribution is symmetric about $N(N^2-1)/24$ when $m = n$, but the symmetry property does not hold for unequal sample sizes. For larger samples, the normal approximation with the moments in (2.2) and (2.3) can be used to find critical values. In all cases, the test may be either one- or two-sided. Under the assumption of normal populations differing only in variance, the asymptotic relative efficiency of the Mood test to the F test is $15/2\pi^2 = 0.76$.

10.3 THE FREUND–ANSARI–BRADLEY–DAVID–BARTON TESTS

In the Mood test of the last section, the deviation of each rank from its average rank was squared to eliminate the problem of positive and negative deviations balancing out. If the absolute values of these deviations are used instead to give equal weight to positive and negative deviations, the linear rank statistic is

$$A_N = \sum_{i=1}^{N} \left| i - \frac{N+1}{2} \right| Z_i = (N+1) \sum_{i=1}^{N} \left| \frac{i}{N+1} - \frac{1}{2} \right| Z_i \tag{3.1}$$

There are several variations of this test statistic in the literature, proposed mainly by Freund and Ansari (1957), Ansari and Bradley (1960), and David and Barton (1958). There seems to be some confusion over which test should be attributed to whom, but they are all essentially equivalent anyway.

The Freund-Ansari-Bradley test can be written as a linear rank statistic in the form

$$F_N = \sum_{i=1}^{N} \left(\frac{N+1}{2} - \left| i - \frac{N+1}{2} \right| \right) Z_i = \frac{m(N+1)}{2} - A_N \tag{3.2}$$

or

$$F_N = \sum_{i=1}^{[(N+1)/2]} iZ_i + \sum_{i=[(N+1)/2]+1}^{N} (N-i+1)Z_i \qquad (3.3)$$

where $[x]$ denotes the largest integer not exceeding the value of x. Specifically the weights assigned then are 1 to both the smallest and largest observations in the combined sample, 2 to the next smallest and next largest, etc., $N/2$ to the two middle observations if N is even, and $(N+1)/2$ to the one middle observation if N is odd. Since the smaller weights are at the two extremes here, which is the reverse of the assignment for the Mood statistic, a small value of F_N would suggest that the X population has larger dispersion. The appropriate rejection regions for the scale-model alternative

$$H_S: F_{Y-M}(x) = F_{X-M}(\theta x) \qquad \text{for all } x \text{ and some } \theta > 0, \theta \neq 1$$

are then

Subclass of Alternatives	Rejection Region
$\theta > 1$	$F_N \leq k_1$
$\theta < 1$	$F_N \geq k_2$
$\theta \neq 1$	$F_N \leq k_3$ or $F_N \geq k_4$

The fact that this test is consistent for these subclasses of alternatives will be shown later in Sec. 10.7.

To determine the critical values for rejection, the exact null distribution of F_N could be found by enumeration. From Theorem 8.3.6, we note that the null distribution of F_N is symmetric about its mean if N is even. A recursion relation may be used to generate the null distribution systematically. For a sequence of $m + n = N$ letters occurring in a particular order, let $r_{m,n}(f)$ denote the number of distinguishable arrangements of m X and n Y letters such that the value of the F_N statistic is the number f, and let $p_{m,n}(f)$ denote the corresponding probability. A sequence of N letters is formed by adding a letter to each sequence of $N - 1$ letters. If $N - 1$ is even (N odd), the extra score will be $(N+1)/2$, so that f will be increased by $(N+1)/2$ if the new letter is an X and be unchanged if a Y. If $N - 1$ is odd, the extra score will be $N/2$. Therefore we have the relations

$$N \text{ odd:} \qquad r_{m,n}(f) = r_{m-1,n}\left(f - \frac{N+1}{2}\right) + r_{m,n-1}(f)$$

$$N \text{ even:} \qquad r_{m,n}(f) = r_{m-1,n}\left(f - \frac{N}{2}\right) + r_{m,n-1}(f)$$

These can be combined in the single recurrence relation

$$r_{m,n}(f) = r_{m-1,n}(f-k) + r_{m,n-1}(f) \qquad \text{for } k = \left[\frac{N+1}{2}\right]$$

Then in terms of the probabilities, the result is

$$p_{m,n}(f) = \frac{r_{m,n}(f)}{\binom{m+n}{m}}$$

$$(m+n)p_{m,n}(f) = mp_{m-1,n}(f-k) + np_{m,n-1}(f)$$

which is the same form as (7.5.14) and (9.2.2) for the Mann-Whitney and Wilcoxon tests. Tables of the null probability distribution for $N \leq 20$ are available in Ansari and Bradley (1960).

For larger samples the normal approximation to the distribution of F_N can be used. The exact mean and variance are easily found by applying the results of Theorem 8.3.2 to F_N in the forms of (3.3) and (3.2) as follows, where $x = (N+1)/2$.

$$NE(F_N) = m\left[\sum_{i=1}^{[x]} i + \sum_{i=[x]+1}^{N} (N-i+1)\right] = m\left(\sum_{i=1}^{[x]} i + \sum_{j=1}^{N-[x]} j\right)$$

N even:
$$E(F_N) = 2m\sum_{i=1}^{N/2} \frac{i}{N} = \frac{m(N+2)}{4}$$

N odd:
$$E(F_N) = \frac{m\left(2\sum_{i=1}^{(N-1)/2} i + \frac{N+1}{2}\right)}{N} = \frac{m(N+1)^2}{4N}$$

$$\text{var}(F_N) = \text{var}(A_N) = \frac{mn}{N^2(N-1)}\left[N\sum_{i=1}^{N}\left(i - \frac{N+1}{2}\right)^2 - \left(\sum_{i=1}^{N}\left|i - \frac{N+1}{2}\right|\right)^2\right]$$

$$= \frac{mn}{N^2(N-1)}\left\{\frac{N^2(N^2-1)}{12} - \left[\frac{N}{m}E(A_N)\right]^2\right\}$$

$$= \frac{mn}{N^2(N-1)}\left\{\frac{N^2(N^2-1)}{12} - \left[\frac{N(N+1)}{2} - \frac{N}{m}E(F_N)\right]^2\right\}$$

N even:
$$\text{var}(F_N) = \frac{mn}{N^2(N-1)}\left[\frac{N^2(N^2-1)}{12} - \left(\frac{N^2}{4}\right)^2\right]$$

$$= \frac{mn(N^2-4)}{48(N-1)}$$

$$N \text{ odd:} \quad \text{var}(F_N) = \frac{mn}{N^2(N-1)}\left[\frac{N^2(N^2-1)}{12} - \left(\frac{N^2-1}{4}\right)^2\right]$$
$$= \frac{mn(N+1)(N^2+3)}{48N^2}$$

Collecting these results, we have

N even	N odd	
$E(F_N) = \dfrac{m(N+2)}{4}$	$E(F_N) = \dfrac{m(N+1)^2}{4N}$	
$\text{var}(F_N) = \dfrac{mn(N^2-4)}{48(N-1)}$	$\text{var}(F_N) = \dfrac{mn(N+1)(N^2+3)}{48N^2}$	(3.4)

Another test which is almost identical is generally attributed to David and Barton (1958). This test also assigns symmetric integer weights but in the reverse order. That is, scores are given starting from the middle with 1 for N even, and 0 for N odd, and going out in both directions. The David-Barton test can be written as a linear rank statistic as

$$B_N = \sum_{i=1}^{[(N+1)/2]} \left(\left[\frac{N+2}{2}\right] - i\right) Z_i + \sum_{i=[(N+1)/2]+1}^{N} \left(i - \left[\frac{N+1}{2}\right]\right) Z_i$$
(3.5)

For N even, B_N and F_N have the exact same set of weights (but rear-ranged), and therefore the means and variances are equal. But for N odd this is not true because of the difference in relative assignment of the one "odd" weight, i.e., the middle observation. B_N assigns a weight of 0 to this observation, while F_N scores it as $(N+1)/2$. The following results are easily verified from Theorem 8.3.2:

N even	N odd	
$E(B_N) = \dfrac{m(N+2)}{4}$	$E(B_N) = \dfrac{m(N^2-1)}{4N}$	
$\text{var}(B_N) = \dfrac{mn(N^2-4)}{48(N-1)}$	$\text{var}(B_N) = \dfrac{mn(N+1)(N^2+3)}{48N^2}$	(3.6)

The exact relationship between B_N and F_N is

$$F_N + B_N = m\left[\frac{N+2}{2}\right]$$
(3.7)

Since the relation is linear, the tests are equivalent in properties. Tables of the null distribution of B_N are given in David and Barton (1958) for $m = n \leq 8$.

Since these three tests—F_N, B_N, and A_N—are all linearly related, they all have equivalent properties. All are consistent against the same alternatives. The asymptotic relative efficiency of each to the F test is $6/\pi^2 = 0.608$ for normal populations differing only in scale.

10.4 THE SIEGEL-TUKEY TEST

Even simpler than the use of positive integer weights symmetric about the middle would be some arrangement of the first N integers. Since these are the weights of the well-known Wilcoxon test W_N for location, tables of the null probability distribution would then be readily available. Siegel and Tukey (1960) proposed a rearrangement of the first N positive integers as weights which does provide a statistic sensitive to differences in scale. This rearrangement is

Here N is assumed to be even.

i	1	2	3	4	5	$\cdots$	$\dfrac{N}{2}$†	$\cdots$	$N-4$	$N-3$	$N-2$	$N-1$	N
a_i	1	4	5	8	9	$\cdots$	N	$\cdots$	10	7	6	3	2

† If $N/2$ is odd, $i = (N/2) + 1$ here.

for N even, and if N is odd, the middle observation in the array is thrown out and the same weights used for the reduced N. This rearrangement achieves the desired symmetry in terms of sums of pairs of adjacent weights, although the weights themselves are not exactly symmetric. Since the weights are smaller at the extremes, we should reject the null hypothesis in favor of an alternative that the X's have the greater variability when the linear rank statistic S_N is small.

In the symbolic form of a linear rank statistic, the Siegel-Tukey statistic is

$$S_N = \sum_{i=1}^{N} a_i Z_i$$

(N is assumed to be even)

See pp. 152-153

where

$$a_i = \begin{cases} 2i & \text{for } i \text{ even, } 1 < i \le \dfrac{N}{2} \\[2mm] 2i - 1 & \text{for } i \text{ odd, } 1 \le i \le \dfrac{N}{2} \\[2mm] 2(N - i) + 2 & \text{for } i \text{ even, } \dfrac{N}{2} < i \le N \\[2mm] 2(N - i) + 1 & \text{for } i \text{ odd, } \dfrac{N}{2} < i < N \end{cases} \qquad (4.1)$$

See Hájek, A Course in Nonparametric Statistics (1969), p. 10, Theorem 3D.

Since the probability distribution of S_N is the same as the Wilcoxon W_N, the moments are also the same

$$E(S_N) = \frac{m(N+1)}{2} \qquad \text{var}(S_N) = \frac{mn(N+1)}{12} \qquad (4.2)$$

To find critical values of S_N, tables of the distribution of W_N may be used, but they are also given in Siegel and Tukey's original article.

The asymptotic relative efficiency of the Siegel-Tukey test is equivalent to that of the tests F_N, B_N, and A_N, because of the following relations. With N even, let S'_N be a test with weights constructed in the same manner as for S_N but starting at the right-hand end of the array, as displayed in Table 4.1 for $N/2$ even. If $N/2$ is odd, the weights $a_{N/2}$ and $a'_{N/2}$ are interchanged, as are $a_{(N/2)+1}$ and $a'_{(N/2)+1}$. In either case, the weights $(a_i + a'_i + 1)/4$ are equal to the set of weights for F_N when N is even, and therefore the following complete cycle of relations is established for N even:

$$S''_N = F_N = m\left(\frac{N}{2}+1\right) - B_N = \frac{m(N+1)}{2} - A_N \qquad (4.3)$$

10.5 THE KLOTZ NORMAL-SCORES TEST

The Klotz (1962) normal-scores test for scale uses the same idea as the Mood test in that it employs as weights the squares of the weights used in the inverse-normal-scores test for location [van der Waerden test of Eq.

Table 4.1

Test	Weights	i						
		1	2	3	4	5	$\cdots$	$N/2$
S_N	a_i	1	4	5	8	9	$\cdots$	N
S'_N	a'_i	2	3	6	7	10	$\cdots$	$N-1$
$S_N + S'_N$	$a_i + a'_i$	3	7	11	15	19	$\cdots$	$2N-1$
S''_N	$(a_i + a'_i + 1)/4$	1	2	3	4	5	$\cdots$	$N/2$

Test	Weights	i					
		$(N/2)+1$	$\cdots$ $N-4$	$N-3$	$N-2$	$N-1$	N
S_N	a_i	$N-1$	$\cdots$ 10	7	6	3	2
S'_N	a'_i	N	$\cdots$ 9	8	5	4	1
$S_N + S'_N$	$a_i + a'_i$	$2N-1$	$\cdots$ 19	15	11	7	3
S''_N	$(a_i + a'_i + 1)/4$	$N/2$	$\cdots$ 5	4	3	2	1

p-169

(9.3.2)]. Symbolically, the test statistic is

$$K_N = \sum_{i=1}^{N} \left[\Phi^{-1} \left(\frac{i}{N+1} \right) \right]^2 Z_i \tag{5.1}$$

where $\Phi(x)$ is the cumulative standard normal probability distribution. Since the larger weights are at the extremes, we again reject H_0 for large K_N for the alternative that the X population has the larger spread. Tables of critical values for $N \le 20$ are given in Klotz (1962). The moments are *by Theorem 8.3.2 (p 154)*

$$E(K_N) = \frac{m}{N} \sum_{i=1}^{N} \left[\Phi^{-1} \left(\frac{i}{N+1} \right) \right]^2$$

$$\text{var}(K_N) = \frac{mn}{N(N-1)} \sum_{i=1}^{N} \left[\Phi^{-1} \left(\frac{i}{N+1} \right) \right]^4 - \frac{n}{m(N-1)} [E(K_N)]^2$$

Since this is an asymptotically optimum test against the alternative of normal distributions differing only in variance, its ARE relative to the F test equals 1.

An asymptotically equivalent test might be formed using the squares of the expected normal scores as weights, or

$$\sum_{i=1}^{N} [E(\xi_{(i)})]^2 Z_i \tag{5.2}$$

The Capon normal scores test.
See Jack Capon, Asymptotic efficiency of certain locally most powerful rank test. Ann. Math. Statist. 1961, 32, 88-100.

where $\xi_{(i)}$ is the ith order statistic from a standard normal distribution. This test then is the scale analog of the Terry-Hoeffding test for location given in (9.3.1). The equivalence between (5.1) and (5.2) follows from the fact that $E(\xi_{(i)})$ converges in probability to $\Phi^{-1}[(i/N+1)]$, as discussed in Sec. 9.3.

p-169

10.6 THE PERCENTILE MODIFIED RANK TESTS FOR SCALE

If the T_p and B_r statistics defined in (9.3.3) are added instead of subtracted, the desired symmetry of weights to detect scale differences is achieved. When N is even and $P = R = N/2$, $T + B$ is equivalent to the David-Barton type of test. The mean and variance of the statistic for N even and $P = R$ are

$$E(T_p + B_r) = \frac{mP^2}{N} \qquad \text{var}(T_p + B_r) = \frac{mnP(4NP^2 - N - 6P^3)}{6N^2(N-1)}$$

The null distribution is symmetric for $P = R$ when $m = n$. Tables for $m = n \le 6$ are given in Gibbons and Gastwirth (1966), and, as for the location problem, the normal approximation to critical values may be used for $m = n \ge 6$.

This scale test has a higher asymptotic relative efficiency than its full-sample counterparts for all choices of $p = r < 0.50$. The maximum ARE (with respect to p) is 0.850, which occurs for normal alternatives when $p = r = \frac{1}{8}$. This result is well above the ARE of 0.76 for Mood's test and the 0.608 value for the tests of Secs. 3 and 4. Thus asymptotically at least, in the normal case, a test based on only the 25 percent of the sample at the extremes is more efficient than a comparable test using the entire sample. The normal-scores tests of Sec. 5 have a higher ARE, of course, but they are more difficult to use because of the complicated sets of scores.

10.7 THE SUKHATME TEST

A number of other tests have been proposed for the scale problem. The only other one we shall discuss in detail here is the Sukhatme test statistic. Although it is less useful in applications than the others, this test has some nice theoretical properties. The test procedure also has the advantage of being easily adapted to the construction of confidence intervals for the unknown scale parameter.

When the X and Y populations have or can be adjusted to have equal medians, we can assume without loss of generality that this common median is zero. If the Y's have a larger spread than the X's, those X observations which are negative should be larger than most of the negative Y observations, and the positive observations should be arranged so that most of the Y's are larger than the X's. In other words, most of the negative Y's should precede negative X's, and most of the positive Y's should follow positive X's. Using the same type of indicator variables as for the Mann-Whitney statistic of (7.5.2), we define

$$D_{ij} = \begin{cases} 1 & \text{if } Y_j < X_i < 0 \text{ or } 0 < X_i < Y_j \\ 0 & \text{otherwise} \end{cases}$$

and the Sukhatme test statistic [Sukhatme (1957)] is

$$T = \sum_{i=1}^{m} \sum_{j=1}^{n} D_{ij} \tag{7.1}$$

The parameter relevant here is

$$\pi = P(Y < X < 0 \text{ or } 0 < X < Y) = \int_{-\infty}^{0} \int_{-\infty}^{x} f_Y(y) f_X(x) \, dy \, dx$$
$$+ \int_{0}^{\infty} \int_{x}^{\infty} f_Y(y) f_X(x) \, dy \, dx$$
$$= \int_{-\infty}^{0} F_Y(x) f_X(x) \, dx + \int_{0}^{\infty} [1 - F_Y(x)] f_X(x) \, dx$$
$$= \int_{-\infty}^{0} [F_Y(x) - F_X(x)] f_X(x) \, dx$$
$$+ \int_{0}^{\infty} [F_X(x) - F_Y(x)] f_X(x) \, dx + \frac{1}{4} \tag{7.2}$$

Then the null hypothesis of identical populations has been parameterized to $H_0: \pi = \frac{1}{4}$, and T/mn is an unbiased estimator of π since

$$E(T) = mn\pi$$

By redefining the parameters π, π_1, and π_2 of the Mann-Whitney statistic as appropriate for the present indicator variables D_{ij}, the variance of T can be expressed as in (7.5.10) and (7.5.11). The probabilities relevant here are

$$\pi_1 = P[(Y_j < X_i < 0 \text{ or } 0 < X_i < Y_j)$$
$$\cap (Y_k < X_i < 0 \text{ or } 0 < X_i < Y_k)]$$
$$= P[(Y_j < X_i < 0) \cap (Y_k < X_i < 0)]$$
$$+ P[(Y_j > X_i > 0) \cap (Y_k > X_i > 0)]$$
$$= \int_{-\infty}^{0} [F_Y(x)]^2 f_X(x) \, dx + \int_{0}^{\infty} [1 - F_Y(x)]^2 f_X(x) \, dx \qquad (7.3)$$

$$\pi_2 = P[(Y_j < X_i < 0 \text{ or } 0 < X_i < Y_j)$$
$$\cap (Y_j < X_h < 0 \text{ or } 0 < X_h < Y_j)]$$
$$= P[(Y_j < X_i < 0) \cap (Y_j < X_h < 0)]$$
$$+ P[(Y_j > X_i > 0) \cap (Y_j > X_h > 0)]$$
$$= \int_{-\infty}^{0} [\tfrac{1}{2} - F_X(y)]^2 f_Y(y) \, dy + \int_{0}^{\infty} [F_X(y) - \tfrac{1}{2}]^2 f_Y(y) \, dy \quad (7.4)$$

Then from (7.5.11), the variance is

$$\text{var}(T) = mn[\pi - \pi^2(N - 1) + (n - 1)\pi_1 + (m - 1)\pi_2] \qquad (7.5)$$

Since $E(T/mn) = \pi$ and $\text{var}(T/mn) \to 0$ as m, n approach infinity, the Sukhatme statistic provides a consistent test for the following cases where ϵ is defined from (7.2) as $\epsilon = \pi - \frac{1}{4}$ or

$$\epsilon = \int_{-\infty}^{0} [F_Y(x) - F_X(x)] f_X(x) \, dx + \int_{0}^{\infty} [F_X(x) - F_Y(x)] f_X(x) \, dx$$
$$(7.6)$$

Subclass of Alternatives	Rejection Region
$\pi < \frac{1}{4}$ $(\epsilon < 0)$ $(\theta > 1)$	$T - \dfrac{mn}{4} \le k_1$
$\pi > \frac{1}{4}$ $(\epsilon > 0)$ $(\theta < 1)$	$T - \dfrac{mn}{4} \ge k_2$ $\qquad (7.7)$
$\pi \ne \frac{1}{4}$ $(\epsilon \ne 0)$ $(\theta \ne 1)$	$\left| T - \dfrac{mn}{4} \right| \ge k_3$

It would be preferable to state these subclasses of alternatives as a simple relationship between $F_Y(x)$ and $F_X(x)$ instead of this integral expression for ϵ. Although (7.6) defines a large subclass, we are particularly interested now in the scale alternative model where $F_Y(x) = F_X(\theta x)$. Then

1. If $\theta < 1$, $F_Y(x) > F_X(x)$ for $x < 0$ and $F_Y(x) < F_X(x)$ for $x > 0$.
2. If $\theta > 1$, $F_Y(x) < F_X(x)$ for $x < 0$ and $F_Y(x) > F_X(x)$ for $x > 0$.

In both cases, the two integrands in (7.6) have the same sign and can therefore be combined to write

$$\epsilon = \pm \int_{-\infty}^{\infty} |F_X(\theta x) - F_X(x)| f_X(x)\, dx \tag{7.8}$$

where the plus sign applies if $\theta < 1$ and the minus if $\theta > 1$. This explains the statements of subclasses in terms of θ given in (7.7).

The exact null distribution of T can be found by enumeration or a recursive method similar to that for the Mann-Whitney test. The null distribution of T is not symmetric for all m, n. However, upper-tailed critical values are still not needed even for two-sided tests. Let us define

$$T' = \sum_{i=1}^{m} \sum_{j=1}^{n} D_{ij}' \qquad \text{where } D_{ij}' = \begin{cases} 1 & \begin{array}{l} \text{if } X_i < Y_j < 0 \\ \text{or } 0 < Y_j < X_i \end{array} \\ 0 & \text{otherwise} \end{cases}$$

Then since X and Y have the same distribution under the null hypothesis, the D_{ij} and D_{ij}' are also identically distributed, and consequently so are T and T'. A two-sided critical region then can be written $T \le c_{\alpha/2}$ or $T' \le c_{\alpha/2}$, where $c_{\alpha/2}$ is that number such that $P(T \le c_{\alpha/2}) = \alpha/2$. The minimum and maximum values of T are zero and $mn/2$, occurring in the extreme cases where the X or Y variables are all clustered. However, for any given set of data, it is not necessarily possible to have a sample configuration which is consistent with these extreme values. Let U and W denote the number of X and Y observations, respectively, which are negative. Then the maximum value of T (or T') is

$$UW + (m - U)(n - W)$$

which also equals the sum $T + T'$ for any set of data. For large samples, U and W converge, respectively, to $m/2$ and $n/2$, so that asymptotically the distribution of T is symmetric about $mn/4$.

In the null case where $F_Y(x) = F_X(x)$ for all x, $\pi = \frac{1}{4}$, so that $E(T) = mn/4$ and $\pi_1 = \pi_2 = \frac{1}{12}$ from (7.3) and (7.4). Substituting these results in (7.5), the variance under H_0 is found to be

$$\text{var}(T) = \frac{mn(N + 7)}{48}$$

For moderate m and n, the distribution of

$$\frac{4\sqrt{3}\,(T - mn/4)}{\sqrt{mn(N + 7)}} \tag{7.9}$$

may be well approximated by the standard normal.

Ties will present a problem for the T test statistic whenever an $X_i = Y_j$ or an $X_i = 0$. The T statistic could be redefined in a manner similar to (7.5.16) so that a correction for ties can be incorporated into the expression for the null variance.

The Sukhatme test has a distinct disadvantage in application inasmuch as it cannot be employed without knowledge of both the individual population medians M_X and M_Y. Even knowledge of the difference $M_Y - M_X$ is not enough to adjust the observations so that both populations have zero medians. Since the sample medians do converge to the respective population medians, the observations might be adjusted by subtracting the X and Y sample medians from each of the X and Y observations, respectively. The test statistic no longer has the same exact distribution, but for large sample sizes the error introduced by this estimating procedure should not be too large.

The Sukhatme test statistic can be written in the form of a linear rank statistic by a development similar to that used in Sec. 9.2 to show the relationship between the Wilcoxon and Mann-Whitney tests. Looking at (7.1) now, we know that for all values of i, $\sum\limits_{j=1}^{n} D_{ij}$ is the sum of two quantities:

1. The number of values of j for which $Y_j < X_i < 0$, which is $[r(X_i) - U_i]$.
2. The number of values of j for which $Y_j > X_i > 0$, which is

   $$[N - r(X_i) + 1 - V_i]$$

   where

   U_i is the number of X's less than or equal to X_i for all $X_i < 0$
   V_i is the number of X's greater than or equal to X_i for all $X_i > 0$.

Therefore we have

$$
\begin{aligned}
T &= \sum_{\substack{i=1 \\ X_i<0}}^{m} [r(X_i) - U_i] + \sum_{\substack{i=1 \\ X_i>0}}^{m} [N - r(X_i) + 1 - V_i] \\
&= \sum_{X<0} iZ_i + \sum_{X>0} (N - i + 1)Z_i - (U_1 + U_2 + \cdots + U_m) \\
&\qquad\qquad\qquad\qquad\qquad - (V_1 + V_2 + \cdots + V_m) \\
&= \sum_{X<0} iZ_i + \sum_{X>0} (N - i + 1)Z_i - \frac{U(U + 1)}{2} - \frac{V(V + 1)}{2}
\end{aligned}
$$

where $\displaystyle\sum_{X<0}$ indicates that the sum is extended over all values of i such that $X_i < 0$, U is the total number of X observations which are less than zero, and V is the number which are greater than zero. From this result, we can see that T is asymptotically equivalent to the Freund-Ansari-Bradley test, since as $N \to \infty$, the combined sample median will converge in probability to zero, the population median, and U and V will both converge to $m/2$, so that T converges to $F_N - m(m + 2)/4$ with F_N defined as in (3.3). The test statistic is therefore asymptotically equivalent to all the tests presented in Secs. 3 and 4, and the large-sample properties are identical, including the ARE of $6/\pi^2$. Note that inasmuch as consistency is a large-sample property, the consistency of these other tests follows also from our analysis for T here.

10.8 CONFIDENCE-INTERVAL PROCEDURES

If the populations from which the X and Y samples are drawn are identical in every respect except scale, the nonparametric model of (1.2) with $M_X = M_Y = M$ is

$$F_{Y-M}(x) = F_{X-M}(\theta x) \qquad \text{for all } x \text{ and some } \theta > 0$$

Since θ is the relevant scale parameter, a procedure for finding a confidence-interval estimate of θ would be desirable. In the above model, we can assume without loss of generality that the common median M is zero. Then for all $\theta > 0$, the random variable $Y' = Y\theta$ has the distribution

$$P(Y' \leq y) = P\left(Y \leq \frac{y}{\theta}\right) = F_Y\left(\frac{y}{\theta}\right) = F_X(y)$$

and Y' and X have identical distributions. The confidence-interval estimate of θ with confidence coefficient $1 - \alpha$ should consist of all values of θ for which the null hypothesis of identical populations will be accepted for the observations X_i and $Y_j\theta$, $i = 1, 2, \ldots, m; j = 1, 2, \ldots, n$. Using the Sukhatme test criterion of (7.1), here T denotes the number of pairs $(x_i, y_j\theta)$ for which either $y_j\theta < x_i < 0$ or $0 < x_i < y_j\theta$, or equivalently the number such that $x_i/y_j < \theta$. Suppose that the rejection region for a two-sided test of size α based on the T test criterion is to reject H_0 for $T \leq k$ or $T' \leq k$. Then, using an argument analogous to that leading to (7.5.17), the appropriate confidence interval is

$$\left(\frac{x_i}{y_j}\right)_{(k+1)} < \theta < \left(\frac{y_j}{x_i}\right)_{(k+1)} \tag{8.1}$$

where $(x_i/y_j)_{(k+1)}$ denotes the $(k + 1)$st smallest of the ratios x/y. Since there are a total of mn ratios x/y and the reciprocals occur in reverse order of magnitude, $(x_i/y_j)_{(k+1)} = (y_j/x_i)_{(mn-k)}$ and an interval equivalent to (8.1) is

$$\left(\frac{x_i}{y_j}\right)_{(k+1)} < \theta < \left(\frac{x_i}{y_j}\right)_{(mn-k)} \tag{8.2}$$

In practice, then, we need only find the $(k + 1)$st from the smallest and $(k + 1)$st from the largest of the ratios x_i/y_j. If the sample sizes are moderate, the number k can be found using the critical value based on the normal approximation given in (7.9).

10.9 OTHER TESTS FOR THE SCALE PROBLEM

All the tests for scale presented so far in this chapter are basically of the Mann-Whitney-Wilcoxon type, and except for the Mood and Klotz tests all are asymptotically equivalent. Other tests have been proposed— some are related to these while others incorporate essentially different ideas. A few will be summarized here even though they do not all fall within the category of linear rank statistics.

A test whose rationale is similar to the two-sample median test can be useful to detect scale differences. In two populations differing only in scale, the expected proportions of the two samples lying between two symmetric quantile points of the combined sample would not be equal. Since the total number of observations lying between the two quantiles is fixed by the order of the quantile, an appropriate test statistic could be the number of X observations lying between these two points. If these quantiles are the first and third quartiles and the sample sizes are large so that the sample quartiles approach the corresponding population parameters in the null case, then the statistic might be considered asymptotically a test for equal population interquartile ranges. The null distribution of the random variable U, the number of X observations within the sample interquartile range, is the hypergeometric distribution, and the appropriate rejection region for the alternative that the X's are more widely dispersed is $U \leq u_\alpha$. If $m + n = N$ is divisible by 4, so that no observations equal the sample quartile values, the distribution is

$$f_U(u) = \frac{\binom{m}{u}\binom{n}{N/2 - u}}{\binom{N}{N/2}} \tag{9.1}$$

This test is usually attributed to Westenberg (1948).

Rosenbaum suggests that the number of observations in the X sample which are either smaller than the smallest Y or larger than the largest Y is a reasonable test criterion for scale under the assumption that the population locations are the same. The null probability that exactly r X values lie outside the extreme values of the Y sample is

$$f_R(r) = n(n-1) \binom{m}{r} B(m+n-1-r, r+2) \qquad (9.2)$$

This result is easily verified by a combinatorial argument (Prob. 10.9). Tables of critical values are given in Rosenbaum (1953).

Another criterion, suggested by Kamat, is based on the pooled sample ranks of the extreme X and Y observations. Let R_m and R_n denote the ranges of the X ranks and Y ranks, respectively, in the combined sample ordering. If the locations are the same, a test statistic is provided by

$$D_{m,n} = R_m - R_n + n \qquad (9.3)$$

Tables of critical values are given in Kamat (1956). It should be noted that when the X sample observations all lie outside the extremes of the Y sample, we have $D_{m,n} = R + n$, where R is Rosenbaum's statistic. The performance of these two tests is discussed in Rosenbaum (1965).

These three tests, as well as the others presented earlier in this chapter, are reasonable approaches to detecting dispersion differences only when the X and Y populations have the same location. If the observations are adjusted before performing a test, say by subtracting the respective sample medians, the tests are no longer exact or even distribution-free. In fact, Moses (1963) shows that no test based on the ranks of the observations will be satisfactory for the dispersion problem without some sort of strong restriction, like equal or known medians, for the two populations. There is one type of approach to testing which avoids this problem. Although strictly speaking it does not qualify as a rank test, rank scores are used. The procedure is to divide each sample into small random subsets of equal size and calculate some measure of dispersion, e.g., the variance, range, average deviation, for each subsample. The measures for both samples can be arranged in a single sequence in order of magnitude, keeping track of which of the X and Y samples produced the measure. A two-sample location test can then be performed on the result. For example, if m and n are both divisible by 2, random pairs could be formed and the Wilcoxon test applied to the $N/2$-derived observations of ranges of the form $|x_i - x_j|$, $|y_i - y_j|$. The test statistic then is an estimate of a linear function of $P(|X_i - X_j| > |Y_i - Y_j|)$. In general, for any samples of dispersion measures denoted by U and V when computed for the X and Y subsamples, respectively, the Wilcoxon test statistic estimates a linear function of $P(U > V)$. Questions such as the best subsample size and the best type of measure of dispersion remain to be answered generally.

PROBLEMS

10.1. Develop by enumeration for $m = n = 3$ the null probability distribution of Mood's statistic M_N.

10.2. Develop by enumeration for $m = n = 3$ the null probability distribution of the Freund-Ansari-Bradley statistic of (3.3).

10.3. Verify the expression given in (2.3) for $\text{var}(M_N)$.

10.4. Apply Theorem 8.3.2 to derive the mean and variance of the statistic A_N defined in (3.1).

10.5. Apply Theorem 8.3.2 to derive the mean and variance of the statistic B_N defined in (3.5).

10.6. Verify the relationship between A_N, B_N, and F_N given in (4.3) for N even.

10.7. Use the relationships in (4.3) and the moments derived for F_N for N even in (3.4) to verify your answers to Probs. 10.4 and 10.5 for N even.

10.8. Use Theorem 8.3.2 to derive the mean and variance of $T_p + B_r$ for N even, $P \neq R$, where $P + R \leq N$.

10.9. Verify the result given in (9.2) for the null probability distribution of Rosenbaum's R statistic.

11

Tests of the Equality of *k* Independent Samples

11.1 INTRODUCTION

The natural extension of the two-sample problem is the k-sample problem, where observations are taken under a variety of different and independent conditions. Assume we have k independent sets of observations, one from each of k continuous populations $F_1(x)$, $F_2(x)$, . . . , $F_k(x)$, where the ith random sample is of size n_i, $i = 1, 2, \ldots, k$, and there are a total of $\sum_{i=1}^{k} n_i = N$ observations. Note that we are again assuming the independence extends across samples in addition to within samples. The extension of the two-sample hypothesis to the k-sample problem is that all k samples are drawn from identical populations

$$H_0: F_1(x) = F_2(x) = \cdots = F_k(x) \qquad \text{for all } x$$

The general alternative is simply that the populations differ in some way.

In classical statistics, the parametric test of this hypothesis is the analysis of variance for a one-way classification, i.e., the F test. The assumptions for this analysis-of-variance model are that the k populations are identical in shape, in fact normal, and with the same variance and therefore may differ only in location. The test of

$$H_0: \mu_1 = \mu_2 = \cdots = \mu_k$$

is, within the context of this model, equivalent to the hypothesis above of k identical populations. Denoting the observations in the ith sample by $X_{i1}, X_{i2}, \ldots, X_{in_i}$, the ith-sample mean by $\bar{X}_i = \sum_{j=1}^{n_i} X_{ij}/n_i$, and the grand mean by $\bar{X} = \sum_{i=1}^{k} \sum_{j=1}^{n_i} \dfrac{X_{ij}}{N}$, the classical test statistic may be written

$$\frac{\displaystyle\sum_{i=1}^{k} \frac{n_i(\bar{X}_i - \bar{X})^2}{k-1}}{\displaystyle\sum_{i=1}^{k} \sum_{j=1}^{n_i} \frac{(X_{ij} - \bar{X}_i)^2}{N-k}}$$

$$= \frac{\displaystyle\sum_{i=1}^{k} \frac{n_i(\bar{X}_i - \bar{X})^2}{k-1}}{\left[\displaystyle\sum_{i=1}^{k} \sum_{j=1}^{n_i} (X_{ij} - \bar{X})^2 - \sum_{i=1}^{k} n_i(\bar{X}_i - \bar{X})^2\right] \bigg/ (N-k)}$$

This test statistic follows the F distribution exactly, with $k-1$ and $N-k$ degrees of freedom, under the parametric assumptions. The F test is robust for equal sample sizes, but it is known to be sensitive to the assumption of equality of variances for unequal n_i.

The nonparametric techniques which have been developed for this k-sample problem require no assumptions beyond continuous populations and therefore are applicable under any circumstances, besides involving only simple calculations. We shall cover here the extension of the two-sample median test, the Kruskal-Wallis analysis-of-variance test, and some other extensions of rank tests from the two-sample problem. Finally, the chi-square test for equality of k proportions will be discussed. This latter test is applicable only to populations where the random variables are dichotomous, often called *count data*, and therefore does not fit within the basic problem of k continuous populations as defined here. However, when appropriate, it is a useful k-sample technique for the hypothesis of identical populations and therefore is included in this chapter.

11.2 EXTENSION OF THE MEDIAN TEST

Under the null hypothesis of identical populations, we have a single random sample of size $\sum_{i=1}^{k} n_i = N$ from the common population. The grand median δ of the pooled samples is an estimate of the median of this common population. Therefore, an observation from any of the k samples is as likely to be above δ as below it. The set of N observations will support the null hypothesis then if, for each of the k samples, about half of the observations in that sample are less than the grand sample median. A test based on this criterion is attributed to Mood [(1950), pp. 398–406].

As in the two-sample case, the grand sample median will be defined as the observation in the pooled ordered sample which has rank $(N + 1)/2$ if N is odd and any number between the two observations with ranks $N/2$ and $(N + 2)/2$ if N is even. Then, for each sample separately, the observations are dichotomized according as they are less than δ or not. Define the random variable U_i as the number of observations in sample number i which are less than δ, and let t denote the total number of observations which are less than δ. Then, by the definition of δ, we have

$$t = \sum_{i=1}^{k} u_i = \begin{cases} \dfrac{N}{2} & \text{if } N \text{ is even} \\[2ex] \dfrac{N-1}{2} & \text{if } N \text{ is odd} \end{cases}$$

Under the null hypothesis, each of the $\binom{N}{t}$ possible sets of t observations is equally likely to be in the less-than-δ category, and the number of dichotomizations with this particular sample outcome is $\prod_{i=1}^{k} \binom{n_i}{u_i}$. Therefore the null probability distribution of the random variables is the multivariate extension of the hypergeometric distribution, or

$$f(u_1, u_2, \ldots, u_k \mid t) = \frac{\binom{n_1}{u_1}\binom{n_2}{u_2} \cdots \binom{n_k}{u_k}}{\binom{N}{t}} \tag{2.1}$$

If any or all of the U_i differ too much from their expected value of $n_i\theta$, where θ denotes the probability an observation from the common population is less than δ, the null hypothesis should be rejected. Generally, it would be impractical to set up joint rejection regions for the test statistics

$U_1, U_2, \ldots, U_k$, because of the large variety of combinations of the sample sizes $n_1, n_2, \ldots, n_k$. If an exact test is desired, it is best to calculate the value of (2.1) for the actual observed and more extreme $u_1, u_2, \ldots, u_k$ and cumulate these point probabilities. If the sum is smaller than the desired significance level, the null hypothesis is rejected.

The test is seldom carried out by this procedure because it is tedious and time-consuming even with tables of binomial coefficients unless the k and n_i are small. Fortunately, we can use another test criterion which, although an approximation, is reasonably accurate even for N as small as 25 if each sample consists of at least five observations. This test statistic can be derived by appealing to the analysis of Chap. 4 for goodness-of-fit tests. Each of the N elements in the pooled sample is classified according to two criteria, sample number and magnitude relative to δ. Let these $2k$ categories be denoted by (i,j), where $i = 1, 2, \ldots, k$ according to the sample number and $j = 1$ if the observation is less than δ and $j = 2$ otherwise. Denote the observed and expected frequencies for the (i,j) category by f_{ij} and e_{ij}, respectively. Then

$$\begin{aligned} f_{i1} &= u_i \\ f_{i2} &= n_i - u_i \end{aligned} \qquad \text{for } i = 1, 2, \ldots, k$$

and the expected frequencies under H_0 are estimated from the data by

$$\begin{aligned} e_{i1} &= \frac{n_i t}{N} \\ e_{i2} &= \frac{n_i(N - t)}{N} \end{aligned} \qquad \text{for } i = 1, 2, \ldots, k$$

The goodness-of-fit test criterion for these $2k$ categories from (4.2.5) is then

$$\begin{aligned} q &= \sum_{i=1}^{k} \sum_{j=1}^{2} \frac{(f_{ij} - e_{ij})^2}{e_{ij}} \\ &= \sum_{i=1}^{k} \frac{(u_i - n_i t/N)^2}{n_i t/N} + \sum_{i=1}^{k} \frac{[n_i - u_i - n_i(N - t)/N]^2}{n_i(N - t)/N} \\ &= N \sum_{i=1}^{k} \frac{(u_i - n_i t/N)^2}{n_i t} + N \sum_{i=1}^{k} \frac{(n_i t/N - u_i)^2}{n_i(N - t)} \\ &= N \sum_{i=1}^{k} \frac{(u_i - n_i t/N)^2}{n_i} \left(\frac{1}{t} + \frac{1}{N - t} \right) \\ &= \frac{N^2}{t(N - t)} \sum_{i=1}^{k} \frac{(u_i - n_i t/N)^2}{n_i} \end{aligned} \qquad (2.2)$$

and Q has approximately the chi-square distribution. The parameters estimated from the data are the $2k$ probabilities that an observation is less than δ for each of the k samples and that it is not less than δ. But for each sample these probabilities sum to 1, and so there are only k independent parameters estimated. The number of degrees of freedom for Q is then $2k - 1 - k$, or $k - 1$. The chi-square approximation to the distribution of Q is somewhat improved by multiplication of Q by the factor $(N - 1)/N$. Then the rejection region is

$$Q \in R \qquad \text{for } \frac{(N - 1)q}{N} \geq \chi^2_{k-1,\alpha}$$

As with the two-sample median test, tied observations do not present a problem unless there is more than one number equal to the median, which can occur only for N odd, or if N is even and the two middle observations are equal. The conservative approach is suggested, whereby the decision is based on that resolution of ties which leads to the smallest value of q.

11.3 THE KRUSKAL–WALLIS ONE–WAY ANOVA TEST

The median test for k samples uses information about the magnitude of each of the N observations relative to a single number which is the median of the pooled samples. Most of the other k-sample tests use more of the available information by considering the relative magnitude of each observation when compared with every other observation. This comparison is effected in terms of ranks.

Since under H_0 we have essentially a single sample of size N from the common population, combine the N observations into a single ordered sequence from smallest to largest, keeping track of which observation is from which sample, and assign the ranks $1, 2, \ldots, N$ to the sequence. If adjacent ranks are well distributed among the k samples, which would be true for a random sample from a single population, the total sum of ranks, $\sum\limits_{i=1}^{N} i = N(N + 1)/2$, would be divided proportionally according to sample size among the k samples. For the ith sample which contains n_i observations, the expected sum of ranks would be

$$\frac{n_i}{N} \frac{N(N + 1)}{2} = \frac{n_i(N + 1)}{2}$$

Equivalently, since the expected rank for any observation is the average rank $(N + 1)/2$, for n_i observations the expected sum of ranks is $n_i(N + 1)/2$. Denote by R_i the actual sum of ranks assigned to the elements in the ith sample. A reasonable test statistic could be based on a function of the deviations between these observed and expected rank sums. Since deviations in either direction indicate disparity between the samples and absolute values are not particularly tractable mathematically, the sum of squares of these deviations can be employed as

$$S = \sum_{i=1}^{k} \left[R_i - \frac{n_i(N+1)}{2} \right]^2 \tag{3.1}$$

The null hypothesis is rejected for large values of S.

In order to determine the null probability distribution of S, consider the ranked sample data recorded in a table with k columns, where the entries in the ith column are the n_i ranks occupied by the elements in the ith sample. R_i is then the ith-column sum. Under H_0, the integers $1, 2, \ldots, N$ are assigned at random to the k columns except for the restriction that there be n_i integers in column i. The total number of ways to make the assignment of ranks then is the number of partitions of N distinct elements into k ordered sets, the ith of size n_i, and this is

$$\frac{N!}{\prod_{i=1}^{k} n_i!}$$

Each of these possibilities must be enumerated and the value of S calculated for each. If $t(s)$ denotes the number of assignments with the particular value s calculated from (3.1), then

$$f_S(s) = t(s) \prod_{i=1}^{k} \frac{n_i!}{N!}$$

Obviously, the calculations required are extremely tedious and therefore will not be illustrated here. Tables of exact probabilities for S are available in Rijkoort (1952) for $k = 3$, 4, and 5, but only for n_i equal and very small. Critical values for some larger equal sample sizes are also given.

A somewhat more useful test criterion is a weighted sum of squares of deviations, with the reciprocals of the respective sample sizes used as weights. Thus the Kruskal-Wallis statistic is defined as

$$H = \frac{12}{N(N+1)} \sum_{i=1}^{k} \frac{1}{n_i} \left[R_i - \frac{n_i(N+1)}{2} \right]^2 \tag{3.2}$$

The consistency of H is investigated in Kruskal (1952). H and S are equivalent test criteria only for all n_i equal. Exact probabilities for H have been tabulated in Kruskal and Wallis (1952) for $k = 3$, all $n_i \leq 5$;

and more extensive tables for $k = 3$ are given in Alexander and Quade (1968).

Since there are practical limitations on the range of tables which can be constructed, some reasonable approximation to the null distribution is required if a test based on the rationale of S is to be useful in application. It is for this reason that the statistic H was introduced.

Under the null hypothesis, the n_i entries in column i were randomly selected from the set $\{1,2, \ldots ,N\}$. They actually constitute a random sample of size n_i drawn without replacement from the finite population consisting of the first N integers. The mean and variance of this population are

$$\mu = \sum_{i=1}^{N} \frac{i}{N} = \frac{N+1}{2}$$

$$\sigma^2 = \sum_{i=1}^{N} \frac{[i - (N+1)/2]^2}{N} = \frac{N^2 - 1}{12}$$

The average rank sum for the ith column, $\bar{R}_i = R_i/n_i$, is the mean of this random sample, and as for any sample mean from a finite population

$$E(\bar{R}_i) = \mu \qquad \text{var}(\bar{R}_i) = \frac{\sigma^2(N - n_i)}{n_i(N - 1)}$$

Here then we have

$$E(\bar{R}_i) = \frac{N+1}{2} \qquad \text{and} \qquad \text{var}(\bar{R}_i) = \frac{(N+1)(N - n_i)}{12n_i}$$

Since $\bar{R}_i$ is a sample mean, if n_i is large, the central-limit theorem allows us to approximate the distribution of

$$Z_i = \frac{\bar{R}_i - (N+1)/2}{\sqrt{(N+1)(N - n_i)/12n_i}}$$

by the standard normal. Consequently Z_i^2 is distributed approximately as chi square with one degree of freedom. This holds for $i = 1, 2, \ldots, k$, but the Z_i are clearly not independent random variables since $\sum_{i=1}^{k} n_i \bar{R}_i = N(N + 1)/2$, a constant. Thus it should not be surprising that if no n_i is very small, the random variable

$$\sum_{i=1}^{k} \frac{N - n_i}{N} Z_i^2 = \sum_{i=1}^{k} \frac{12n_i[\bar{R}_i - (N+1)/2]^2}{N(N+1)} = H \qquad (3.3)$$

is distributed approximately as chi square with $k - 1$ degrees of freedom [Kruskal (1952)]. The statistic H is easier to calculate in the following form, which is algebraically equivalent to (3.3) and (3.2):

$$H = \frac{12}{N(N + 1)} \sum_{i=1}^{k} \frac{R_i^2}{n_i} - 3(N + 1) \tag{3.4}$$

The rejection region is $H \geq \chi^2_{\alpha, k-1}$.

Some other approximations to the null distribution of H are discussed in Alexander and Quade (1968).

The only assumption made initially was that the population was continuous, and this of course was to avoid the problem of ties. When two or more observations are tied within a column, the value of H is the same regardless of the method used to resolve the ties since the rank sum is not affected. When ties occur across columns, the midrank method is generally used. Alternatively, for a conservative test the ties can be broken in the way which is least conducive to rejection of H_0.

11.4 OTHER RANK-TEST STATISTICS

A general form for any k-sample rank-test statistic which follows the rationale of the Kruskal-Wallis statistic can be written as follows. Denote the $\sum_{i=1}^{k} n_i = N$ items in the pooled (not necessarily ordered) sample by $X_1, X_2, \ldots, X_N$ with respective ranks $r(X_1), r(X_2), \ldots, r(X_N)$, and put a subscript on the ranks to indicate of which sample the observation is a member. Thus $r_j(X_i)$ is the rank of X_i, where X_i is from the jth sample, for some $j = 1, 2, \ldots, k$. The rank sum for the jth sample, previously denoted by R_j, would now be denoted by $\sum_i r_j(X_i)$.

Since the $r_j(X_i)$ for fixed j are a random sample of n_j numbers, for every j the sum of any monotone increasing function g of $r_j(X_i)$ should, if the null hypothesis is true, on the average be approximately equal to the average of the function for all N observations multiplied by n_j. The weighted sum of squares of these deviations provides a test criterion. Thus the general k-sample rank statistic can be written as

$$Q = \sum_{j=1}^{k} \frac{\left\{ \sum_i g[r_j(X_i)] - \frac{n_j}{N} \sum_{j=1}^{k} \sum_i g[r_j(X_i)] \right\}^2}{n_j} \tag{4.1}$$

For simplicity, now let us denote the set of all N values of the function

$g[r_j(x_i)]$ by $\{a_1, a_2, \ldots, a_N\}$ and their mean by

$$\bar{a} = \sum_{i=1}^{N} \frac{a_i}{N} = \sum_{j=1}^{k} \sum_{i} \frac{g[r_j(x_i)]}{N}$$

It can be shown [see Hajek and Sidak (1967), pp. 170–172] that as minimum $(n_1, n_2, \ldots, n_k) \to \infty$, under certain regularity conditions the probability distribution of

$$\frac{(N-1)Q}{\displaystyle\sum_{i=1}^{N} (a_i - \bar{a})^2}$$

approaches the chi-square distribution with $k - 1$ degrees of freedom.

Two obvious possibilities for our function g are suggested by the scores in the two-sample location problem for the Terry and van der Waerden test statistics. Since in both these cases the scores are symmetric about zero, $\bar{a}$ is zero and the k-sample analogs are

$$T = \frac{N-1}{\displaystyle\sum_{i=1}^{N} [E(\xi_{(i)})]^2} \sum_{j=1}^{k} \frac{\left[\displaystyle\sum_i E(\xi_{(i)})_j \right]^2}{n_j}$$

$$X = \frac{N-1}{\displaystyle\sum_{i=1}^{N} \left[\Phi^{-1}\left(\frac{i}{N+1}\right) \right]^2} \sum_{j=1}^{k} \frac{\left[\displaystyle\sum_i \Phi^{-1}\left(\frac{i}{N+1}\right)_j \right]^2}{n_j}$$

The T and X tests are asymptotically equivalent as before.

11.5 DISCUSSION

For all the k-sample tests presented here, actual measurements are not required. Therefore they are applicable to qualitative data for which a unique ordering can be effected. For the median test, even less information is required as long as the common grand median can be determined at least positionally.

The median test uses less information than any of the others and is therefore expected to be less powerful. Its asymptotic relative efficiency to the F test in the case of normal populations with equal variances is $2/\pi = 0.637$. The Kruskal-Wallis test has an ARE of $3/\pi = 0.955$. The other two tests should have an ARE of 1 under these circumstances, since they are asymptotically optimum for normal distributions.

All the tests are much quicker and easier to apply than the F test, and as a rule may perform better if the F-test assumptions are not satisfied. Unfortunately, there does not seem to be any method of singling out which populations differ most if the null hypothesis is rejected. All the tests are for differences which are collectively significant. If then two samples are singled out for comparison, the usual problem of unknown overall probabilities for type I and type II errors results.

11.6 THE CHI–SQUARE TEST FOR k PROPORTIONS

There is one further k-sample test which should be mentioned, although it is applicable in a situation different from that of the other tests in this chapter. If the k populations are all Bernoulli distributions, the samples consist of count data. The k populations are all equivalent if the Bernoulli parameters $\theta_1, \theta_2, \ldots, \theta_k$ are all equal. Let $X_1, X_2, \ldots, X_k$ denote the numbers of successes in each of the k samples, respectively. For $i = 1, 2, \ldots, k$, X_i has the binomial distribution with parameter θ_i and number of trials n_i. A test statistic for the null hypothesis

$$H_0: \theta_1 = \theta_2 = \cdots = \theta_k \text{ unspecified}$$

can be derived from the goodness-of-fit test exactly as in the case of the k-sample median test in Sec. 2, except that our classification criterion besides sample number is now simply success and failure for each sample instead of less than δ or not. Using the same notation as before, with θ_0 denoting the estimated common parameter under H_0 (which replaces t/N), we have

$$f_{i1} = x_i$$
$$f_{i2} = n_i - x_i$$
$$e_{i1} = n_i\theta_0$$
$$e_{i2} = n_i(1 - \theta_0)$$

and the test criterion from (2.2) is

$$Q = \sum_{i=1}^{k} \frac{(X_i - n_i\theta_0)^2}{n_i\theta_0(1 - \theta_0)}$$

where $\theta_0 = \sum_{i=1}^{N} \dfrac{X_i}{N}$, which has approximately the chi-square distribution with $k - 1$ degrees of freedom.

PROBLEMS

11.1. Generate by enumeration the exact null probability distribution of the k-sample median test statistic for $k = 3$, $n_1 = 2$, $n_2 = 1$, $n_3 = 1$. If the rejection region consists of those arrangements which are least likely under the null hypothesis, find this region R and exact α. Compute the values of the Q statistic for all arrangements and compare that critical region for the same value of α with the region R.

11.2. Generate the exact distribution of H for the same k and n_i as in Prob. 11.1. Find the critical region which consists of those sets R_1, R_2, R_3 which have the largest value of H and find exact α.

11.3. By enumeration, place the median test criterion (U_1, U_2, U_3) and the H test criterion (R_1, R_2, R_3) in one-to-one correspondence for the same k and n_i as in Probs. 11.1 and 11.2. If the two tests reject for the largest values of Q and H, respectively, which test seems to distinguish better between extreme arrangements?

11.4. Verify that the form of H given in (3.4) is algebraically equivalent to (3.2).

11.5. Show that H with $k = 2$ is exactly equivalent to the large-sample approximation to the two-sample Wilcoxon test statistic mentioned in Sec. 9.2 with $m = n_1$, $n = n_2$.

11.6. Show that H is equivalent to the F test statistic in a one-way analysis-of-variance problem if applied to the ranks of the observations rather than the actual numbers. *Hint:* Express the F ratio as a function of H in the form given in (3.3) or (3.2) to show that

$$F = \left[\frac{k-1}{N-k} \left(\frac{N-1}{H} - 1 \right) \right]^{-1}$$

11.7. Write the k-sample median test statistic given in (2.2) in the form of (4.1) (cf. Prob. 8.2).

11.8. How could the subsampling procedure described in Sec. 10.9 be extended to test the equality of variances in k populations?

12

Measures of Association for Bivariate Samples

12.1 INTRODUCTION: DEFINITION OF MEASURES OF ASSOCIATION IN A BIVARIATE POPULATION

In Chap. 6 we saw that the ordinary sign-test and Wilcoxon signed-rank test procedures, although discussed in terms of inferences in a single-sample problem, could be applied to paired-sample data by basing the statistical analysis on the differences between the pairs of observations. The inferences then must be concerned with the population of differences as opposed to some general relationship between the two dependent random variables. One parameter of this population of differences, the variance, does contain information concerning their relationship, since

$$\text{var}(X - Y) = \text{var}(X) + \text{var}(Y) - 2\,\text{cov}(X,Y)$$

It is this covariance factor and a similar measure with which we shall be concerned in this chapter.

In general, if X and Y are two random variables with a bivariate probability distribution, their covariance, in a certain sense, reflects the direction and amount of association or correspondence between the variables.   The covariance is large and positive if there is a high probability that large (small) values of X are associated with large (small) values of Y. On the other hand, if the correspondence is inverse so that large (small) values of X generally occur in conjunction with small (large) values of Y, their covariance is large and negative.   This comparative type of association is referred to as *concordance* or *agreement*.   The covariance parameter as a measure of association is difficult to interpret because its value depends on the orders of magnitude and units of the random variables concerned.   A nonabsolute or relative measure of association circumvents this difficulty.   Since the Pearson product-moment correlation coefficient, defined as

$$\rho(X,Y) = \frac{\mathrm{cov}(X,Y)}{[\mathrm{var}(X)\ \mathrm{var}(Y)]^{\frac{1}{2}}}$$

is invariant under changes of scale and location in X and Y, in classical statistics this parameter is usually employed as the relative measure of association in a bivariate distribution.   The absolute value of the correlation coefficient does not exceed 1, and its sign is determined by the sign of the covariance.   If X and Y are independent random variables, their correlation is zero, and therefore the magnitude of ρ in some sense measures the degree of association.   Although it is not true in general that a zero correlation implies independence, the bivariate normal distribution is a significant exception, and therefore in the normal-theory model ρ is a good measure of association.   For random variables from other bivariate populations, ρ may not be such a good description of relationship since dependence may be reflected in a wide variety of types of relationships. One can only say in general that ρ is a more descriptive measure of dependence than covariance is since it does not depend on the scales of X and Y.

If the main justification for the use of ρ as a measure of association is that the bivariate normal is such an important distribution in classical statistics and zero correlation is equivalent to independence for that particular population, this reasoning has little significance in nonparametric statistics.   Other population measures of association should be equally acceptable, but the approach to measuring relationships might be analogous, so that interpretations are simplified.   Because ρ is so widely known and accepted, any other measure would preferably emulate its properties.

Suppose we define a "good" relative measure of association as one which satisfies the following criteria:

1. For any two independent pairs (X_i, Y_i) and (X_j, Y_j) of random variables which follow this bivariate distribution, the measure will equal $+1$

if the relationship is direct and perfect in the sense that

$$X_i < X_j \quad \text{whenever } Y_i < Y_j \quad \text{or}$$
$$X_i > X_j \quad \text{whenever } Y_i > Y_j$$

This relation will be referred to as *perfect concordance* (agreement).

2. For any two independent pairs, the measure will equal -1 if the relationship is indirect and perfect in the sense that

$$X_i < X_j \quad \text{whenever } Y_i > Y_j \quad \text{or}$$
$$X_i > X_j \quad \text{whenever } Y_i < Y_j$$

This relation will be referred to as *perfect discordance* (disagreement).

3. If neither criterion 1 nor criterion 2 is true for all pairs, the measure will lie between the two extremes -1 and $+1$.†

4. The measure will equal zero if X and Y are independent.

5. The measure for X and Y will be the same as for Y and X, or $-X$ and $-Y$, or $-Y$ and $-X$.

6. The measure for $-X$ and Y or X and $-Y$ will be the negative of the measure for X and Y.

7. The measure should be invariant under all transformations of X and Y for which order of magnitude is preserved.

The parameter ρ is well known to satisfy the first six of these criteria. It is a type of measure of concordance in the same sense that covariance measures the degree to which the two variables are associated in magnitude. However, although ρ is invariant under positive linear transformations of the random variables, it is not invariant under all order-preserving transformations. This last criterion seems especially desirable in nonparametric statistics, as we have seen that in order to be distribution-free, inferences must usually be determined by relative magnitudes as opposed to absolute magnitudes of the variables under study. Since probabilities of events involving only inequality relations between random variables are invariant under all order-preserving transformations, a measure of association which is a function of the probabilities of concordance and discordance will satisfy the seventh criterion. Perfect direct and indirect association between X and Y are reflected by perfect concordance and perfect discordance, respectively, and in the same spirit as ρ measures a perfect direct and indirect linear relationship between the variables. Thus an appropriate combination of these probabilities will provide a measure of association which will satisfy all seven of these desirable criteria.

† It is also desirable that, in some sense, increasing *degrees of concordance* are reflected by increasing positive values and increasing *degrees of discordance* are reflected by increasing negative values.

For any two independent pairs of random variables (X_i, Y_i) and (X_j, Y_j), we denote by π_c and π_d the probabilities of concordance and discordance, respectively.

$$
\begin{aligned}
\pi_c &= P\{[(X_i < X_j) \cap (Y_i < Y_j)] \cup [(X_i > X_j) \cap (Y_i > Y_j)]\} \\
&= P[(X_j - X_i)(Y_j - Y_i) > 0] \\
&= P[(X_i < X_j) \cap (Y_i < Y_j)] + P[(X_i > X_j) \cap (Y_i > Y_j)] \\
\pi_d &= P[(X_j - X_i)(Y_j - Y_i) < 0] \\
&= P[(X_i < X_j) \cap (Y_i > Y_j)] + P[(X_i > X_j) \cap (Y_i < Y_j)]
\end{aligned}
$$

Perfect association between X and Y is reflected by either perfect concordance or perfect discordance, and thus some combination of these probabilities should provide a measure of association. The Kendall coefficient τ is defined as the difference

$$
\tau = \pi_c - \pi_d
$$

and this measure of association satisfies our symmetry criteria 1 to 6. If the marginal probability distributions of X and Y are continuous, so that the possibility of ties $X_i = X_j$ or $Y_i = Y_j$ within groups is eliminated, we have

$$
\begin{aligned}
\pi_c &= \{P(Y_i < Y_j) - P[(X_i > X_j) \cap (Y_i < Y_j)]\} \\
&\qquad + \{P(Y_i > Y_j) - P[(X_i < X_j) \cap (Y_i > Y_j)]\} \\
&= P(Y_i < Y_j) + P(Y_i > Y_j) - \pi_d \\
&= 1 - \pi_d
\end{aligned}
$$

Thus in this case τ can also be expressed as

$$
\tau = 2\pi_c - 1 = 1 - 2\pi_d
$$

How does τ measure independence? If X and Y are independent and continuous random variables, $P(X_i < X_j) = P(X_i > X_j)$ and further the joint probabilities in π_c or π_d are the product of the individual probabilities. Using these relations, we can write

$$
\begin{aligned}
\pi_c &= P(X_i < X_j)P(Y_i < Y_j) + P(X_i > X_j)P(Y_i > Y_j) \\
&= P(X_i > X_j)P(Y_i < Y_j) + P(X_i < X_j)P(Y_i > Y_j) \\
&= \pi_d
\end{aligned}
$$

and thus $\tau = 0$ for independent continuous random variables. In general, the converse is not true, but this disadvantage is shared by ρ. For the bivariate normal population, however, $\tau = 0$ if and only if $\rho = 0$, that is, if and only if X and Y are independent. This fact follows from the relation

$$
\tau = \frac{2}{\pi} \arcsin \rho
$$

which can be derived as follows. Suppose that X and Y are bivariate normal with variances $\sigma_X{}^2$ and $\sigma_Y{}^2$ and correlation coefficient ρ. Then for any two independent pairs (X_i, Y_i) and (X_j, Y_j) from this population, the differences

$$U = \frac{X_i - X_j}{\sqrt{2}\,\sigma_X} \quad \text{and} \quad V = \frac{Y_i - Y_j}{\sqrt{2}\,\sigma_Y}$$

also have a bivariate normal distribution, with zero means, unit variances, and covariance equal to ρ. Thus $\rho(U, V) = \rho(X, Y)$. Since

$$\pi_c = P(UV > 0)$$

we have

$$\pi_c = \int_{-\infty}^{0} \int_{-\infty}^{0} \varphi(x, y)\,dx\,dy + \int_{0}^{\infty} \int_{0}^{\infty} \varphi(x, y)\,dx\,dy$$
$$= 2 \int_{-\infty}^{0} \int_{-\infty}^{0} \varphi(x, y)\,dx\,dy = 2\Phi(0, 0)$$

where $\varphi(x, y)$ and $\Phi(x, y)$ denote the density and cumulative distributions, respectively, of a standardized bivariate normal probability distribution. Since it can be shown that

$$\Phi(0, 0) = \frac{1}{4} + \frac{1}{2\pi} \arcsin \rho$$

we see that for the bivariate normal

$$\pi_c = \frac{1}{2} + \frac{1}{\pi} \arcsin \rho$$

and

$$\tau = \frac{2}{\pi} \arcsin \rho$$

In this chapter, the problem of point estimation of these two population measures of association, ρ and τ, will be considered. We shall find estimates which are distribution-free and discuss their individual properties and procedures for hypothesis testing, and the relationship between the two estimates will be determined. Another measure of association will be discussed briefly.

12.2 KENDALL'S TAU COEFFICIENT

In Sec. 1 Kendall's tau, a measure of association between random variables from any bivariate population, was defined as

$$\tau = \pi_c - \pi_d \tag{2.1}$$

where, for any two independent pairs of observations (X_i, Y_i), (X_j, Y_j) from the population,

$$\begin{aligned} \pi_c &= P[(X_j - X_i)(Y_j - Y_i) > 0] \\ \pi_d &= P[(X_j - X_i)(Y_j - Y_i) < 0] \end{aligned} \tag{2.2}$$

In order to estimate the parameter τ from a random sample of n pairs

$$(X_1, Y_1), (X_2, Y_2), \ldots, (X_n, Y_n)$$

drawn from this bivariate population, we must find point estimates of the probabilities π_c and π_d. For each set of pairs (X_i, Y_i), (X_j, Y_j) of sample observations, define the indicator variables

$$A_{ij} = \operatorname{sgn}(X_j - X_i) \operatorname{sgn}(Y_j - Y_i) \tag{2.3}$$

where

$$\operatorname{sgn}(u) = \begin{cases} -1 & \text{if } u < 0 \\ 0 & \text{if } u = 0 \\ 1 & \text{if } u > 0 \end{cases}$$

Then the values assumed by A_{ij} are

$$a_{ij} = \begin{cases} 1 & \text{if these pairs are concordant} \\ -1 & \text{if these pairs are discordant} \\ 0 & \text{if these pairs are neither concordant nor discordant} \\ & \text{because of a tie in either component} \end{cases}$$

The marginal probability distribution of these indicator variables is

$$f_{A_{ij}}(a_{ij}) = \begin{cases} \pi_c & \text{if } a_{ij} = 1 \\ \pi_d & \text{if } a_{ij} = -1 \\ 1 - \pi_c - \pi_d & \text{if } a_{ij} = 0 \end{cases} \tag{2.4}$$

and the expected value is

$$E(A_{ij}) = 1\pi_c + (-1)\pi_d = \pi_c - \pi_d = \tau \tag{2.5}$$

Since obviously we have $a_{ij} = a_{ji}$ and $a_{ii} = 0$, there are only $\binom{n}{2}$ sets of pairs which need be considered. An unbiased estimator of τ is therefore provided by

$$T = \sum\sum_{1 \le i < j \le n} \frac{A_{ij}}{\binom{n}{2}} = 2 \sum\sum_{1 \le i < j \le n} \frac{A_{ij}}{n(n-1)} \tag{2.6}$$

This measure of the association in the paired-sample observations is called *Kendall's sample tau coefficient*.

The reader should note that with the definition of A_{ij} in (2.3) that allows for tied observations, no assumption regarding the continuity of the population was necessary, and thus T is an unbiased estimator of the parameter τ in *any* bivariate distribution. Since the variance of T approaches zero as the sample size approaches infinity, T is also a consistent estimator of τ for any bivariate distribution.

In order to determine the variance of T, the variances and covariances of the A_{ij} must be evaluated since T is a linear combination of these indicator random variables. From (2.6), we have

$$n^2(n-1)^2 \operatorname{var}(T) = 4\left[\sum_{1\leq i<j\leq n}\sum \operatorname{var}(A_{ij}) + \sum_{\substack{1\leq i<j\leq n \\ 1\leq h<k\leq n \\ i\neq h \text{ or } j\neq k}}\sum\sum\sum \operatorname{cov}(A_{ij},A_{hk})\right]$$

$$(2.7)$$

Since the A_{ij} are identically distributed for all $i < j$, and A_{ij} and A_{hk} are independent for all $i \neq h$ and $j \neq k$ (no pairs in common), (2.7) can be written as

$$n^2(n-1)^2 \operatorname{var}(T) = 4\left[\binom{n}{2}\operatorname*{var}_{i<j}(A_{ij}) + \sum_{i=1}^{n-1}\sum_{j=i+1}^{n}\sum_{\substack{k=i+1 \\ j\neq k}}^{n} \operatorname{cov}(A_{ij},A_{ik})\right.$$

$$+ \sum_{j=2}^{n}\sum_{i=1}^{j-1}\sum_{\substack{k=1 \\ i\neq k}}^{j-1} \operatorname{cov}(A_{ij},A_{kj})$$

$$\left. + \sum_{j=2}^{n}\sum_{i=1}^{j-1}\sum_{\substack{k=j+1 \\ i\neq k}}^{n} \operatorname{cov}(A_{ij},A_{jk}) + \sum_{i=2}^{n-1}\sum_{j=i+1}^{n}\sum_{\substack{k=1 \\ j\neq k}}^{i-1} \operatorname{cov}(A_{ij},A_{ki})\right] \quad (2.8)$$

By symmetry, all the covariance terms in (2.8) are equal. They are grouped together according to which of the (X,Y) pairs are common to the (A_{ij},A_{hk}) in order to facilitate counting the number of terms in each summation set. Within the first set we have two distinct permutations, (A_{ij},A_{ik}) and (A_{ik},A_{ij}), for each of the $\binom{n}{2}$ choices of $i \neq j \neq k$, and similarly for the second set. But the third and fourth sets do not allow for reversal of the A_{ij} and A_{hk} terms since this makes a different (X,Y) pair in common, and so there are only $\binom{n}{3}$ covariance terms in each of these summations. The total number of distinguishable covariance terms then is $(2+2+1+1)\binom{n}{3} = 6\binom{n}{3}$, and (2.8) can be written as simply

$$n^2(n-1)^2 \operatorname{var}(T) = 4\left[\binom{n}{2}\operatorname{var}(A_{ij}) + 6\binom{n}{3}\operatorname{cov}(A_{ij},A_{ik})\right]$$

or

$$n(n-1)\operatorname{var}(T) = 2\operatorname{var}(A_{ij}) + 4(n-2)\operatorname{cov}(A_{ij},A_{ik}) \quad (2.9)$$

for any

$$i < j \qquad i < k \qquad j \neq k \qquad \begin{aligned} i &= 1, 2, \ldots, n - 1 \\ j &= 2, 3, \ldots, n \\ k &= 2, 3, \ldots, n \end{aligned}$$

Using the marginal probability distribution of A_{ij} given in (2.4), the variance of A_{ij} is easily evaluated as follows:

$$E(A_{ij}^2) = 1\pi_c + (-1)^2\pi_d = \pi_c + \pi_d$$
$$\text{var}(A_{ij}) = (\pi_c + \pi_d) - (\pi_c - \pi_d)^2 \tag{2.10}$$

The covariance expression, however, requires knowledge of the joint distribution of A_{ij} and A_{ik}, which can be expressed as

$$f_{A_{ij},A_{ik}}(a_{ij},a_{ik}) = \begin{cases} \pi_{cc} & \text{if } a_{ij} = a_{ik} = 1 \\ \pi_{dd} & \text{if } a_{ij} = a_{ik} = -1 \\ \pi_{cd} & \text{if } a_{ij} = 1, a_{ik} = -1 \text{ or} \\ & \quad a_{ij} = -1, a_{ik} = 1 \\ 1 - \pi_{cc} - \pi_{dd} - 2\pi_{cd} & \text{if } a_{ij} = 0, a_{ik} = -1, 0, 1 \\ & \quad \text{or } a_{ij} = -1, 0, 1, a_{ik} = 0 \\ 0 & \text{otherwise} \end{cases}$$

$$\tag{2.11}$$

for all $i < j$, $i < k$, $j \neq k$, $i = 1, 2, \ldots, n$, and some $0 \leq \pi_{cc}, \pi_{dd}, \pi_{cd} \leq 1$. Thus we can evaluate

$$E(A_{ij}A_{ik}) = 1^2\pi_{cc} + (-1)^2\pi_{dd} + 2(-1)\pi_{cd}$$
$$\text{cov}(A_{ij},A_{ik}) = \pi_{cc} + \pi_{dd} - 2\pi_{cd} - (\pi_c - \pi_d)^2 \tag{2.12}$$

Substitution of (2.10) and (2.12) in (2.9) gives

$$n(n - 1)\,\text{var}(T) = 2(\pi_c + \pi_d) + 4(n - 2)(\pi_{cc} + \pi_{dd} - 2\pi_{cd})$$
$$- 2(2n - 3)(\pi_c - \pi_d)^2 \tag{2.13}$$

so that the variance of T is of order $1/n$ and therefore approaches zero as n approaches infinity.

The results obtained so far are completely general, applying to all random variables. If the marginal distributions of X and Y are continuous, $P(A_{ij} = 0) = 0$ and the resulting identities

$$\pi_c + \pi_d = 1 \qquad \text{and} \qquad \pi_{cc} + \pi_{dd} + 2\pi_{cd} = 1$$

allow us to simplify (2.13) to a function of, say, π_c and π_{cd} only:

$$n(n - 1)\,\text{var}(T) = 2 - 2(2n - 3)(2\pi_c - 1)^2 + 4(n - 2)(1 - 4\pi_{cd})$$
$$= 8(2n - 3)\pi_c(1 - \pi_c) - 16(n - 2)\pi_{cd} \tag{2.14}$$

Since for X and Y continuous we also have

$$\pi_{cd} = P(A_{ij} = 1 \cap A_{ik} = -1)$$
$$= P(A_{ij} = 1) - P(A_{ij} = 1 \cap A_{ik} = 1)$$
$$= \pi_c - \pi_{cc}$$

another expression equivalent to (2.14) is

$$n(n-1)\,\text{var}(T) = 8(2n-3)\pi_c(1-\pi_c) - 16(n-2)(\pi_c - \pi_{cc})$$
$$= 8\pi_c(1-\pi_c) + 16(n-2)(\pi_{cc} - \pi_c^2) \qquad (2.15)$$

We have already interpreted π_c as the probability that the pair (X_i, Y_i) is concordant with (X_j, Y_j). Since the parameter π_{cc} is

$$\pi_{cc} = P(A_{ij} = 1 \cap A_{ik} = 1)$$
$$= P[(X_j - X_i)(Y_j - Y_i) > 0 \cap (X_k - X_i)(Y_k - Y_i) > 0] \qquad (2.16)$$

for all $i < j$, $i < k$, $j \neq k$, $i = 1, 2, \ldots, n$, we interpret π_{cc} as the probability the pair (X_i, Y_i) is concordant with both (X_j, Y_j) and (X_k, Y_k).

Integral expressions can be obtained as follows for the probabilities π_c and π_{cc} for random variables X and Y from any continuous bivariate population $F_{X,Y}(x,y)$.

$$\pi_c = P[(X_i < X_j) \cap (Y_i < Y_j)] + P[(X_i > X_j) \cap (Y_i > Y_j)]$$
$$= \int_{-\infty}^{\infty} \int_{-\infty}^{\infty} P[(X_i < x_j) \cap (Y_i < y_j)]f_{X_j,Y_j}(x_j,y_j)\,dx_j\,dy_j$$
$$+ \int_{-\infty}^{\infty} \int_{-\infty}^{\infty} P[(X_j < x_i) \cap (Y_j < y_i)]f_{X_i,Y_i}(x_i,y_i)\,dx_i\,dy_i$$
$$= 2\int_{-\infty}^{\infty} \int_{-\infty}^{\infty} F_{X,Y}(x,y)f_{X,Y}(x,y)\,dx\,dy \qquad (2.17)$$

$$\pi_{cc} = P(\{[(X_i < X_j) \cap (Y_i < Y_j)] \cup [(X_i > X_j) \cap (Y_i > Y_j)]\}$$
$$\cap \{[(X_i < X_k) \cap (Y_i < Y_k)] \cup [(X_i > X_k) \cap (Y_i > Y_k)]\})$$
$$= P[(A \cup B) \cap (C \cup D)] \qquad \text{say}$$
$$= P[(A \cap C) \cup (B \cap D) \cup (A \cap D) \cup (B \cap C)]$$
$$= P(A \cap C) + P(B \cap D) + 2P(A \cap D)$$
$$= \int_{-\infty}^{\infty} \int_{-\infty}^{\infty} \{P[(X_j > x_i) \cap (Y_j > y_i) \cap (X_k > x_i) \cap (Y_k > y_i)]$$
$$+ P[(X_j < x_i) \cap (Y_j < y_i) \cap (X_k < x_i) \cap (Y_k < y_i)]$$
$$+ 2P[(X_j > x_i) \cap (Y_j > y_i) \cap (X_k < x_i) \cap (Y_k < y_i)]\}$$
$$f_{X_i,Y_i}(x_i,y_i)\,dx_i\,dy_i$$
$$= \int_{-\infty}^{\infty} \int_{-\infty}^{\infty} (\{P[(X > x) \cap (Y > y)]\}^2 + \{P[(X < x)$$
$$\cap (Y < y)]\}^2 + 2P[(X > x) \cap (Y > y)]P[(X < x)$$
$$\cap (Y < y)])f_{X,Y}(x,y)\,dx\,dy$$
$$= \int_{-\infty}^{\infty} \int_{-\infty}^{\infty} \{P[(X > x) \cap (Y > y)]$$
$$+ P[(X < x) \cap (Y < y)]\}^2 f_{X,Y}(x,y)\,dx\,dy$$
$$= \int_{-\infty}^{\infty} \int_{-\infty}^{\infty} [1 - F_X(x) - F_Y(y) + 2F_{X,Y}(x,y)]^2 f_{X,Y}(x,y)\,dx\,dy \qquad (2.18)$$

Although T as given in (2.6) is perhaps the simplest form for deriving theoretical properties, the coefficient can be written in a number of other ways. In terms of all n^2 pairs for which A_{ij} is defined, (2.6) can be written as

$$T = \sum_{i=1}^{n} \sum_{j=1}^{n} \frac{A_{ij}}{n(n-1)} \tag{2.19}$$

Now we introduce the notation

$$U_{ij} = \text{sgn}(X_j - X_i) \qquad \text{and} \qquad V_{ij} = \text{sgn}(Y_j - Y_i)$$

so that $A_{ij} = U_{ij}V_{ij}$ for all i, j. Assuming that $X_i \neq X_j$ and $Y_i \neq Y_j$ for all $i \neq j$, we have

$$\sum_{i=1}^{n} \sum_{j=1}^{n} U_{ij}{}^2 = \sum_{i=1}^{n} \sum_{j=1}^{n} V_{ij}{}^2 = n(n-1)$$

and (2.19) can be written in a form resembling an ordinary sample correlation coefficient as

$$T = \frac{\sum\limits_{i=1}^{n} \sum\limits_{j=1}^{n} U_{ij}V_{ij}}{\left[\left(\sum\limits_{i=1}^{n} \sum\limits_{j=1}^{n} U_{ij}{}^2 \right) \left(\sum\limits_{i=1}^{n} \sum\limits_{j=1}^{n} V_{ij}{}^2 \right) \right]^{\frac{1}{2}}} \tag{2.20}$$

Kendall (1962) often uses T in still another form, which arises by simply classifying sets of differences according to the resulting sign of A_{ij}. If P and Q denote the number of positive and negative A_{ij} for $1 \leq i < j \leq n$, respectively, and the total sum is $S = P - Q$, we have

$$T = \frac{P - Q}{\binom{n}{2}} = \frac{S}{\binom{n}{2}} \tag{2.21}$$

If there are no ties within either the X or Y groups, that is, $A_{ij} \neq 0$ for $i \neq j$, $P + Q = \binom{n}{2}$ and (2.21) can be written as

$$T = \frac{2P}{\binom{n}{2}} - 1 = 1 - \frac{2Q}{\binom{n}{2}} \tag{2.22}$$

These two forms in (2.22) are analogous to the expressions in Sec. 1 for the parameter

$$\tau = 2\pi_c - 1 = 1 - 2\pi_d$$

and $P \Big/ \binom{n}{2}$ and $Q \Big/ \binom{n}{2}$ are obviously unbiased estimators for π_c and

π_d, respectively. The quantity P is perhaps the simplest to calculate for a given sample of n pairs. Assuming that the pairs are written from smallest to largest according to the value of the X component, P is simply the number of values of $1 \leq i < j \leq n$ for which $Y_j - Y_i > 0$, since only then shall we have $a_{ij} = 1$.

Another interpretation of T is as a coefficient of disarray, since it can be shown [see Kendall (1962), pp. 24–26] that the total number of interchanges between two consecutive Y observations required to transform the Y arrangement into the natural ordering from smallest to largest, i.e., to transform the Y arrangement into the X arrangement, is equal to Q, or $\binom{n}{2}(1 - T)/2$.

NULL DISTRIBUTION OF T

Suppose we wish to test the null hypothesis that the X and Y random variables are independent. Since $\tau = 0$ for independent variables, the null distribution of T is symmetric about the origin. For a general alternative of nonindependence, the rejection region of size α then should be

$$T \in R \qquad \text{for } |t| \geq t_{\alpha/2}$$

where $t_{\alpha/2}$ is chosen so that

$$P(|T| \geq t_{\alpha/2} \mid H_0) = \alpha$$

For an alternative of positive dependence, a similar one-sided critical region is appropriate.

We must now determine the random sampling distribution of T under the assumption of independence. For this purpose, it will be more convenient, but not necessary, to assume that the X and Y sample observations have both been ordered from smallest to largest and assigned positive integer ranks. The data then consist of n sets of pairs of ranks. The justification for this assumption is that, like τ, T is invariant under all order-preserving transformations. Its numerical value then depends only on the relative magnitudes of the observations and is the same whether calculated for variate values or ranks. For samples with no ties, the $n!$ distinguishable pairings of ranks are all equally likely under the null hypothesis. The value of T is completely determined by the value of P or S because of the expressions in (2.21) and (2.22), and it is more convenient to work with P. Denote by $u(n,p)$ the number of pairings of n ranks which result in exactly p positive a_{ij}, $1 \leq i < j \leq n$. Then

$$P(P = p) = \frac{u(n,p)}{n!} \tag{2.23}$$

and

$$f_T(t) = P(T = t) = P\left[P = \binom{n}{2}\frac{t+1}{2} \right] \qquad (2.24)$$

We shall now find a recursive relation to generate the values of $u(n + 1, p)$ from knowledge of the values of $u(n,p)$ for some n and all p. Assuming that the observations are written in order of magnitude of the X component, the value of P depends only on the resulting permutation of the Y ranks. If s_i denotes the rank of the Y observation which is paired with the rank i in the X sample, for $i = 1, 2, \ldots, n$, p equals the number of integers greater than s_1, plus the number of integers greater than s_2 excluding s_1, plus the number exceeding s_3 excluding s_1 and s_2, etc. For any given permutation of n integers which has this sum p, we need only consider what insertion of the number $n + 1$ in any of the $n + 1$ possible positions of the permutation $(s_1, s_2, \ldots, s_n)$ does to the value of p. If $n + 1$ is in the first position, p is clearly unchanged. If $n + 1$ is in the second position, there is one additional integer greater than s_1, so that p is increased by 1. If in the third position, there is one additional integer greater than both s_1 and s_2, so that p is increased by 2. In general, if $n + 1$ is in the kth position, p is increased by $k - 1$ for all $k = 1, 2, \ldots, n + 1$. Therefore the desired recursive relation is

$$u(n + 1, p) = u(n,p) + u(n, p - 1) + u(n, p - 2) + \cdots$$
$$+ u(n, p - n) \quad (2.25)$$

In terms of s, since for a set of n pairs

$$s = 2p - \frac{n(n-1)}{2} \qquad (2.26)$$

insertion of $n + 1$ in the kth position increases p by $k - 1$, the new value s' of s for $n + 1$ pairs will be

$$s' = 2p' - \frac{n(n+1)}{2} = 2(p + k - 1) - \frac{n(n+1)}{2}$$

$$= 2p - \frac{n(n-1)}{2} + 2(k - 1) - n$$

$$= s + 2(k - 1) - n$$

In other words, s is increased by $2(k - 1) - n$ for $k = 1, 2, \ldots, n + 1$, and corresponding to (2.25) we have

$$u(n + 1, s) = u(n, s + n) + u(n, s + n - 2) + u(n, s + n - 4)$$
$$+ \cdots + u(n, s - n + 2) + u(n, s - n) \quad (2.27)$$

The distribution of S is symmetrical about zero, and from (2.26) it is clear that S for n pairs is an even or odd integer according as $n(n-1)/2$ is even or odd. Because of this symmetry, tables are most easily constructed for S (or T) rather than P or Q. The null distribution of S is given in Kendall [(1962), p. 173] for $4 \leq n \leq 10$.

A simple example will suffice to illustrate the use of (2.25) or (2.27) to set up tables of these probability distributions. When $n = 3$, the 3! permutations of the Y ranks and the corresponding values of P and S are:

Permutation	123	132	213	231	312	321
p	3	2	2	1	1	0
s	3	1	1	-1	-1	-3

The frequencies then are:

p	0	1	2	3
s	-3	-1	1	3
$u(3,p)$ or $u(3,s)$	1	2	2	1

For P, using (2.25), $u(4,p) = \sum_{i=0}^{3} u(3,\, p - i)$, or

$u(4,0) = u(3,0) = 1$
$u(4,1) = u(3,1) + u(3,0) = 3$
$u(4,2) = u(3,2) + u(3,1) + u(3,0) = 5$
$u(4,3) = u(3,3) + u(3,2) + u(3,1) + u(3,0) = 6$
$u(4,4) = u(3,3) + u(3,2) + u(3,1) = 5$
$u(4,5) = u(3,3) + u(3,2) = 3$
$u(4,6) = u(3,3) = 1$

Alternatively, we could use (2.27), or $u(4,s) = \sum_{i=0}^{3} u(3,\, s + 3 - 2i)$. Therefore the probability distributions for $n = 4$ are:

p	0	1	2	3	4	5	6
s	-6	-4	-2	0	2	4	6
t	-1	$-\frac{2}{3}$	$-\frac{1}{3}$	0	$\frac{1}{3}$	$\frac{2}{3}$	1
$f(p, s, \text{or } t)$	$\frac{1}{24}$	$\frac{3}{24}$	$\frac{5}{24}$	$\frac{6}{24}$	$\frac{5}{24}$	$\frac{3}{24}$	$\frac{1}{24}$

The way in which the $u(n, s,$ or $p)$ are built up by cumulative sums indicates that simple schemes for their generation may be easily worked out [see, for example, Kendall (1962), p. 68].

The exact null distribution is thus easily found for moderate n. Since T is a sum of random variables, it can be shown using general limit theorems for dependent variables that the distribution of a standardized T approaches the standard normal distribution as n approaches infinity. To use this fact to facilitate inferences concerning independence in large samples, we must determine the null mean and variance of T. Since T was defined to be an unbiased estimator of τ for any bivariate population and we showed in Sec. 1 that $\tau = 0$ for independent, continuous random variables, the mean is $E(T|H_0) = 0$. In order to find $\text{var}(T|H_0)$ for X and Y continuous, (2.15) is used with the appropriate π_c and π_{cc} under H_0. Because of the probability-integral transformation, under the assumption that X and Y have continuous marginal distributions and are independent, they can be assumed to be identically distributed according to the uniform distribution over the interval $(0,1)$. Then, in (2.17) and (2.18), we have

$$
\begin{aligned}
\pi_c &= 2 \int_0^1 \int_0^1 xy \, dx \, dy = \tfrac{1}{2} \\
\pi_{cc} &= \int_0^1 \int_0^1 (1 - x - y + 2xy)^2 \, dx \, dy = \tfrac{5}{18}
\end{aligned}
\tag{2.28}
$$

Substituting these results in (2.15), we obtain

$$
n(n - 1) \, \text{var}(T) = 2 + \frac{16(n - 2)}{36}
$$

$$
\text{var}(T) = \frac{2(2n + 5)}{9n(n - 1)}
\tag{2.29}
$$

For large n, the random variable

$$
Z = \frac{3 \sqrt{n(n - 1)} \, T}{\sqrt{2(2n + 5)}}
\tag{2.30}
$$

can be treated as a standard normal variable with distribution $\varphi(z)$.

If the null hypothesis of independence of X and Y is accepted, we can of course infer that the population parameter τ is zero. However, if the hypothesis is rejected, this implies dependence between the random variables but not necessarily that $\tau \neq 0$.

THE LARGE-SAMPLE NONNULL DISTRIBUTION OF KENDALL'S STATISTIC

The probability distribution of T is asymptotically normal for sample pairs from any bivariate population. Therefore, if the general mean and variance of T could be determined, T would be useful in large samples for

other inferences relating to population characteristics besides independence. Since $E(T) = \tau$ for any distribution, T is particularly relevant in
inferences concerning the value of τ. The expressions previously found
for var(T) in (2.13) for any distribution and (2.15) for continuous distributions depend on unknown probabilities. Unless the hypothesis under
consideration somehow determines π_c, π_d, π_{cc}, π_{dd}, and π_{cd} (or simply π_c and
π_{cc} for the continuous case), the exact variance cannot be found without
some information about $f_{X,Y}(x,y)$. However, unbiased and consistent
estimates of these probabilities can be found from the sample data to
provide a consistent estimate $\hat{\sigma}^2(T)$ of the variance of T. The asymptotic
distribution of $(T - \tau)/\hat{\sigma}(T)$ then remains standard normal.

Such estimates will be found here for paired samples containing no
tied observations. We observed before that $P \left/ \binom{n}{2} \right.$ is an unbiased and
consistent estimator of π_c. However, for the purpose of finding estimates
for all the probabilities involved, it will be more convenient now to introduce a different notation. As before, we can assume without loss of
generality that the n pairs are arranged in natural order according to the x
component and that s_i is the rank of that y which is paired with the ith
smallest x for $i = 1, 2, \ldots, n$, so that the data are $(s_1, s_2, \ldots, s_n)$.
Define

a_i = number of integers to the left of s_i and less than s_i
b_i = number of integers to the right of s_i and greater than s_i

Then

$c_i = a_i + b_i$ = number of values of $j = 1, 2, \ldots, n$ such that
        (x_i, y_i) is concordant with (x_j, y_j)

There are $n(n - 1)$ distinguishable sets of pairs, of which $\sum\limits_{i=1}^{n} c_i$ are concordant. An unbiased estimate of π_c then is

$$p_c = \sum_{i=1}^{n} \frac{c_i}{n(n - 1)} \tag{2.31}$$

Similarly, we define

a_i' = number of integers to the left of s_i and greater than s_i
b_i' = number of integers to the right of s_i and less than s_i

and

$d_i = a_i' + b_i'$ = number of values of $j = 1, 2, \ldots, n$ such that
        (x_i, y_i) is discordant with (x_j, y_j)

Then

$$p_d = \sum_{i=1}^{n} \frac{d_i}{n(n-1)} \qquad (2.32)$$

gives an unbiased estimate of π_d.

An unbiased and consistent estimate of π_{cc} is the number of sets of three pairs (x_i,y_i), (x_j,y_j), (x_k,y_k), for all $i \neq j \neq k$, for which both the products $(x_i - x_j)(y_i - y_j)$ and $(x_i - x_k)(y_i - y_k)$ are positive, divided by the number of distinguishable sets $n(n-1)(n-2)$. Denote by c_{ii} the number of values of j and k, $i \neq j \neq k$, $1 \leq j$, $k \leq n$, such that (x_i,y_i) is concordant with both (x_j,y_j) and (x_k,y_k), so that

$$p_{cc} = \sum_{i=1}^{n} \frac{c_{ii}}{n(n-1)(n-2)}$$

The pair (x_i,y_i) is concordant with both (x_j,y_j) and (x_k,y_k) if:

Group 1:	$s_j < s_i < s_k$	for $j < i < k$
	$s_k < s_i < s_j$	for $k < i < j$
Group 2:	$s_i < s_j < s_k$	for $i < j < k$
	$s_i < s_k < s_j$	for $i < k < j$
Group 3:	$s_j < s_k < s_i$	for $j < k < i$
	$s_k < s_j < s_i$	for $k < j < i$

Therefore c_{ii} is twice the sum of the following three corresponding numbers:

1. The number of unordered pairs of integers, one to the left and one to the right of s_i, such that the one to the left is less than s_i and the one to the right is greater than s_i.
2. The number of unordered pairs of integers, both to the right of s_i, such that both are greater than s_i.
3. The number of unordered pairs of integers, both to the left of s_i, such that both are less than s_i.

Then, employing the same notation as before, we have

$$c_{ii} = 2\left[\binom{a_i}{1}\binom{b_i}{1} + \binom{b_i}{2} + \binom{a_i}{2} \right] = (a_i + b_i)^2 - (a_i + b_i)$$
$$= c_i^2 - c_i = c_i(c_i - 1)$$

and

$$p_{cc} = \sum_{i=1}^{n} \frac{c_i(c_i - 1)}{n(n-1)(n-2)} \qquad (2.33)$$

Similarly we can obtain

$$p_{dd} = \sum_{i=1}^{n} \frac{d_i(d_i - 1)}{n(n-1)(n-2)} \tag{2.34}$$

$$p_{cd} = \sum_{i=1}^{n} \frac{a_i b_i' + a_i a_i' + b_i a_i' + b_i b_i'}{n(n-1)(n-2)}$$

$$= \sum_{i=1}^{n} \frac{c_i d_i}{n(n-1)(n-2)} \tag{2.35}$$

Substituting the results (2.31) and (2.33) in the form (2.15), the estimated variance of T in samples for continuous variables is

$$n(n-1)\hat{\sigma}^2(T) = 8p_c - 8p_c^2(2n-3) + 16(n-2)p_{cc}$$

$$n^2(n-1)^2\hat{\sigma}^2(T) = 8\left[2\sum_{i=1}^{n} c_i^2 - \frac{2n-3}{n(n-1)}\left(\sum_{i=1}^{n} c_i\right)^2 - \sum_{i=1}^{n} c_i\right] \tag{2.36}$$

In order to obviate any confusion regarding the calculation of the c_i and c_{ii} to estimate the variance from (2.36) in the case of no tied observations, a simple example is provided below for achievement tests in Mathematics and English administered to a group of six randomly chosen students.

Student	A	B	C	D	E	F
Math score	91	52	69	99	72	78
English score	89	72	69	96	66	67

The two sets of scores ranked and rearranged in order of increasing Mathematics scores are:

Student	B	C	E	F	A	D
Math rank	1	2	3	4	5	6
English rank	4	3	1	2	5	6

The numbers $c_i = a_i + b_i$ are

$$c_1 = 0 + 2 \qquad c_2 = 0 + 2 \qquad c_3 = 0 + 3 \qquad c_4 = 1 + 2$$
$$c_5 = 4 + 1 \qquad c_6 = 5 + 0$$

$$\Sigma c_i = 20 \qquad \Sigma c_i{}^2 = 76 \qquad n = 6$$

$$p_c = \frac{20}{6(5)} = \frac{2}{3}$$

$$p_{cc} = \frac{76 - 20}{6(5)(4)} = \frac{7}{15}$$

$$t = 2(\tfrac{2}{3}) - 1 = \tfrac{1}{3}$$

$$30^2\hat{\sigma}^2(T) = 8\left[2(76) - 20 - \frac{9}{6(5)}20^2\right] = 96$$

$$\hat{\sigma}^2(T) = 0.1067 \qquad \hat{\sigma}(T) = 0.33$$

If we wish to count the c_{ii} directly, we have for $c_{ii} = 2(\text{group } 1 + \text{group } 2 + \text{group } 3)$, the pairs relevant to c_{44}, say, are

Group 1: (1,5)(1,6)
Group 2: (5,6)
Group 3: None

so that $c_{44} = 2(3) = 6 = c_4(c_4 - 1)$.

On the other hand, suppose the English scores corresponding to increasing Math scores were ranked as

y:	3	1	4	2	6	5

Then we can calculate

$$c_1 = c_4 = 3 \qquad c_2 = c_3 = c_5 = c_6 = 4$$
$$p_c = {}^{11}\!/_{15} \qquad p_{cc} = \tfrac{1}{2} \qquad t = {}^{7}\!/_{15} \qquad \hat{\sigma}^2(T) = -{}^{32}\!/_{1125}$$

and the estimated variance is negative! A negative variance from (2.15) of course cannot occur, but when the parameters π are replaced by estimates p and combined, the result can be negative. Since the probability estimates are consistent, the estimated variance of T will be positive for n sufficiently large.

Two applications of this asymptotic approximation to the nonnull distribution of T in nonparametric inference for large samples are:

1. An approximate $(1 - \alpha)100$ percent confidence-interval estimate of the population Kendall tau coefficient is

$$t - z_{\alpha/2}\hat{\sigma}(T) < \tau < t + z_{\alpha/2}\hat{\sigma}(T)$$

2. An approximate test of

$$H_0: \tau = \tau_0 \qquad \text{versus} \qquad H_1: \tau \neq \tau_0$$

with significance level α is to reject H_0 when

$$\frac{|t - \tau_0|}{\hat{\sigma}(T)} \geq z_{\alpha/2}$$

A one-sided alternative can also be tested.

THE PROBLEM OF TIED OBSERVATIONS

Whether or not the marginal distributions of X and Y are assumed continuous, tied observations can occur within either or both samples. Ties across samples do not present any problem of course. Since the definition of A_{ij} in (2.3) assigned a value of zero to a_{ij} if a tie occurs in the (i,j) set of pairs for either the x or y sample values, T as defined before allows for, and essentially ignores, all zero differences. With τ defined as the difference $\pi_c - \pi_d$, T as calculated from (2.6), (2.19), or (2.21) is an unbiased estimator of τ with variance as given in (2.13) even in the presence of ties. If the occurrence of ties in the sample is attributed to a lack of precision in measurement as opposed to discrete marginal distributions, the simplified expression for $\text{var}(T)$ in (2.15) may still be used. If there are sample ties, however, the expressions (2.20) and (2.22) are no longer equivalent to (2.6), (2.19), or (2.21).

For small samples with a small number of tied observations, the exact null distribution of T (or S) conditional on the observed ties can be determined by enumeration. There will be mw pairings of the two samples, each occurring with equal probability $1/mw$, if there are m and w distinguishable permutations of the x and y sample observations, respectively. For larger samples, the normal approximation to the distribution of T can still be used but with corrected moments. Conditional upon the observed ties, the parameters π_c, π_d, π_{cc}, π_{dd}, and π_{cd} must have a slightly different interpretation. For example, $\pi_c + \pi_d$ here would be the probability that we select two pairs (x_i,y_i) and (x_j,y_j) which do not have a tie in either coordinate, and under the assumption of independence this is

$$\left[1 - \frac{\Sigma u(u - 1)}{n(n - 1)} \right] \left[1 - \frac{\Sigma v(v - 1)}{n(n - 1)} \right]$$

where u denotes the multiplicity of a tie in the x set and the sum is extended over all ties and v has the same interpretation for the y set. These parameters in the conditional distribution can be determined and substituted in (2.13) to find the conditional variance [see, for example, Noether (1967), pp. 76–77]. The conditional mean of T, however, is unchanged, since even for the new parameters $\pi_c = \pi_d$ for independent samples.

Conditional on the observed ties, however, there are no longer $\binom{n}{2}$ distinguishable sets of pairs to check for concordance, and thus if T is calculated in the ordinary way, it cannot equal 1 even for perfect agreement. Therefore an alternative definition of T in the presence of ties is to replace the $n(n - 1)$ in the denominator of (2.6), (2.19), or (2.21) by a smaller quantity. To obtain a result still analogous to a correlation coefficient, we might take (2.20) as the definition of T in general. Since $\sum_{i=1}^{n} \sum_{j=1}^{n} U_{ij}{}^2$ is the number of nonzero differences $X_j - X_i$ for all (i,j), the sum is the total number of distinguishable differences less the number involving tied observations, or $n(n - 1) - \Sigma u(u - 1)$. Similarly for the Y observations. Therefore our modified T from (2.20) is

$$
T = \frac{\sum_{i=1}^{n} \sum_{j=1}^{n} U_{ij} V_{ij}}{\{[n(n - 1) - \Sigma u(u - 1)][n(n - 1) - \Sigma v(v - 1)]\}^{\frac{1}{2}}} \tag{2.37}
$$

which reduces to all previously given forms if there are no ties. This expression for T still may not equal 1 even for perfect agreement, but it is always greater than a coefficient calculated using (2.19) when ties are present.

A RELATED MEASURE OF ASSOCIATION FOR DISCRETE POPULATIONS

In Sec. 1 we stated the criterion that a good measure of association between two random variables would equal $+1$ for a perfect direct relationship and -1 for a perfect indirect relationship. In terms of the probability parameters, perfect concordance requires $\pi_c = 1$, and perfect discordance requires $\pi_d = 1$. With Kendall's coefficient defined as $\tau = \pi_c - \pi_d$, the criterion is satisfied if and only if $\pi_c + \pi_d = 1$. But if the marginal distributions of X and Y are not continuous,

$$
\begin{aligned}
\pi_c + \pi_d &= P[(X_j - X_i)(Y_j - Y_i) > 0] + P[(X_j - X_i)(Y_j - Y_i) < 0] \\
&= 1 - P[(X_j - X_i)(Y_j - Y_i) = 0] \\
&= 1 - P[(X_i = X_j) \cup (Y_i = Y_j)] = 1 - \pi_t
\end{aligned}
$$

where π_t denotes the probability that a pair is neither concordant nor discordant. Thus τ cannot be considered a "good" measure of association if $\pi_t \neq 0$.

However, a modified parameter which does satisfy the criteria for all distributions can easily be defined as

$$\tau^* = \frac{\tau}{1 - \pi_t} = \pi_c^* - \pi_d^*$$

where π_c^* and π_d^* are, respectively, the conditional probabilities of concordance and discordance given that there are no ties

$$\pi_c^* = \frac{\pi_c}{1 - \pi_t} = \frac{P[(X_j - X_i)(Y_j - Y_i) > 0]}{P[(X_j - X_i)(Y_j - Y_i) \neq 0]}$$

Since τ^* is a linear function of τ, an estimate is provided by

$$T^* = \frac{T}{1 - p_t} = \frac{p_c - p_d}{p_c + p_d}$$

with p_c and p_d defined as before in (2.31) and (2.32). Since p_c and p_d are consistent estimators, the asymptotic distribution of $T/(p_c + p_d)$ is equivalent to the asymptotic distribution of $T/(\pi_c + \pi_d)$, which we know to be the normal distribution. Therefore for large samples inferences concerning τ^* can be made [see, for example, Goodman and Kruskal (1954, 1959, 1963)].

USE OF KENDALL'S STATISTIC TO TEST AGAINST TREND

In Chap. 3 regarding tests for randomness, we observed that the arrangement of relative magnitudes of a single sequence of time-ordered observations can indicate some sort of trend. When the theory of runs up and down was used to test a hypothesis of randomness, the magnitude of each observation relative to its immediately preceding value was considered, and a long run of plus (minus) signs or a sequence with a large predominance of plus (minus) signs was considered indicative of an upward (downward) trend. If time is treated as an X variable, say, and a set of time-ordered observations as the Y variable, an association between X and Y might be considered indicative of a trend. Thus the degree of concordance between such X and Y observations would be a measure of trend, and Kendall's tau statistic becomes a measure of trend. Unlike the case of runs up and down, however, the tau coefficient considers the relative magnitude of each observation relative to *every* preceding observation.

A hypothesis of randomness in a single set of n time-ordered observations is the same as a hypothesis of independence between these observations when paired with the numbers $1, 2, \ldots, n$. Therefore, assuming that $x_i = i$ for $i = 1, 2, \ldots, n$, the indicator variables A_{ij} defined in (2.3) become

$$A_{ij} = \operatorname{sgn}(j - i)\,\operatorname{sgn}(Y_j - Y_i)$$

and (2.6) can be written as

$$\binom{n}{2} T = \sum\sum_{1 \leq i < j \leq n} \operatorname{sgn}(Y_j - Y_i)$$

The exact null distribution of T is the same as before. If the alternative is an upward trend, the rejection region consists of large positive values of T, and T can be considered an unbiased estimate of τ, a relative measure of population trend.

12.3 SPEARMAN'S COEFFICIENT OF RANK CORRELATION

A random sample of n pairs

$$(X_1, Y_1), (X_2, Y_2), \ldots, (X_n, Y_n)$$

is drawn from a bivariate population with Pearson product-moment correlation coefficient ρ. In classical statistics, the estimate commonly used for ρ is the sample correlation coefficient defined as

$$R = \frac{\displaystyle\sum_{i=1}^{n} (X_i - \bar{X})(Y_i - \bar{Y})}{\left[\displaystyle\sum_{i=1}^{n} (X_i - \bar{X})^2 \sum_{i=1}^{n} (Y_i - \bar{Y})^2\right]^{\frac{1}{2}}} \tag{3.1}$$

In general, of course, the sampling distribution of R depends upon the form of the bivariate population from which the sample of pairs is drawn. However, suppose the X observations are ranked from smallest to largest using the integers $1, 2, \ldots, n$, and the Y observations are ranked separately using the same ranking scheme. In other words, each observation is assigned a rank according to its magnitude relative to the others in its own group. If the marginal distributions of X and Y are assumed continuous, unique sets of rankings exist theoretically. The data then consist of n sets of paired ranks from which R as defined in (3.1) can be calculated. The resulting statistic is then called *Spearman's coefficient of rank correlation*. It measures the degree of correspondence between rankings, instead of between actual variate values, but it can still be considered a measure of association between the samples and an estimate of the association between X and Y in the continuous bivariate population. It is difficult to interpret exactly what R is estimating in the population from which these samples were drawn and ranks obtained, but the measure has intuitive appeal anyway. The problem of interpretation will be treated in Sec. 4.

The fact that we know the numerical values of the derived observations from which Spearman's R is computed, if not their scheme of pairing, means that the expression in (3.1) can be considerably simplified. Denoting the respective ranks of the random variables in the samples by

$$R_i = \text{rank}(X_i) \qquad \text{and} \qquad S_i = \text{rank}(Y_i)$$

the derived sample observations of n pairs are

$$(r_1, s_1), (r_2, s_2), \ldots, (r_n, s_n)$$

$r_i, s_i = 1, 2, \ldots, n$, for $i = 1, 2, \ldots, n$. Since addition is commutative, we have the constant values for all samples

$$\sum_{i=1}^{n} r_i = \sum_{i=1}^{n} s_i = \sum_{i=1}^{n} i = \frac{n(n+1)}{2} \qquad \bar{r} = \bar{s} = \frac{n+1}{2} \qquad (3.2)$$

$$\sum_{i=1}^{n} (r_i - \bar{r})^2 = \sum_{i=1}^{n} (s_i - \bar{s})^2 = \sum_{i=1}^{n} \left(i - \frac{n+1}{2}\right)^2 = \frac{n(n^2-1)}{12} \tag{3.3}$$

Substituting these constants in (3.1), the following equivalent forms of R are obtained:

$$R = \frac{12 \sum_{i=1}^{n} (R_i - \bar{R})(S_i - \bar{S})}{n(n^2-1)} \tag{3.4}$$

$$R = \frac{12 \left[\sum_{i=1}^{n} R_i S_i - \frac{1}{4} n(n+1)^2 \right]}{n(n^2-1)} \tag{3.5}$$

$$R = \frac{12 \sum_{i=1}^{n} R_i S_i}{n(n^2-1)} - \frac{3(n+1)}{n-1} \tag{3.6}$$

Another useful form of R is in terms of the differences

$$D_i = R_i - S_i = (R_i - \bar{R}) - (S_i - \bar{S})$$

Substituting (3.3) in the expression

$$\sum_{i=1}^{n} D_i^2 = \sum_{i=1}^{n} (R_i - \bar{R})^2 + \sum_{i=1}^{n} (S_i - \bar{S})^2 - 2 \sum_{i=1}^{n} (R_i - \bar{R})(S_i - \bar{S})$$

and using the result back in (3.4), the most common form of the Spearman coefficient of rank correlation is obtained as

$$R = 1 - \frac{6 \sum_{i=1}^{n} D_i^2}{n(n^2 - 1)} \tag{3.7}$$

We can assume without loss of generality that the n sample pairs are labeled in accordance with increasing magnitude of the X component, so that $R_i = i$ for $i = 1, 2, \ldots, n$. Then S_i is the rank of the Y observation that is paired with the rank i in the X sample, and $D_i = i - S_i$.

In Sec. 1, criteria were defined for a "good" relative measure of association between two random variables. Although the parameter analogous to R has not been specifically defined, we can easily verify that Spearman's R does satisfy the corresponding criteria of a good measure of association between sample ranks.

1. For any two sets of paired ranks (i,S_i) and (j,S_j) of random variables in a sample from any continuous bivariate distribution, in order to have perfect concordance between ranks, the Y component must also be increasing, or, equivalently, $s_i = i$ and $d_i = 0$ for $i = 1, 2, \ldots, n$ so that r equals 1.
2. For perfect discordance between ranks, the Y arrangement must be the reverse of the X arrangement to have decreasing Y components, so that $s_i = n - i + 1$ and

$$\sum_{i=1}^{n} d_i^2 = \sum_{i=1}^{n} [i - (n - i + 1)]^2$$
$$= 4 \sum_{i=1}^{n} \left(i - \frac{n+1}{2} \right)^2 = \frac{n(n^2 - 1)}{3}$$

from (3.3). Substituting this in (3.7), we find $r = -1$.
3 to 6. Since R in, say, (3.7) is algebraically equivalent to (3.1) and the value of (3.1) is in the interval $[-1,1]$ for all sets of numerical pairs, the same bounds apply here. Further, R is commutative and symmetric about zero and has expectation zero when the X and Y observations are independent. These properties will be shown later in this section.
7. Since ranks are preserved under all order-preserving transformations, the measure R based on ranks is invariant.

EXACT NULL DISTRIBUTION OF R

If the X and Y random variables from which these n pairs of ranks (R_i,S_i) are derived are independent, R is a distribution-free statistic since each of the $n!$ distinguishable sets of pairings of n ranks is equally likely. There-

<sup>... the
... If
... for the</sup>

The null d... ... symmetric about the origin, since the random variable $D = \sum_{i=1}^{n} D_i^2$ is symmetric about $n(n^2 - 1)/6$. This property is the result of the fact that for any set of pairs

$$(1,s_1),(2,s_2), \ldots , (n,s_n)$$

with

$$\sum_{i=1}^{n} d_i^2 = \sum_{i=1}^{n} (i - s_i)^2$$

there exists a conjugate set of pairs

$$(1,s_n), (2,s_{n-1}), \ldots , (n,s_1)$$

with

$$\sum_{i=1}^{n} d_i'^2 = \sum_{i=1}^{n} (i - s_{n-i+1})^2 = \sum_{i=1}^{n} (n - i + 1 - s_i)^2$$

The sums of squares of the respective sum and difference of rank differences are

$$\sum_{i=1}^{n} (d_i + d_i')^2 = \sum_{i=1}^{n} (n + 1 - 2s_i)^2$$

$$= 4 \sum_{i=1}^{n} \left(s_i - \frac{n + 1}{2}\right)^2 = \frac{n(n^2 - 1)}{3}$$

$$\sum_{i=1}^{n} (d_i - d_i')^2 = \sum_{i=1}^{n} (2i - n - 1)^2$$

$$= 4 \sum_{i=1}^{n} \left(i - \frac{n + 1}{2}\right)^2 = \frac{n(n^2 - 1)}{3}$$

Substituting these results in the relation

$$\sum_{i=1}^{n} [(d_i + d_i') + (d_i - d_i')]^2 = 4 \sum_{i=1}^{n} d_i^2$$

$$= \sum_{i=1}^{n} (d_i + d_i')^2 + \sum_{i=1}^{n} (d_i - d_i')^2 + 2 \sum_{i=1}^{n} (d_i^2 - d_i'^2)$$

we obtain

$$4 \sum_{i=1}^{n} d_i^2 = \frac{2n(n^2 - 1)}{3} + 2 \sum_{i=1}^{n} d_i^2 - 2 \sum_{i=1}^{n} d_i'^2$$

or

$$\sum_{i=1}^{n} d_i^2 + \sum_{i=1}^{n} d_i'^2 = \frac{n(n^2 - 1)}{3} = \text{const}$$

Further, r cannot equal zero unless n is even, since $\sum_{i=1}^{n} d_i^2$ is always even because $\sum_{i=1}^{n} d_i = 0$, an even number.

The direct approach to determining u_r, of course, is by enumeration, which is probably least tedious for r in the form of (3.6). Because of the symmetry property, only $n!/2$ cases need be considered. For $n = 3$, for example, we list the following sets (s_1, s_2, s_3) which may be paired with $(1,2,3)$, and the resulting values of r:

(s_1, s_2, s_3)	$\sum_{i=1}^{n} i s_i$	r
1,2,3	14	1.0
1,3,2	13	0.5
2,1,3	13	0.5

The complete probability distribution then is

r	-1.0	-0.5	0.5	1.0
$f_R(r)$	$\frac{1}{6}$	$\frac{2}{6}$	$\frac{2}{6}$	$\frac{1}{6}$

This method of generating the distribution is time-consuming, even for moderate n. Although there are more efficient methods of enumeration [see, for example, Kendall (1962), pp. 74–75], no recursive relation has been found. The exact complete probability distribution of $\sum_{i=1}^{n} d_i^2$ for $n \leq 10$ is tabulated in Kendall [(1962), pp. 174–175].

Although the general null probability distribution of R requires enumeration, the marginal and joint distributions of any number of the individual ranks of a single random sample of size n are easily determined from combinatorial theory. For example, for the Y sample, we have

$$f_{S_i}(s_i) = \frac{1}{n} \qquad s_i = 1, 2, \ldots, n \tag{3.8}$$

$$f_{S_i,S_j}(s_i,s_j) = \frac{1}{n(n-1)} \qquad s_i, s_j = 1, 2, \ldots, n$$

$$s_i \neq s_j \tag{3.9}$$

Thus, using (3.2) and (3.3),

$$E(S_i) = \frac{n+1}{2} \qquad \text{var}(S_i) = \frac{n^2-1}{12}$$

For the covariance, we have for all $i \neq j$,

$$\text{cov}(S_i,S_j) = E(S_iS_j) - E(S_i)E(S_j)$$

$$= \frac{1}{n(n-1)} \sum_{\substack{i=1 \\ i \neq j}}^{n} \sum_{j=1}^{n} ij - \frac{1}{n^2}\left(\sum_{i=1}^{n} i\right)^2$$

$$= \frac{1}{n^2(n-1)}\left[n\left(\sum_{i=1}^{n} i\right)^2 - n\sum_{i=1}^{n} i^2 - (n-1)\left(\sum_{i=1}^{n} i\right)^2\right]$$

$$= \frac{-1}{n^2(n-1)}\left[\frac{n^2(n+1)(2n+1)}{6} - \frac{n^2(n+1)^2}{4}\right]$$

$$= -\frac{n+1}{12} \tag{3.10}$$

The same results hold for the ranks R_i of the X sample. Under the null hypothesis that the X and Y samples are independent, the ranks R_i and S_j are independent for all i, j, and the null mean and variance of R are easily found as follows:

$$E\left(\sum_{i=1}^{n} R_iS_i\right) = nE(R_i)E(S_i) = \frac{n(n+1)^2}{4} \tag{3.11}$$

$$\text{var}\left(\sum_{i=1}^{n} R_iS_i\right) = n\,\text{var}(R_i)\,\text{var}(S_i)$$

$$+ n(n-1)\,\text{cov}(R_i,R_j)\,\text{cov}(S_i,S_j)$$

$$= \frac{n(n^2-1)^2 + n(n-1)(n+1)^2}{144}$$

$$= \frac{n^2(n-1)(n+1)^2}{144} \tag{3.12}$$

Then using the form of R in (3.6)

$$E(R \mid H_0) = 0 \qquad \text{var}(R \mid H_0) = \frac{1}{n-1} \tag{3.13}$$

ASYMPTOTIC NULL DISTRIBUTION OF R

Considering R in the form of (3.6), and as before assuming S_i denotes the rank of the Y observation paired with the ith smallest X observation, we see that the distribution of R depends only on the random variable $\sum_{i=1}^{n} iS_i$. This quantity is a linear combination of random variables, which can be shown to be asymptotically normally distributed [see, for example, Fraser (1957), pp. 247–248]. The mean and variance are given in (3.11) and (3.12). The standardized normal variable used for an approximate test of independence then is

$$Z = \left(12 \sum_{i=1}^{n} iS_i - 3n^3\right) n^{-5/2}$$

or, equivalently,

$$Z = R \sqrt{n-1}$$

This approximation is good even for n as small as 10.

TESTING THE NULL HYPOTHESIS

Since R has mean zero for independent random variables, the appropriate rejection region of size α is large absolute values of R for a general alternative of nonindependence and large positive values of R for alternatives of positive dependence. As in the case of Kendall's coefficient, if the null hypothesis of independence is accepted, we can infer that $\rho(X,Y)$ equals zero, but dependence between the variables does not necessarily imply that $\rho(X,Y) \neq 0$. Besides, the coefficient of rank correlation is measuring association between ranks, not variate values. Since the distribution of R was derived only under the assumption of independence, these results cannot be used to construct confidence-interval estimates of $\rho(X,Y)$ or $E(R)$.

THE PROBLEM OF TIED OBSERVATIONS

In all of the foregoing discussion we assumed that the data to be analyzed consisted of n sets of paired integer ranks. These integer ranks may be obtained by ordering observations from two continuous populations, but the theory is also equally applicable to any two sets of n pairs which can be placed separately in a unique preferential order. In the first case, ties can still occur within either or both sets of sample measurements, and in the second case it is possible that no preference can be made between two or more of the individuals in either group. Thus, for practical purposes, the problem of ties within a set of ranks must be considered.

If within each set of tied observations the ranks they would have if distinguishable are assigned at random, nothing is changed since we still have the requisite type of data to be analyzed. However, such an approach has little intuitive appeal, and besides an additional element of chance is introduced. The most common practice for dealing with tied observations here, as in most other nonparametric procedures, is to assign equal ranks to indistinguishable observations. If that rank is the midrank in every case, the sum of the ranks for each sample is still $n(n + 1)/2$, but the sum of squares of the ranks is changed, so that the expressions in (3.4) to (3.7) are no longer equivalent to (3.1). Assuming that the spirit of the rank correlation coefficient is unchanged, the expression in (3.1) can be calculated directly from the ranks assigned. However, a form analogous to (3.7) which is equivalent to (3.1) can still be found for use in the presence of ties.

We shall investigate what happens to the sum of squares

$$\sum_{i=1}^{n} (s_i - \bar{s})^2 = \sum_{i=1}^{n} s_i{}^2 - \frac{n(n + 1)^2}{2}$$

when there are one or more groups of u tied observations within the Y sample and each is assigned the appropriate midrank. In each group of u tied observations which, if not tied, would be assigned the ranks $p_k + 1, p_k + 2, \ldots, p_k + u$, the rank assigned to all is

$$\sum_{i=1}^{u} \frac{p_k + i}{u} = p_k + \frac{u + 1}{2}$$

The sum of squares for these tied ranks then is

$$u \left(p_k + \frac{u + 1}{2} \right)^2 = u \left[p_k{}^2 + p_k(u + 1) + \frac{(u + 1)^2}{4} \right] \qquad (3.14)$$

and the corresponding sum in the absence of ties would be

$$\sum_{i=1}^{u} (p_k + i)^2 = u p_k{}^2 + p_k u(u + 1) + \frac{u(u + 1)(2u + 1)}{6} \qquad (3.15)$$

This particular group of u tied observations then decreases the sum of squares by the difference between (3.15) and (3.14) or

$$\frac{u(u + 1)(2u + 1)}{6} - \frac{u(u + 1)^2}{4} = \frac{u(u^2 - 1)}{12}$$

Since this is true for each group of u tied observations, the sum of squares in the presence of ties is

$$\sum_{i=1}^{n} (s_i - \bar{s})^2 = \frac{n(n^2 - 1)}{12} - u' \tag{3.16}$$

where $u' = \Sigma u(u^2 - 1)/12$ and the summation is extended over all sets of u tied ranks in the Y sample. Letting t' denote the corresponding sum for the X sample, we obtain the alternative forms of (3.1) as

$$R = \frac{12 \left[\sum_{i=1}^{n} R_i S_i - \frac{1}{4} n(n + 1)^2 \right]}{\{[n(n^2 - 1) - 12t'][n(n^2 - 1) - 12u']\}^{\frac{1}{2}}} \tag{3.17}$$

or

$$R = \frac{n(n^2 - 1) - 6 \sum_{i=1}^{n} D_i^2 - 6(t' + u')}{\{[n(n^2 - 1) - 12t'][n(n^2 - 1) - 12u']\}^{\frac{1}{2}}}$$

analogous to (3.5) and (3.7), respectively, since here

$$\sum_{i=1}^{n} D_i^2 = \frac{n(n^2 - 1)}{6} - t' - u' - 2 \sum_{i=1}^{n} (R_i - \bar{R})(S_i - \bar{S})$$

Assuming this to be our definition of the sample coefficient of rank correlation in the presence of ties, its probability distribution under the null hypothesis of independence is clearly not the same as the null distribution discussed before for n distinct ranks. For small n, it is possible again to obtain the exact null distribution conditional upon a given set of ties by enumeration. This of course is very tedious. The asymptotic distribution of our R as modified for ties is also normal since it is still a linear combination of the S_i random variables. Since the total sum of ranks is unchanged when tied ranks are assigned by the midrank method, $E(S_i)$ is unchanged and $E(R \mid H_0)$ is obviously still zero. The fact that the variance of modified R is also unchanged in the presence of ties is not so obvious. The marginal and joint distributions of the ranks of the Y sample in the presence of ties can still be written in the forms (3.8) and (3.9) except that the domain is now n numbers, not all distinct, which we can write as $s_1', s_2', \ldots , s_n'$. Then using (3.16)

$$\text{var}(S_i) = \sum_{i=1}^{n} \frac{(s_i' - \bar{s})^2}{n} = \frac{n(n^2 - 1) - 12u'}{12n}$$

For the covariance, proceeding as in the steps leading to (3.10)

$$\mathrm{cov}(S_i, S_j) = \frac{1}{n(n-1)} \sum_{\substack{i=1 \\ i \neq j}}^{n} \sum_{j=1}^{n} s_i' s_j' - \bar{s}^2$$

$$= - \frac{\displaystyle\sum_{i=1}^{n} (s_i' - \bar{s})^2}{n(n-1)}$$

$$= - \frac{n(n^2 - 1) - 12u'}{12n(n-1)}$$

Similar results hold for the X ranks. Now using R in the form of (3.17), we have

$$\mathrm{var}(R \mid H_0) = \frac{144[n\,\mathrm{var}(R_i)\,\mathrm{var}(S_i) + n(n-1)\,\mathrm{cov}(R_i, R_j)\,\mathrm{cov}(S_i, S_j)]}{[n(n^2-1) - 12t'][n(n^2-1) - 12u']}$$

and substitution of the appropriate variances and covariances gives as before

$$\mathrm{var}(R \mid H_0) = \frac{1}{n-1}$$

Thus for large samples with ties, a modified $R\sqrt{n-1}$ can still be treated as a standard normal variable for testing a hypothesis of independence. However, unless the ties are extremely extensive, they will have little effect on the value of R. In practice, the common expression given in (3.7) is often used without corrections for ties. It should be noted that the effect of the correction factors is to decrease the value of R. This of course means that a negative R is closer to -1, not to zero.

USE OF SPEARMAN'S R TO TEST AGAINST TREND

As with Kendall's T, R can be considered a measure of trend in a single sequence of time-ordered observations and used to test a hypothesis of no trend.

12.4 THE RELATIONS BETWEEN R AND T; $E(R)$, τ, AND ρ

In Sec. 1 we defined the parameters τ and ρ as two different measures of association in a bivariate population, one in terms of concordances and the other as a function of covariance, but noted that concordance and covariance measure relationship in the same spirit at least. The sample estimate of τ was found to have exactly the same numerical value and theoretical properties whether calculated in terms of actual variate values or ranks, since the parameter τ and its estimate are both invariant under all order-preserving transformations. However, this is not true for the parameter ρ or for a sample estimate calculated from (3.1) with variate values. The Pearson product-moment correlation coefficient is invariant under linear transformations only, and ranks usually cannot be generated using only linear transformations.

The coefficient of rank correlation is certainly a measure of association between ranks. It has a certain intuitive appeal as an estimate of ρ, but it is not a direct sample analog of this parameter. Nor can it be considered a direct sample analog of a "population coefficient of rank correlation" if the marginal distributions of our random variables are continuous, since theoretically continuous random variables cannot be ranked. If an infinite number of values can be assumed by a random variable, the values cannot be enumerated and therefore cannot be ordered. However, we still would like some conception, however nebulous, of a population parameter which is the analog of the Spearman coefficient of rank correlation in a random sample of pairs from a continuous bivariate population. Since probabilities of order properties are population parameters and these probabilities are the same for either ranks or variate values, if R can be defined in terms of sample proportions of types of concordance, as T was, we shall be able to define a population parameter other than ρ for which the coefficient of rank correlation is an unbiased estimate.

For this purpose, we shall first investigate the relationship between R and T for samples with no ties from any continuous bivariate population. In (2.20), T was written in a form resembling R as

$$T = \sum_{i=1}^{n} \sum_{j=1}^{n} \frac{U_{ij}V_{ij}}{n(n-1)} \tag{4.1}$$

where

$$U_{ij} = \text{sgn}(X_j - X_i) \qquad \text{and} \qquad V_{ij} = \text{sgn}(Y_j - Y_i) \tag{4.2}$$

To complete the similarity, we must determine the general relation between U_{ij} and R_i, V_{ij}, and S_i. In (5.1.1), a functional definition of R_i was given as

$$R_i = 1 + \sum_{\substack{j=1 \\ j \neq i}}^{n} S(X_i - X_j) \tag{4.3}$$

where

$$S(u) = \begin{cases} 0 & \text{if } u < 0 \\ 1 & \text{if } u \geq 0 \end{cases}$$

In general, then, the relation is

$$\text{sgn}(X_j - X_i) = 1 - 2S(X_i - X_j) \qquad \text{for all } 1 \le i \ne j \le n$$
$$\text{sgn}(X_i - X_i) = 0 \tag{4.4}$$

Substituting this form back in (4.3), we have

$$R_i = \frac{n+1}{2} - \frac{1}{2} \sum_{j=1}^{n} \text{sgn}(X_j - X_i)$$

or

$$R_i - \bar{R} = - \sum_{j=1}^{n} \frac{U_{ij}}{2}$$

Using R in the form (3.4), by substitution we have

$$n(n^2 - 1)R = 12 \sum_{i=1}^{n} (R_i - \bar{R})(S_i - \bar{S})$$

$$= 3 \sum_{i=1}^{n} \left(\sum_{j=1}^{n} U_{ij} \sum_{k=1}^{n} V_{ik} \right)$$

$$= 3 \sum_{i=1}^{n} \sum_{j=1}^{n} U_{ij}V_{ij} + 3 \sum_{i=1}^{n} \sum_{j=1}^{n} \sum_{\substack{k=1 \\ k \ne j}}^{n} U_{ij}V_{ik}$$

or from (4.1)

$$R = \frac{3}{n+1} T + \frac{6}{n(n^2-1)} \sum_{\substack{i=1 \\ i<j}}^{n} \sum_{j=1}^{n} \sum_{\substack{k=1 \\ k \ne j}}^{n} U_{ij}V_{ik} \tag{4.5}$$

Before, we defined two pairs (X_i, Y_i) and (X_j, Y_j) as being concordant if $U_{ij}V_{ij} > 0$, with π_c denoting the probability of concordance and p_c the corresponding sample estimate, the number of concordant sample pairs divided by $n(n-1)$, the number of distinguishable pairs, and found $T = 2p_c - 1$. To complete a definition of R in terms of concordances, because of the last term in (4.5), we must now define another type of concordance, this time involving three pairs. We shall say that the three pairs (X_i, Y_i), (X_j, Y_j), and (X_k, Y_k) exhibit a concordance of the second order if

$$X_i < X_j \qquad \text{whenever } Y_i < Y_k$$

or

$$X_i > X_j \qquad \text{whenever } Y_i > Y_k$$

or, equivalently, if

$$(X_j - X_i)(Y_k - Y_i) = U_{ij}V_{ik} > 0$$

The probability of a second-order concordance is

$$\pi_{c_2} = P[(X_j - X_i)(Y_k - Y_i) > 0]$$

and the corresponding sample estimate p_{c_2} is the number of sets of three pairs with the product $U_{ij}V_{ik} > 0$ for $i < j, k \neq j$, divided by $\binom{n}{2}(n-2)$, the number of distinguishable sets of three pairs. The triple sum in (4.5) is the totality of all these products, whether positive or negative, and therefore equals

$$\binom{n}{2}(n-2)[p_{c_2} - (1 - p_{c_2})] = \frac{n(n-1)(n-2)(2p_{c_2} - 1)}{2}$$

In terms of sample concordances, then, (4.5) can be written as

$$R = \frac{3}{n+1}(2p_c - 1) + \frac{3(n-2)}{n+1}(2p_{c_2} - 1) \tag{4.6}$$

and the population parameter for which R is an unbiased estimator is

$$E(R) = \frac{3[\tau + (n-2)(2\pi_{c_2} - 1)]}{n+1} \tag{4.7}$$

We shall now express π_{c_2} for any continuous bivariate population $F_{X,Y}(x,y)$ in a form analogous to (2.17) for π_c:

$$\begin{aligned}
\pi_{c_2} &= P[(X_i < X_j) \cap (Y_i < Y_k)] + P[(X_i > X_j) \cap (Y_i > Y_k)] \\
&= \int_{-\infty}^{\infty} \int_{-\infty}^{\infty} \{P[(X_i < x_j) \cap (Y_i < y_k)] \\
&\qquad\qquad + P[(X_i > x_j) \cap (Y_i > y_k)]\} f_{X_j,Y_k}(x_j,y_k)\, dx_j\, dy_k \\
&= \int_{-\infty}^{\infty} \int_{-\infty}^{\infty} [F_{X,Y}(x,y) + 1 - F_X(x) - F_Y(y) \\
&\qquad\qquad\qquad\qquad + F_{X,Y}(x,y)] f_X(x) f_Y(y)\, dx\, dy \\
&= 1 + 2 \int_{-\infty}^{\infty} \int_{-\infty}^{\infty} F_{X,Y}(x,y) f_X(x) f_Y(y)\, dx\, dy \\
&\qquad\qquad\qquad\qquad\qquad - 2 \int_{-\infty}^{\infty} F_X(x) f_X(x)\, dx \\
&= 2 \int_{-\infty}^{\infty} \int_{-\infty}^{\infty} F_{X,Y}(x,y) f_X(x) f_Y(y)\, dx\, dy \tag{4.8}
\end{aligned}$$

A similar development yields another equivalent form

$$\pi_{c_2} = 2 \int_{-\infty}^{\infty} \int_{-\infty}^{\infty} F_X(x) F_Y(y) f_{X,Y}(x,y)\, dx\, dy \tag{4.9}$$

If X and Y are independent, of course, a comparison of these expressions with (2.17) shows that $\pi_{c_2} = \pi_c = \frac{1}{2}$. Unlike π_c, however, which ranges between 0 and 1, π_{c_2} ranges only between $\frac{1}{3}$ and $\frac{2}{3}$, with the extreme values obtained for perfect indirect and direct linear relationships, respectively. This result can be shown easily. For the upper limit, since for all x, y

$$2F_X(x)F_Y(y) \leq F_X{}^2(x) + F_Y{}^2(y)$$

we have from (4.9)

$$\pi_{c_2} \leq 2 \int_{-\infty}^{\infty} \int_{-\infty}^{\infty} F_X{}^2(x)f_{X,Y}(x,y)\, dx\, dy = \tfrac{2}{3}$$

Similarly, for all x, y

$$2F_X(x)F_Y(y) = [F_X(x) + F_Y(y)]^2 - F_X{}^2(x) - F_Y{}^2(y)$$

so that from (4.9)

$$\pi_{c_2} = \int_{-\infty}^{\infty} \int_{-\infty}^{\infty} [F_X(x) + F_Y(y)]^2 f_{X,Y}(x,y)\, dx\, dy - \tfrac{2}{3}$$

$$\geq \left[\int_{-\infty}^{\infty} \int_{-\infty}^{\infty} [F_X(x) + F_Y(y)]f_{X,Y}(x,y)\, dx\, dy \right]^2 - \tfrac{2}{3} = \tfrac{1}{3}$$

Now if X and Y have a perfect direct linear relationship, we can assume without loss of generality that $X = Y$, so that

$$F_{X,Y}(x,y) = \begin{cases} F_X(x) & \text{if } x \leq y \\ F_X(y) & \text{if } x > y \end{cases}$$

Then from (4.8)

$$\pi_{c_2} = 2(2) \int_{-\infty}^{\infty} \int_{-\infty}^{y} F_X(x)f_X(x)f_X(y)\, dx\, dy = \tfrac{2}{3}$$

For a perfect indirect relationship, we assume $X = -Y$, so that

$$F_{X,Y}(x,y) = \begin{cases} F_X(x) - F_X(-y) & \text{if } x \geq -y \\ 0 & \text{if } x < -y \end{cases}$$

and

$$\pi_{c_2} = 2 \int_{-\infty}^{\infty} \int_{-y}^{\infty} [F_X(x) - F_X(-y)]f_X(x)f_X(-y)\, dx\, dy$$

$$= \int_{-\infty}^{\infty} \{1 - F_X{}^2(-y) - 2[1 - F_X(-y)]F_X(-y)\}f_X(-y)\, dy$$

$$= \int_{-\infty}^{\infty} [1 - F_X(-y)]^2 f_X(-y)\, dy = \tfrac{1}{3}$$

Substitution of these extreme values in (4.7) shows that for any continuous population, ρ, τ, and $E(R)$ all have the same value for the following cases:

X, Y relation	$\rho = \tau = E(R)$
Indirect linear dependence	-1
Independence	0
Direct linear dependence	1

Although strictly speaking we cannot talk about a parameter for a bivariate distribution which is a coefficient of rank correlation, it seems natural to define the pseudo rank-correlation parameter, say ρ_2, as that constant for which R is an unbiased estimator in large samples. Then from (4.7), we have the definition

$$\rho_2 = \lim_{n \to \infty} E(R) = 3(2\pi_{c_2} - 1) \tag{4.10}$$

and for a sample of size n, the relation between $E(R)$, ρ_2, and τ is

$$E(R) = \frac{3\tau + (n - 2)\rho_2}{n + 1} \tag{4.11}$$

The relation between ρ_2 (for ranks) and ρ (for variate values) depends on the relation between π_{c_2} and covariance. From (4.9), we see that

$$\begin{aligned}
\pi_{c_2} &= 2E[F_X(X)F_Y(Y)] \\
&= 2 \, \text{cov}[F_X(X), F_Y(Y)] + 2E[F_X(X)]E[F_Y(Y)] \\
&= 2 \, \text{cov}[F_X(X), F_Y(Y)] + \tfrac{1}{2}
\end{aligned}$$

Since

$$\text{var}[F_X(X)] = \text{var}[F_Y(Y)] = \tfrac{1}{12}$$

we have

$$6\pi_{c_2} = \rho[F_X(X), F_Y(Y)] + 3$$

and we see from (4.10) that

$$\rho_2 = \rho[F_X(X), F_Y(Y)]$$

Therefore ρ_2 is sometimes called the *grade correlation coefficient*, since the grade of a number x is usually defined as the cumulative probability $F_X(x)$.

12.5 ANOTHER MEASURE OF ASSOCIATION

Another nonparametric type of measure of association for paired samples which is related to the Pearson product-moment correlation coefficient has been investigated by Fieller, Hartley, Pearson, and others. This is the ordinary Pearson sample correlation coefficient of (3.1) calculated using expected normal scores in place of ranks or variate values. That is, if $\xi_i = E(U_{(i)})$, where $U_{(i)}$ is the ith order statistic in a sample of n from the standard normal population and S_i denotes the rank of the Y observation which is paired with the ith smallest X observation, the random sample of pairs of ranks

$$(1, s_1), (2, s_2), \ldots, (n, s_n)$$

is replaced by the derived sample of pairs

$$(\xi_1, \xi_{s_1}), (\xi_2, \xi_{s_2}), \ldots, (\xi_n, \xi_{s_n})$$

and the correlation coefficient for these pairs is

$$R_F = \frac{\sum\limits_{i=1}^{n} \xi_i \xi_{s_i}}{\sum\limits_{i=1}^{n} \xi_i^2}$$

This coefficient is discussed in Fieller, Hartley, and Pearson (1957) and Fieller and Pearson (1961). The authors show that the transformed random variable

$$Z_F = \tanh^{-1} R_F$$

is approximately normally distributed with moments

$$E(Z_F) = \tanh^{-1}\left[\rho\left(1 - \frac{0.6}{n + 8}\right)\right]$$

$$\text{var}(Z_F) = \frac{1}{n - 3}$$

where ρ is the correlation coefficient in the bivariate population from which the sample is drawn.

The authors also show that analogous transformations on R and T,

$$Z_R = \tanh^{-1} R$$
$$Z_T = \tanh^{-1} T$$

produce approximately normally distributed random variables, but in the nonnull case the approximation for Z_F is best.

PROBLEMS

12.1. A beauty contest has eight contestants. The two judges are each asked to rank the contestants in a preferential order of pulchritude. The results are shown in the table. Answer parts (a) and (b) using (i) the Kendall tau-coefficient procedures and

	Contestant							
Judge	A	B	C	D	E	F	G	H
1	2	1	3	5	4	8	7	6
2	1	2	4	5	7	6	8	3

(*ii*) the Spearman rank-correlation-coefficient procedures:

    (*a*) Calculate the measure of association.

    (*b*) Test the null hypothesis that the judges ranked the contestants independently (use tables).

    (*c*) Find a 95 percent confidence-interval estimate of τ.

12.2. Verify the result given in (4.9).

12.3. Two independent random samples of sizes m and n contain no ties. A set of $m + n$ paired observations can be derived from these data by arranging the combined samples in ascending order of magnitude and (*a*) assigning ranks, (*b*) assigning sample indicators. Show that Kendall's tau, calculated for these pairs without a correction for ties, is linearly related to the Mann-Whitney U statistic for these data, and find the relation if the sample indicators are (*i*) sample numbers 1 and 2, (*ii*) 1 for the first sample and 0 for the second sample as in the Z vector of Chap. 8.

13
Measures of Association in Multiple Classifications

13.1 INTRODUCTION

Suppose we have a set of data presented in the form of a complete two-way layout of I rows and J columns, with one entry in each of the IJ cells. In the sampling situation of Chap. 11, if the independent samples drawn from each of I univariate populations were all of the same size J, we would have a complete layout of IJ cells. However, this would be called a one-way layout since only one factor is involved, the populations. Under the null hypothesis of identical populations, the data can be considered a single random sample of size IJ from the common population. The parallel to this problem in classical statistics is the one-way analysis of variance. In this chapter we shall study some nonparametric analogs of the two-way analysis-of-variance problem, all parallel in the sense that the data are presented in the form of a two-way layout which cannot be considered a single random sample because of certain relationships among elements.

Let us first review the techniques of the analysis-of-variance approach to testing the null hypothesis that the column effects are all the same. The mixed-effects model† is usually written

$$X_{ij} = \mu + \beta_i + T_j + E_{ij} \qquad \text{for } i = 1, 2, \ldots, I$$
$$j = 1, 2, \ldots, J$$

where $\sum_{i=1}^{I} \beta_i = 0$. The β_i and T_j are known as the fixed-row and random-column effects, respectively. In the normal-theory model, the errors E_{ij} are independent, normally distributed random variables with mean zero and variance σ_E^2, and are independent of the T_j, which have the J-variate normal distribution with means zero, variances σ_T^2, and covariances $\rho\sigma_T^2$. The test statistic for the null hypothesis of equal column effects or, equivalently,

$$H_0: \sigma_T^2 = 0$$

is the ratio

$$\frac{(I - 1)I \sum_{j=1}^{J} (\bar{x}_j - \bar{x})^2}{\sum_{i=1}^{I} \sum_{j=1}^{J} (x_{ij} - \bar{x}_i - \bar{x}_j + \bar{x})^2}$$

where

$$\bar{x}_i = \sum_{j=1}^{J} \frac{x_{ij}}{J} \qquad \bar{x}_j = \sum_{i=1}^{I} \frac{x_{ij}}{I} \qquad \bar{x} = \sum_{i=1}^{I} \sum_{j=1}^{J} \frac{x_{ij}}{IJ}$$

If all the assumptions of the model are met, this test statistic has the F distribution with $J - 1$ and $(I - 1)(J - 1)$ degrees of freedom.

The first two parallels of this design which we shall consider are the k-related or k-matched sample problems. The matching can arise in two different ways, but both are somewhat analogous to the randomized-block design of a two-way layout. In this design, IJ experimental units are grouped into I blocks, each containing J units. A set of J treatments is assigned at random to the units within each block in such a way that all $J!$ assignments are equally likely, and the assignments in different blocks are independent. The scheme of grouping into blocks is important, since the purpose of such a design is to minimize the differences between units in the same block. If the design is successful, an estimate of experimental error can be obtained which is not inflated by differences between blocks. This model is often appropriate in agricultural field experimentation since the effects of a possible fertility gradient can be reduced. Dividing the field into I blocks, the plots within each block can be kept in close proximity. Any differences between plots within the same block can be attributed to differences between treatments and the block effect eliminated from the estimate of experimental error.

† The mixed-effects model is given here because it provides a closer analogy to the two-way layouts presented in this chapter, where the columns will be dependent.

The first related-sample problem arises where IJ subjects are grouped into I blocks each containing J-matched subjects, and within each block J treatments are assigned randomly to the matched subjects. The effects of the treatments are observed, and we let X_{ij} denote the observation in block i of treatment number j, $i = 1, 2, \ldots, I$, $j = 1, 2, \ldots, J$. Since the observations in different blocks are independent, the collection of entries in column number j are independent. In order to determine whether the treatment (column) effects are all the same, the analysis-of-variance test is appropriate if the requisite assumptions are justified. If the observations in each row $X_{i1}, X_{i2}, \ldots, X_{iJ}$ are replaced by their ranking within that row, a nonparametric test involving the column sums of this $I \times J$ table, called *Friedman's two-way analysis of variance by ranks*, can be used to test the same hypothesis. This is a k-related sample problem when $J = k$.

The second related-sample problem arises by considering a single group of J subjects, each of which is observed under I different conditions. The matching here is by condition rather than subject, and the observation X_{ij} denotes the effect of condition i on subject number j, $i = 1, 2, \ldots, I$, $j = 1, 2, \ldots, J$. We have here a random sample of size J from an I-variate population. Under the null hypothesis that the I variates are independent, the expected sum of the I observations on subject number j is the same for all $j = 1, 2, \ldots, J$. In order to determine whether the column effects are all the same, the analysis-of-variance test may be appropriate. Testing the independence of the I variates involves a comparison of J column totals, so that in a sense the roles of treatments and blocks have been reversed in terms of which factor is of interest. This is a k-related sample problem when $I = k$. If the observations in each row are ranked as before, Friedman's two-way analysis of variance will provide a nonparametric test of independence of the k variates. Thus, in order to effect consistency of results as opposed to consistency of sampling situations, the presentation here in both cases will be for a table containing k rows and n columns, where each row is a set of positive integer ranks.

In this second related-sample problem, particularly if the null hypothesis of the independence of the k variates is rejected, a measure of the association between the k variates would be desirable. In fact, this sampling situation is the direct extension of the paired-sample problem of Chap. 12 to the k-related sample case. A measure of the overall agreement between the k sets of rankings, called *Kendall's coefficient of concordance*, can be determined. This statistic can also be used to test the null hypothesis of independence, but the test is equivalent to Friedman's test for n treatments and k blocks. An analogous measure of concordance will be found for k sets of incomplete rankings, which relate to the balanced incomplete-blocks design.

Two additional subjects will be covered in this chapter, not because of their similarity to analysis-of-variance techniques but because they also relate to testing hypotheses concerning independence or measuring association when data are presented in the form of a two-way layout. In contingency tables, where the test procedure can also be extended to higher-way layouts, the analysis is for count data, since the table entries are the numbers of subjects in each of the multiple classifications. A test of independence between classifications may be performed, and a measure of association, called the *contingency coefficient*, obtained. The other topic to be treated briefly is a nonparametric approach to finding a measure of partial correlation or correlation between two variables when a third is held constant.

13.2 FRIEDMAN'S TWO-WAY ANALYSIS OF VARIANCE BY RANKS IN A $k \times n$ TABLE

As suggested in Sec. 1, in the first related-sample problem we have data presented in the form of a two-way layout of k rows and n columns. The rows indicate block, subject, or sample numbers, and the columns are treatment numbers. The observations in different rows are independent, but the columns are not because of some unit of association. In order to avoid making the assumptions requisite for the usual analysis-of-variance test that the n treatments are the same, Friedman (1937, 1940) suggested replacing each treatment observation within the ith block by a number from the set $\{1, 2, \ldots, n\}$ which represents that treatment's magnitude relative to the other observations in the same block. We denote the ranked observations by R_{ij}, $i = 1, 2, \ldots, k, j = 1, 2, \ldots, n$, so that R_{ij} is the rank of treatment number j when observed in block number i. Then $R_{i1}, R_{i2}, \ldots, R_{in}$ is a permutation of the first n integers, and $R_{1j}, R_{2j}, \ldots, R_{kj}$ is the set of ranks given to treatment number j in all blocks. We represent the data in tabular form as follows:

$$
\begin{array}{c}
\\
\\
Blocks
\end{array}
\begin{array}{c}
Treatments \\
\begin{array}{cccc}
1 & 2 & \cdots & n
\end{array}
\end{array}
$$

		Treatments 1	2	$\cdots$	n	Row totals
	1	R_{11}	R_{12}	$\cdots$	R_{1n}	$\dfrac{n(n+1)}{2}$
	2	R_{21}	R_{22}	$\cdots$	R_{2n}	$\dfrac{n(n+1)}{2}$
Blocks	.					.
	.					.
	k	R_{k1}	R_{k2}	$\cdots$	R_{kn}	$\dfrac{n(n+1)}{2}$
Column totals		R_1	R_2	$\cdots$	R_n	$\dfrac{kn(n+1)}{2}$

$$(2.1)$$

The row totals are of course constant, but the column totals are affected by differences between treatments. If the treatments are all the same, each expected column total is the same and equals the average column total $k(n + 1)/2$. The sum of deviations of observed column totals around this mean is zero, but the sum of squares of these deviations will be indicative of the differences in treatment effects. Therefore we shall consider the sampling distribution of the random variable

$$S = \sum_{j=1}^{n} \left[R_j - \frac{k(n + 1)}{2} \right]^2 = \sum_{j=1}^{n} \left[\sum_{i=1}^{k} \left(R_{ij} - \frac{n + 1}{2} \right) \right]^2 \qquad (2.2)$$

under the hypothesis of no difference between the n treatments. For this null case, in the ith block the ranks are assigned completely at random, and each row in the two-way layout constitutes a random permutation of the first n integers if there are no ties. There are then a total of $(n!)^k$ distinguishable sets of entries in the $k \times n$ table, and each is equally likely. These possibilities can be enumerated and the value of S calculated for each. The probability distribution of S then is

$$f_S(s) = \frac{u_s}{(n!)^k}$$

where u_s is the number of those assignments which yield s as the sum of squares of column total deviations. A systematic method of generating the values of u_s for n, k from the values of u_s for $n, k - 1$ can be employed [see Kendall (1962), pp. 113–114]. Tables of the distribution of S for $n = 3, 4, 5$, and k small are available in Kendall [(1962), pp. 184–187]. More extensive tables for the distribution of F, a linear function of S to be defined later in (2.8), are given in Owen (1962) for $k = 3, n \leq 15$ and $k = 4, n \leq 8$. However, the calculations are considerable even using the systematic approach. Therefore, outside the range of existing tables, an approximation to the null distribution is generally used for tests of significance.

Using the symbol μ to denote $(n + 1)/2$, (2.2) can be written as

$$\begin{aligned}
S &= \sum_{j=1}^{n} \sum_{i=1}^{k} (R_{ij} - \mu)^2 + 2 \sum_{j=1}^{n} \sum_{1 \leq i < p \leq k} (R_{ij} - \mu)(R_{pj} - \mu) \\
&= k \sum_{j=1}^{n} (j - \mu)^2 + 2U \\
&= \frac{kn(n^2 - 1)}{12} + 2U
\end{aligned} \qquad (2.3)$$

The moments of S then are determined by the moments of U, which can be found using the following relations from (3.2), (3.3), and (3.10) of Chap. 12:

$$E(R_{ij}) = \frac{n+1}{2} \qquad \text{var}(R_{ij}) = \frac{n^2 - 1}{12}$$

$$\text{cov}(R_{ij}, R_{iq}) = -\frac{n+1}{12}$$

Furthermore, by the design assumptions, observations in different rows are independent, so that for all $i \neq p$ the expected value of a product of functions of R_{ij} and R_{pq} is the product of the expected values and $\text{cov}(R_{ij}, R_{pq}) = 0$. Then

$$E(U) = n \binom{k}{2} \text{cov}(R_{ij}, R_{pj}) = 0$$

so that $\text{var}(U) = E(U^2)$, where

$$U^2 = \sum_{j=1}^{n} \sum_{1 \leq i < p \leq k} (R_{ij} - \mu)^2 (R_{pj} - \mu)^2 + 2 \sum\sum_{1 \leq j < q \leq n} \sum\sum_{1 \leq i < p \leq k} \sum\sum_{1 \leq r < s \leq k}$$
$$(R_{ij} - \mu)(R_{pj} - \mu)(R_{rq} - \mu)(R_{sq} - \mu) \quad (2.4)$$

Since R_{ij} and R_{pq} are independent whenever $i \neq p$, we have

$$E(U^2) = \sum_{j=1}^{n} \sum_{1 \leq i < p \leq k} \text{var}(R_{ij})\, \text{var}(R_{pj})$$

$$+ 2 \sum\sum_{1 \leq j < q \leq n} \binom{k}{2} \text{cov}(R_{ij}, R_{iq})\, \text{cov}(R_{pj}, R_{pq}) \quad (2.5)$$

$$E(U^2) = n \binom{k}{2} \frac{(n^2 - 1)^2}{144} + 2 \binom{n}{2}\binom{k}{2} \frac{(n+1)^2}{144}$$

$$= n^2 \binom{k}{2} (n+1)^2 \frac{n-1}{144} \quad (2.6)$$

Using these results back in (2.3), we find

$$E(S) = \frac{kn(n^2 - 1)}{12} \qquad \text{var}(S) = \frac{n^2 k(k-1)(n-1)(n+1)^2}{72} \quad (2.7)$$

A linear function of the random variable defined as

$$F = \frac{12S}{kn(n+1)} = \frac{12 \sum_{j=1}^{n} R_j^2}{kn(n+1)} - 3k(n+1) \quad (2.8)$$

has moments $E(F) = n - 1$, $\text{var}(F) = 2(n-1)(k-1)/k \approx 2(n-1)$, which are the first two moments of a chi-square distribution with $n-1$

degrees of freedom.   The higher moments of F are also closely approximated by corresponding higher moments of the chi square for k large. For all practical purposes then, F can be treated as a chi-square variable with $n - 1$ degrees of freedom.   Numerical comparisons have shown this to be a good approximation as long as $n > 7$.   The rejection region for a test of equal treatment effects with significance level approximately α is

$$F \in R \qquad \text{for } f \geq \chi^2_{n-1,\alpha}$$

From classical statistics, we are accustomed to thinking of an analysis-of-variance test statistic as the ratio of two estimated variances or mean squares of deviations.   The total sum of squares of deviations of all nk ranks around the average rank is

$$s_t = \sum_{i=1}^{k} \sum_{j=1}^{n} (r_{ij} - \bar{r})^2 = k \sum_{j=1}^{n} \left(j - \frac{n+1}{2} \right)^2 = kn \frac{n^2 - 1}{12}$$

and thus we could write Friedman's test statistic in (2.8) as

$$F = \frac{(n-1)S}{s_t}$$

Even though s_t is a constant, as in classical analysis-of-variance problems, it can be partitioned into a sum of squares of deviations between columns plus a residual sum of squares.   Denoting the grand mean and column means by

$$\bar{r} = \sum_{i=1}^{k} \sum_{j=1}^{n} \frac{r_{ij}}{nk} = \frac{n+1}{2} \qquad \bar{r}_j = \frac{r_j}{k} = \sum_{i=1}^{k} \frac{r_{ij}}{k}$$

we have

$$s_t = \sum_{i=1}^{k} \sum_{j=1}^{n} (r_{ij} - \bar{r})^2 = \sum_{i=1}^{k} \sum_{j=1}^{n} (r_{ij} - \bar{r}_j + \bar{r}_j - \bar{r})^2$$

$$= \sum_{i=1}^{k} \sum_{j=1}^{n} (r_{ij} - \bar{r}_j)^2 + k \sum_{j=1}^{n} (\bar{r}_j - \bar{r})^2 + 2 \sum_{j=1}^{n} (\bar{r}_j - \bar{r}) \sum_{i=1}^{k} (r_{ij} - \bar{r}_j)$$

$$= \sum_{i=1}^{k} \sum_{j=1}^{n} (r_{ij} - \bar{r}_j)^2 + \sum_{j=1}^{n} \frac{\{r_j - [k(n+1)/2]\}^2}{k}$$

or

$$s_t = \sum_{i=1}^{k} \sum_{j=1}^{n} (r_{ij} - \bar{r}_j)^2 + \frac{s}{k} = kn \frac{n^2 - 1}{12} \tag{2.9}$$

An analogy to the classical analysis-of-variance table is given in Table 2.1.

Table 2.1

Source of variation	Degrees of freedom	Sum of squares
Between columns	$n - 1$	s/k
Between rows†	$k - 1$	0
Residual	$(n - 1)(k - 1)$	$s_t - s/k$
Total	$nk - 1$	s_t

† There is no variation between rows here since the row sums are all equal.

The usual F statistic for equal column effects, with $n - 1$ and $(n - 1)$ $(k - 1)$ degrees of freedom, would be the ratio of the column and residual mean squares, or

$$\frac{(k - 1)s}{ks_t - s} \tag{2.10}$$

13.3 THE COEFFICIENT OF CONCORDANCE OF k SETS OF RANKINGS OF n OBJECTS

The second k-related sample problem mentioned in Sec. 1 involves k sets of rankings of n subjects, where we are interested both in testing the hypothesis that the k sets are independent and in finding a measure of the relationship between rankings. In the common parlance of this type of sample problem, the k conditions are called *observers*, each of whom is presented with the same set of n objects to be ranked. The measure of relationship then will describe the agreement or concordance between observers in their judgments on the n objects.

Since the situation here is an extension of the paired-sample problem of Chap. 12, one possibility for a measure of agreement is to select one of the measures for paired samples and apply it to each of the $\binom{k}{2}$ sets of pairs of rankings of n objects. However, if $\binom{k}{2}$ tests of the null hypothesis of independence are then made using the sampling distribution appropriate for the measure employed, the tests are not independent and the overall probability of a type I error is difficult to determine but necessarily increased. Such a method of hypothesis testing is then statistically undesirable. We need a single measure of overall association which will provide a single test statistic designed to detect overall dependence between samples with a specified significance level. If we could somehow combine measures obtained for each of the $\binom{k}{2}$ pairs, this would provide a single coefficient of overall association which can be used to test the null hypothesis of independence if its sampling distribution can be determined.

The coefficient of concordance is such an approach to the problem of relationship between k sets of rankings. It is a linear function of the average of the coefficients of rank correlation for all pairs of rankings, as will be shown later in this section. However, the rationale of the measure will be developed independently of the procedures of the last chapter so that the analogy to analysis-of-variance techniques will be more apparent.

For the purpose of this parallel, then, we visualize the data as presented in the form of a two-way layout of dimension $k \times n$ as in (2.1), with row and column labels now designating observers and objects instead of blocks and treatments. The table entries R_{ij} denote the rank given by the ith observer to the jth object. Then the ith row is a permutation of the numbers $1, 2, \ldots, n$, and the jth column is the collection of ranks given to object number j by all observers. The ranks in each column are then indicative of the agreement between observers, since if the jth object has the same magnitude relative to all other objects in the opinion of each of the k observers, all ranks in the jth column will be identical. If this is true for every column, the observers agree perfectly and the respective column totals $(R_1, R_2, \ldots, R_n)$ will be some permutation of the numbers

$$1k, 2k, 3k, \ldots, nk$$

Since the average column total is $k(n + 1)/2$, for perfect agreement between rankings the sum of squares of deviations of column totals from the average column total will be a constant

$$\sum_{j=1}^{n} \left[jk - \frac{k(n + 1)}{2} \right]^2 = k^2 \sum_{j=1}^{n} \left(j - \frac{n + 1}{2} \right)^2$$
$$= k^2 n \frac{(n^2 - 1)}{12} \tag{3.1}$$

The actual observed sum of squares of these deviations is

$$S = \sum_{j=1}^{n} \left[R_j - \frac{k(n + 1)}{2} \right]^2 \tag{3.2}$$

We found in (2.9) that

$$ks_t = \frac{k^2 n(n^2 - 1)}{12} = s + k \sum_{i=1}^{k} \sum_{j=1}^{n} (r_{ij} - \bar{r}_j)^2 \tag{3.3}$$

where s_t is the total sum of squares of deviations of all ranks around the average rank. In terms of this situation, however, we see from (3.1) that

ks_t is the sum of squares of column total deviations when there is perfect agreement. Therefore the value of s for any set of k rankings ranges between zero and $k^2n(n^2 - 1)/12$, with the maximum value attained when $r_j = jk$ for all j, that is, when there is perfect agreement, and the minimum value attained when $r_j = k(n + 1)/2$ for all j, that is, when each observer's rankings are assigned completely at random so that there is no agreement between observers. If the observers are called samples, no agreement between observers is equivalent to independence of the k samples.

The ratio of S to its maximum value

$$W = \frac{S}{ks_t} = \frac{12S}{k^2n(n^2 - 1)} \tag{3.4}$$

therefore provides a measure of agreement between observers, or concordance between sample rankings, or dependence of the samples. This measure is called Kendall's coefficient of concordance. It ranges between 0 and 1, with 1 designating perfect concordance, and 0 indicating no agreement or independence of samples. As W increases, the set of ranks given to each object must become more similar because in the error term of (3.3), $\sum_{i=1}^{k} (r_{ij} - \bar{r}_j)^2$ becomes smaller for all j, and thus there is greater agreement between observers. In order to have the interpretation of this k-sample coefficient be consistent with a two-sample measure of association, one might think some measure which ranges from -1 to $+1$ with -1 designating perfect discordance would be preferable. However, for more than two samples, there is no such thing as perfect disagreement between rankings, and thus concordance and discordance are not symmetrical opposites. Therefore the range 0 to 1 is really more appropriate for a k-sample measure of association.

RELATIONSHIP BETWEEN W AND RANK CORRELATION

We shall now show that the statistic W is related to the average of the $\binom{k}{2}$ coefficients of rank correlation which can be calculated for the $\binom{k}{2}$ pairings of sample rankings. The average value is

$$r_{\text{av}} = \sum\sum_{1 \leq i < m \leq k} r_{i,m} \Big/ \binom{k}{2} = \sum_{i=1}^{k} \sum_{\substack{m=1 \\ i \neq m}}^{k} \frac{r_{i,m}}{k(k-1)} \tag{3.5}$$

where

$$r_{i,m} = \frac{12}{n(n^2 - 1)} \sum_{j=1}^{n} \left(r_{ij} - \frac{n+1}{2} \right) \left(r_{mj} - \frac{n+1}{2} \right) \qquad \text{for all } i \neq m$$

Denoting the average rank $(n + 1)/2$ by μ, we have

$$
r_{\text{av}} = 12 \sum_{j=1}^{n} \sum_{\substack{i=1 \\ i \neq m}}^{k} \sum_{m=1}^{k} \frac{(r_{ij} - \mu)(r_{mj} - \mu)}{kn(k-1)(n^2 - 1)}
$$

$$
= 12 \sum_{j=1}^{n} \frac{\left[\sum_{i=1}^{k} (r_{ij} - \mu) \right]^2 - \sum_{i=1}^{k} (r_{ij} - \mu)^2}{kn(k-1)(n^2 - 1)}
$$

$$
= \frac{\sum_{j=1}^{n} (r_j - k\mu)^2 - s_t}{(k-1)s_t}
$$

$$
= \frac{s - s_t}{(k-1)s_t} = \frac{kw - 1}{k - 1} \tag{3.6}
$$

or

$$
w = r_{\text{av}} + \frac{1 - r_{\text{av}}}{k} = \frac{r_{\text{av}}(k-1) + 1}{k} \tag{3.7}
$$

From this relation, we see that $w = 1$ when $r_{\text{av}} = 1$, which can only occur when $r_{i,m}$ equals 1 for all sets (i,m) of two samples, since always $r_{i,m} \leq 1$. It is impossible to have $r_{\text{av}} = -1$, since $r_{i,m} = -1$ cannot occur for all sets (i,m) simultaneously. Since we have already shown that the minimum value of w is zero, the smallest possible value of r_{av} is $-1/(k-1)$.

TESTS OF SIGNIFICANCE BASED ON W

Suppose we consider each column in our $k \times n$ table as the ranks of observations from a k-variate population. With n columns, we can say that $(R_{1j}, R_{2j}, \ldots, R_{kj}), j = 1, 2, \ldots, n$, constitute ranks of a random sample of size n from a k-variate population. We wish to test the null hypothesis that the variates are independent. The coefficient of concordance W is an overall measure of the association between the ranks of the k variates or the k sets of rankings of n objects, which in turn estimates some measure of the relationship between the k variates in the population. If the variates are independent, there is no association and W is zero, and for complete dependence there is perfect agreement and W equals 1. Therefore the statistic W may be used to test the null hypothesis that the variates are independent. The appropriate rejection region is large values of W.

In the null case, the ranks assigned to the n observations are completely random for each of the k variates, and the $(n!)^k$ assignments are all equally likely. The random sampling distribution of S (or W) then is exactly the same as in Sec. 2. The tables in Kendall (1962) or Owen

(1962) can be used for small samples, and for k large the distribution of

$$\frac{12S}{kn(n+1)} = k(n-1)W$$

may be approximated by the chi-square distribution with $n-1$ degrees of freedom.

Other approximations are also occasionally employed for tests of significance. Although the mean and variance of W are easily found using the moments already obtained for S in (2.7), it will be more instructive to determine the null moments of W directly by using its relationship with R_{av} given in (3.7). From (12.3.13), for $R_{i,m}$ the rank-correlation coefficient of any pairing of independent sets of ranks,

$$E(R_{i,m}) = 0 \qquad \text{var}(R_{i,m}) = \frac{1}{n-1} \qquad \text{for all } 1 \le i < m \le k$$

For any two independent sets of pairings of independent ranks, say (i,m) and (p,j), where $1 \le i < m \le k$, $1 \le p < j \le k$,

$$\text{cov}(R_{i,m}, R_{p,j}) = 0 \qquad \text{unless } i = p \text{ and } m = j$$

Therefore, from the definition of R_{av} in (3.5),

$$E(R_{av}) = 0$$
$$\binom{k}{2}^2 \text{var}(R_{av}) = \sum_{1 \le i < m \le k} \sum \text{var}(R_{i,m}) + \sum_{\substack{1 \le i < m \le k \\ }} \sum \sum_{\substack{1 \le p < j \le k \\ i \ne p \text{ or } m \ne j}} \sum \text{cov}(R_{i,m}, R_{p,j})$$
$$= \frac{\binom{k}{2}}{n-1}$$

and

$$\text{var}(R_{av}) = \frac{2}{k(k-1)(n-1)}$$

Now using (3.7),

$$E(W) = \frac{1}{k} \qquad \text{var}(W) = \frac{2(k-1)}{k^3(n-1)} \tag{3.8}$$

These are the exact first two moments of the beta distribution with parameters

$$\frac{k(n-1)-2}{2k} \qquad \text{and} \qquad \frac{(k-1)[k(n-1)-2]}{2k}$$

An investigation of the higher moments of W shows that they are approximately equal to the corresponding higher moments of the beta distribution unless $k(n - 1)$ is small. Thus an approximation to the distribution of W is the beta distribution, for which tables are available. However, if any random variable, say X, has the beta distribution with parameters α and β, the transformation $Y = \beta X/[\alpha(1 - X)]$ produces a random variable with Snedecor's F distribution with parameters $\nu_1 = 2\alpha$ and $\nu_2 = 2\beta$, and the transformed variable $Z = (\log Y)/2$ has Fisher's z distribution with the same parameters. Applying these transformations here, we find the approximate distributions

1. $(k - 1)W/(1 - W)$ is Snedecor's F
2. $\log[(k - 1)W/(1 - W)]/2$ is Fisher's z $\Bigg\}$ $\nu_1 = n - 1 - \dfrac{2}{k}, \nu_2 = (k - 1)\nu_1$

in addition to our previous approximation

3. $k(n - 1) W$ is chi square with $n - 1$ degrees of freedom.

Approximation 1 is not surprising, as we found in (2.10) that the random variable

$$\frac{(k - 1)S}{ks_t - S} = \frac{(k - 1)W}{1 - W}$$

was the ratio of mean squares analogous to the analysis-of-variance test statistic with $n - 1$ and $(n - 1)(k - 1)$ degrees of freedom.

ESTIMATION OF THE TRUE PREFERENTIAL ORDER OF OBJECTS

Assume that the coefficient of concordance has been found for some k sets of rankings of n objects and the null hypothesis of no agreement is rejected. The magnitude of this relative measure of agreement implies that not all these particular observers ranked the objects strictly randomly and independently. This might be interpreted to mean that there is some agreement among these observers and that perhaps some unique ordering of these objects exists in their estimation. Suppose we call this the true preferential ordering. If there were perfect agreement, we would know which object is least preferred, which is next, etc., by the agreed-upon ranks. Object number j would have the position m in the true preferential ordering if the sum of ranks given object j is km. In our $k \times n$ table of ranks, the ordering corresponds to the ranks of the column sums. In a case of less-than-perfect agreement, then, the true preferential ordering might be estimated by assigning ranks to the objects in accordance with the magnitudes of the column sums.

This estimate is best in the sense that if the coefficient of rank correlation is calculated between this estimated ranking and each of the k-

observed rankings, the average of these k correlation coefficients is a maximum. To show this, we let $r_{e1}, r_{e2}, \ldots r_{en}$ be any estimate of the true preferential ordering, where r_{ej} is the estimated rank of object number j. If $r_{e,i}$ denotes the rank-correlation coefficient between this estimated ranking and the ranking assigned by the ith observer, the average rank correlation is

$$\sum_{i=1}^{k} \frac{r_{e,i}}{k} = 12 \sum_{i=1}^{k} \sum_{j=1}^{n} \frac{(r_{ej} - \mu)(r_{ij} - \mu)}{kn(n^2 - 1)}$$

$$= 12 \sum_{j=1}^{n} \frac{(r_{ej} - \mu)(r_j - k\mu)}{kn(n^2 - 1)}$$

$$= \frac{12 \sum\limits_{j=1}^{n} r_{ej} r_j}{kn(n^2 - 1)} - \frac{3(n + 1)}{(n - 1)}$$

where $\mu = (n + 1)/2$ and r_j is the jth column sum as before. This average then is a maximum when $\sum\limits_{j=1}^{n} r_{ej} r_j$ is a maximum, i.e., when the r_{ej} are in the same relative order of magnitude as the r_j.

This estimate is also best in a least-squares sense. If r_{ej} is any estimated rank of object j and the estimate is the true preferential rank, the jth column sum would equal kr_{ej}. A measure of the error in this estimate then is the sum of squares of deviations

$$\sum_{j=1}^{n} (r_j - kr_{ej})^2 = \sum_{j=1}^{n} r_j^2 + k^2 \sum_{j=1}^{n} r_{ej}^2 - 2k \sum_{j=1}^{n} r_j r_{ej}$$

$$= \sum_{j=1}^{n} r_j^2 + k^2 \sum_{i=1}^{n} i^2 - 2k \sum_{j=1}^{n} r_j r_{ej}$$

$$= c - 2k \sum_{j=1}^{n} r_j r_{ej}$$

where c is a constant. The error is thus minimized when $\sum\limits_{j=1}^{n} r_j r_{ej}$ is a maximum, and the r_{ej} should be chosen as before.

THE PROBLEM OF TIED OBSERVATIONS

Up to now we have assumed that each row of our $k \times n$ table is a permutation of the first n integers. If an observer cannot express any preference between two or more objects, or if the objects are actually indistinguishable, we may wish to allow him to assign equal ranks. If these numbers are the midranks of the positions each set of tied objects would occupy if a preference could be expressed, the average rank of any object and the average column sum are not changed. However, the sum of squares of deviations of any set of n ranks is reduced if there are ties. As we found in (12.3.16), for any $i = 1, 2, \ldots, k$,

$$\sum_{j=1}^{n} \left[r_{ij} - \frac{n+1}{2} \right]^2 = \frac{n(n^2 - 1) - \Sigma t(t^2 - 1)}{12}$$

The maximum value of s/k, as in (2.9), is then reduced to

$$s_t' = \sum_{i=1}^{k} \sum_{j=1}^{n} \left[r_{ij} - \frac{n+1}{2} \right]^2 = \frac{kn(n^2 - 1) - \Sigma\Sigma t(t^2 - 1)}{12}$$

where the sum is extended over all sets of t tied ranks in each of the k rows. The relative measure of agreement in the presence of ties then is $W' = S/ks_t'$. The significance of the corrected coefficient W' can be tested using any of the previously mentioned approximations for W.

13.4 THE COEFFICIENT OF CONCORDANCE OF k SETS OF INCOMPLETE RANKINGS

As an extension of the sampling situation of Sec. 3, suppose that we have n objects to be ranked and a fixed number of observers to rank them but each observer ranks only some subset of the n objects. This situation could arise for reasons of economy or practicality. In the case of human observers particularly, the ability to rank objects effectively and reliably may be a function of the number of comparative judgments to be made. For example, after 10 different brands of bourbon have been tasted, the discriminatory powers of the observers may legitimately be questioned.

We shall assume that the experimental design in this situation is such that the rankings are incomplete in the same symmetrical way as in the balanced incomplete-blocks design which is used effectively in agricultural field experiments. In terms of our situation, this means that:

1. Each observer will rank the same number m of objects for some $m < n$.
2. Every object will be ranked exactly the same total number k of times.
3. Each pair of objects will be presented together to some observer a total of exactly λ times, $\lambda \geq 1$, a constant for all pairs.

These specifications then ensure that all comparisons are made with the same frequency.

In order to visualize the design, imagine a two-way layout of p rows and n columns, where the entry δ_{ij} in the (i,j) cell equals 1 if object j is presented to observer i and 0 otherwise. The design specifications then can be written symbolically as

1. $\displaystyle\sum_{j=1}^{n} \delta_{ij} = m$ for $i = 1, 2, \ldots, p$

2. $\displaystyle\sum_{i=1}^{p} \delta_{ij} = k$ for $j = 1, 2, \ldots, n$

3. $\displaystyle\sum_{i=1}^{p} \delta_{ij}\delta_{ir} = \lambda$ for all $r \neq j = 1, 2, \ldots, n$

Summing on the other subscript in specifications 1 and 2, we obtain

$$\sum_{i=1}^{p} \sum_{j=1}^{n} \delta_{ij} = mp = kn$$

which implies that the number of observers is fixed by the design to be $p = kn/m$. Now using specification 3, we have

$$\sum_{i=1}^{p} \left(\sum_{j=1}^{n} \delta_{ij} \right)^2 = \sum_{i=1}^{p} \left(\sum_{j=1}^{n} \delta_{ij}^{2} + \sum_{\substack{j=1 \\ j \neq r}}^{n} \sum_{r=1}^{n} \delta_{ij}\delta_{ir} \right) = mp + \lambda n(n - 1)$$

and from specification 1, this same sum equals pm^2. This requires

$$\lambda = \frac{pm(m - 1)}{n(n - 1)} = \frac{k(m - 1)}{n - 1}$$

Since p and λ must both be positive integers, m must be a factor of kn and $n - 1$ must be a factor of $k(m - 1)$. Designs of this type are called *Youden squares* or *incomplete Latin squares*. Such plans have been tabulated [for example, in Cochran and Cox (1957), pp. 520–544]. An example of this design for $n = 7$, $\lambda = 1$, $m = k = 3$, where the objects are designated by A, B, C, D, E, F, and G is:

Observer	1	2	3	4	5	6	7
Objects presented	A	B	C	D	E	F	G
for ranking	B	C	D	E	F	G	A
	D	E	F	G	A	B	C

We are interested in determining a single measure of the overall concordance or agreement between the kn/m observers in their relative comparisons of the objects. For simplification, suppose there is some natural ordering of all n objects and the objects labeled accordingly. In other words, object number r would receive rank r by all observers if each observer was presented with all n objects and the observers agreed perfectly in their evaluation of the objects. For perfect agreement in a balanced incomplete ranking then where each observer assigns ranks $1, 2, \ldots, m$ to the subset presented to him, object 1 will receive rank 1 whenever it is presented; object 2 will receive rank 2 whenever it is presented along with object 1, and rank 1 otherwise; object 3 will receive rank 3 when presented along with both objects 1 and 2, rank 2 when with either objects 1 or 2 but not both, and rank 1 otherwise, etc. In general, then, the rank of object j when presented to observer i is 1 more than the number of objects presented to that observer from the subset of objects $\{1, 2, \ldots, j - 1\}$, for all $2 \leq j \leq n$. Symbolically, using the δ notation of before, the rank of object j when presented to observer i is 1 for $j = 1$ and

$$1 + \sum_{r=1}^{j-1} \delta_{ir} \qquad \text{for all } 2 \leq j \leq n$$

The sum of the ranks assigned to object j by all p observers in the case of perfect agreement then is

$$\sum_{i=1}^{p} \left(1 + \sum_{r=1}^{j-1} \delta_{ir} \right) \delta_{ij} = \sum_{i=1}^{p} \delta_{ij} + \sum_{r=1}^{j-1} \sum_{i=1}^{p} \delta_{ir} \delta_{ij}$$
$$= k + \lambda(j - 1) \qquad \text{for } j = 1, 2, \ldots, n$$

as a result of the design specifications 2 and 3.

Since each object is ranked a fixed number, k, of times, the observed data for an experiment of this type can easily be presented in a two-way layout of k rows and n columns, where the jth column contains the collection of ranks assigned to object j by those observers to whom object j was presented. The rows no longer have any significance, but the column sums can be used to measure concordance. The sum of all ranks in the table is $\dfrac{m(m + 1)}{2} \dfrac{kn}{m} = kn(m + 1)/2$, and thus the average column sum is $k(m + 1)/2$. In the case of perfect concordance, the column sums are some permutation of the numbers

$$k, k + \lambda, k + 2\lambda, \ldots, k + (n - 1)\lambda$$

and the sums of squares of deviations of column sums around their mean is

$$\sum_{j=0}^{n-1} \left[(k + j\lambda) - \frac{k(m + 1)}{2} \right]^2 = \frac{\lambda^2 n(n^2 - 1)}{12}$$

Let R_j denote the actual sum of ranks in the jth column. A relative measure of concordance between observers may be defined here as

$$W = \frac{12 \sum_{j=1}^{n} \left[R_j - \frac{k(m+1)}{2} \right]^2}{\lambda^2 n(n^2 - 1)} \qquad (4.1)$$

$$= \frac{12 \sum_{j=1}^{n} R_j^2 - 3k^2 n(m+1)^2}{\lambda^2 n(n^2 - 1)}$$

If $m = n$ and $\lambda = k$ so that each observer ranks all n objects, (4.1) is equivalent to (3.4), as it should be.

This coefficient of concordance also varies between 0 and 1 with larger values reflecting greater agreement between observers. If there is no agreement, the column sums would all tend to be equal to the average column sum and W would be zero.

TESTS OF SIGNIFICANCE BASED ON W

For testing the null hypothesis that the ranks are allotted randomly by each observer to the subset of objects presented to him so that there is no concordance, the appropriate rejection region is large values of W.

The exact sampling distribution of W could be determined only by an extensive enumeration process. For k large an approximation to the null distribution may be employed for tests of significance. We shall first determine the exact null mean and variance of W using an approach analogous to the steps leading to (2.7). Let $R_{ij}, i = 1, 2, \ldots, k$, denote the collection of ranks allotted to object number j by the k observers to whom it was presented. From (12.3.2), (12.3.3), and (12.3.10), in the null case then for all i, j and $q \neq j$

$$E(R_{ij}) = \frac{m+1}{2} \qquad \text{var}(R_{ij}) = \frac{m^2 - 1}{12} \qquad \text{cov}(R_{ij}, R_{iq}) = -\frac{m+1}{12}$$

and R_{ij} and R_{hj} are independent for all j where $i \neq h$. Denoting $(m+1)/2$ by μ, the numerator of W in (4.1) may be written as

$$12 \sum_{j=1}^{n} \left(\sum_{i=1}^{k} R_{ij} - k\mu \right)^2 = 12 \sum_{j=1}^{n} \left[\sum_{i=1}^{k} (R_{ij} - \mu) \right]^2$$

$$= 12 \sum_{j=1}^{n} \sum_{i=1}^{k} (R_{ij} - \mu)^2$$

$$+ 24 \sum_{j=1}^{n} \sum_{1 \leq i < h \leq k} (R_{ij} - \mu)(R_{hj} - \mu)$$

$$= pm(m^2 - 1) + 24U$$

$$= \lambda^2 n(n^2 - 1)W \qquad (4.2)$$

Since $\text{cov}(R_{ij}, R_{hj}) = 0$ for all $i < h$, $E(U) = 0$. Squaring the sum represented by U, we have

$$U^2 = \sum_{j=1}^{n} \sum_{1 \leq i < h \leq k} (R_{ij} - \mu)^2 (R_{hj} - \mu)^2$$
$$+ 2 \sum_{1 \leq j < q \leq n} \sum_{1 \leq i < h \leq k} \sum_{1 \leq r < s \leq k} (R_{ij} - \mu)(R_{hj} - \mu)(R_{rq} - \mu)(R_{sq} - \mu)$$

and

$$E(U^2) = \sum_{j=1}^{n} \sum_{1 \leq i < h \leq k} \text{var}(R_{ij}) \, \text{var}(R_{hj})$$
$$+ 2 \sum_{1 \leq j < q \leq n} \binom{\lambda}{2} \text{cov}(R_{ij}, R_{iq}) \, \text{cov}(R_{hj}, R_{hq})$$

since objects j and q are presented together to both observers i and h a total of $\binom{\lambda}{2}$ times in the experiment. Substituting the respective variances and covariances, we obtain

$$\text{var}(U) = E(U^2) = \frac{n \binom{k}{2}(m^2 - 1)^2 + 2\binom{n}{2}\binom{\lambda}{2}(m + 1)^2}{144}$$
$$= nk(m + 1)^2 (m - 1) \frac{(m - 1)(k - 1) + (\lambda - 1)}{288}$$

From (4.2), the moments of W are

$$E(W) = \frac{m + 1}{\lambda(n + 1)}$$
$$\text{var}(W) = 2(m + 1)^2 \frac{(m - 1)(k - 1) + (\lambda - 1)}{nk\lambda^2(m - 1)(n + 1)^2}$$

As in the case of complete rankings, a linear function of W has moments approximately equal to the corresponding moments of the chi-square distribution with $n - 1$ degrees of freedom if k is large. This function is

$$V = \frac{\lambda(n^2 - 1)W}{m + 1}$$

and its exact mean and variance are

$$E(V) = n - 1$$
$$\text{var}(V) = 2(n - 1)\left[1 - \frac{m(n - 1)}{nk(m - 1)}\right] \approx 2(n - 1)\left(1 - \frac{1}{k}\right)$$

The rejection region for large k and significance level α then is

$$V \in R \qquad \text{for } v \geq \chi^2_{n-1, \alpha}$$

THE PROBLEM OF TIED OBSERVATIONS

Unlike the case of complete rankings, no simple correction factor can be introduced to account for the reduction in total sum of squares of deviations of column totals around their mean when the midrank method is used to handle ties. If there are only a few ties, the null distribution of W should not be seriously altered, and thus the statistic can be computed as usual with midranks assigned. Alternatively, any of the other methods of handling ties discussed in Sec. 5.3 (except omission of tied observations) may be adopted.

13.5 CONTINGENCY TABLES

In the two-way classifications considered up to now in this chapter, each cell entry was the actual measurement or ranking of the experimental unit having the properties of that particular cross-classification. Assuming that only the first of these properties is of interest because the other property was introduced solely for the purpose of obtaining good experimental design (the blocks or the objects to be ranked), we were able to make inferences about relationships between the categories of that first property (the treatments or the observers).

Now suppose that there are two or more properties of interest, say $A, B, C, \ldots$, which are called *sets* or *families of attributes*, each of which has two or more categories, say $A_1, A_2, \ldots$, for family A, etc. These attributes may be measurable or not, as long as the categories are clearly defined, mutually exclusive, and exhaustive. Some number n of experimental units are observed, and each is classified into exactly one category of each family. The number of units classified into each cross-category constitute the sample data. Such a layout is called a *contingency table* of order $r_1 \times r_2 \times r_3 \times \cdots$ if there are r_1 categories of family A, r_2 of family B, etc. We are interested in a measure of association between the families, or in testing the null hypothesis that the families are completely independent, or that one particular family is independent of the others. In general, a group of families of events is defined to be completely independent if

$$P(A_i \cap B_j \cap C_k \cap \cdots) = P(A_i)P(B_j)P(C_k) \cdots$$

for all $A_i \subset A$, $B_j \subset B$, $C_k \subset C$, etc. For subgroup independence, say that family A is independent of all others, the requirement is

$$P(A_i \cap B_j \cap C_k \cap \cdots) = P(A_i)P(B_j \cap C_k \cap \cdots)$$

For example, in a public-opinion survey concerning a proposed bond issue, the results of each interview or questionnaire may be classified according to the attributes of sex, education, and opinion. Along with the two categories of sex, we might have three categories of opinion, e.g., favor, oppose, and undecided, and five categories of education according to highest level of formal schooling completed. The data may be presented in a 2 × 3 × 5 contingency table of 30 cells. A tally is placed in the appropriate cell for each interviewee, and these count data may be used to determine whether sex and educational level have any observable relationship to opinion or to find some relative measure of their association.

For convenience, we shall restrict our analysis to a two-way classification for two families of attributes, as the extension to higher-way layouts will be obvious. Suppose there are r categories of the type A attribute and k categories of the B attribute, and each of n experimental units is classified into exactly one of the rk cross-categories. In an $r \times k$ layout, the entry in the (i,j) cell, denoted by X_{ij}, is the number of items having the cross-classification $A_i \cap B_j$. The contingency table is written in the following form:

$$
\begin{array}{c}
\text{B family} \\
\begin{array}{cccccl}
 & B_1 & B_2 & \cdots & B_k & \text{Row total} \\
A_1 & X_{11} & X_{12} & \cdots & X_{1k} & X_{1.} \\
A_2 & X_{21} & X_{22} & \cdots & X_{2k} & X_{2.} \\
\vdots & & & & & \vdots \\
A_r & X_{r1} & X_{r2} & \cdots & X_{rk} & X_{r.} \\
\text{Column total} & X_{.1} & X_{.2} & \cdots & X_{.k} & X_{..} = n
\end{array}
\end{array}
$$

The total numbers of items classified respectively into the categories A_i and B_j then are the row and column totals $X_{i.}$ and $X_{.j}$, where

$$X_{i.} = \sum_{j=1}^{k} X_{ij} \quad \text{and} \quad X_{.j} = \sum_{i=1}^{r} X_{ij}$$

Without making any additional assumptions, we know that the rk random variables $X_{11}, X_{12}, \ldots, X_{rk}$ have the multinomial probability distribution with parameters

$$\theta_{ij} = P(A_i \cap B_j) \quad \text{where} \quad \sum_{i=1}^{r} \sum_{j=1}^{k} \theta_{ij} = 1$$

so that the likelihood function of the sample is

$$\prod_{i=1}^{r} \prod_{j=1}^{k} (\theta_{ij})^{x_{ij}}$$

The null hypothesis that the A and B classifications are independent affects only the allowable values of these parameters θ_{ij}. In view of the definition of independence of families, the hypothesis can be stated simply as

$$H_0: \theta_{ij} = \theta_{i.}\theta_{.j} \qquad \text{for all } i = 1, 2, \ldots, r$$
$$j = 1, 2, \ldots, k$$

where

$$\theta_{i.} = \sum_{j=1}^{k} \theta_{ij} = P(A_i) \qquad \theta_{.j} = \sum_{i=1}^{r} \theta_{ij} = P(B_j)$$

If these $\theta_{i.}$ and $\theta_{.j}$ were all specified under the null hypothesis, this would reduce to an ordinary goodness-of-fit test of a simple hypothesis with rk groups. However, the probability distribution is not completely specified under H_0, since only a particular relation between the parameters need be stated for the independence criterion to be satisfied. The chi-square goodness-of-fit test for composite hypotheses discussed in Sec. 4.2 is appropriate here. The unspecified parameters must be estimated by the method of maximum likelihood and the degrees of freedom for the test statistic reduced by the number of independent parameters estimated. The maximum-likelihood estimates of the $(r-1) + (k-1)$ unknown independent parameters are those sample functions which maximize the likelihood function under H_0, which is

$$L(\theta_{1.},\theta_{2.}, \ldots, \theta_{r.},\theta_{.1},\theta_{.2}, \ldots, \theta_{.k}) = \prod_{i=1}^{r} \prod_{j=1}^{k} (\theta_{i.}\theta_{.j})^{x_{ij}} \qquad (5.1)$$

subject to the restrictions

$$\sum_{i=1}^{r} \theta_{i.} = \sum_{j=1}^{k} \theta_{.j} = 1$$

Using Lagrange multipliers or some other method, the maximum-likelihood estimates are easily found to be the corresponding observed proportions, or

$$\hat{\theta}_{i.} = \frac{X_{i.}}{n} \qquad \text{and} \qquad \hat{\theta}_{.j} = \frac{X_{.j}}{n} \qquad \text{for } i = 1, 2, \ldots, r$$
$$j = 1, 2, \ldots, k$$

so that the rk estimated expected cell frequencies under H_0 are

$$n\hat{\theta}_{ij} = \frac{X_{i.}X_{.j}}{n}$$

By the results of Sec. 4.2, the test statistic then is

$$Q = \sum_{i=1}^{r} \sum_{j=1}^{k} \frac{(X_{ij} - X_{i.}X_{.j}/n)^2}{X_{i.}X_{.j}/n} = \sum_{i=1}^{r} \sum_{j=1}^{k} \frac{(nX_{ij} - X_{i.}X_{.j})^2}{nX_{i.}X_{.j}} \qquad (5.2)$$

which under H_0 has approximately the chi-square distribution with degrees of freedom $rk - 1 - (r - 1 + k - 1) = (r - 1)(k - 1)$. Since nonindependence is reflected by lack of agreement between the observed and expected cell frequencies, the rejection region with significance level α is

$$Q \in R \qquad \text{for } q \geq \chi^2_{(r-1)(k-1),\alpha}$$

As before, if any expected cell frequency is too small, say less than 5, the chi-square approximation is improved by combining cells and reducing the degrees of freedom accordingly.

The extension of this to higher-order contingency tables is obvious. For an $r_1 \times r_2 \times r_3$ table, for example, for the hypothesis of complete independence

$$H_0: \theta_{ijk} = \theta_{i..}\theta_{.j.}\theta_{..k} \qquad \text{for all } i = 1, 2, \ldots, r_1$$
$$j = 1, 2, \ldots, r_2$$
$$k = 1, 2, \ldots, r_3$$

the estimates of expected call frequencies are

$$n\hat{\theta}_{ijk} = \frac{X_{i..}X_{.j.}X_{..k}}{n^2}$$

and the chi-square test statistic is

$$\sum_{i=1}^{r_1} \sum_{j=1}^{r_2} \sum_{k=1}^{r_3} \frac{(n^2 X_{ijk} - X_{i..}X_{.j.}X_{..k})^2}{n^2 X_{i..}X_{.j.}X_{..k}}$$

with

$$r_1 r_2 r_3 - 1 - (r_1 - 1 + r_2 - 1 + r_3 - 1)$$
$$= r_1 r_2 r_3 - r_1 - r_2 - r_3 + 2$$

degrees of freedom. For the hypothesis that family A is independent of B and C,

$$H_0: \theta_{ijk} = \theta_{i..}\theta_{.jk}$$

the estimated expected cell frequencies are

$$n\hat{\theta}_{ijk} = \frac{X_{i..}X_{.jk}}{n}$$

and the chi-square test statistic has

$$r_1 r_2 r_3 - 1 - (r_1 - 1 + r_2 r_3 - 1) = (r_1 - 1)(r_2 r_3 - 1)$$

degrees of freedom.

If the experimental situation is such that one set of totals is fixed by the experimenter in advance, say the row totals in an $r \times k$ contingency table, the test statistic for a hypothesis of independence is exactly the same as for completely random totals, although the reasoning is somewhat different. The entries in the ith row of the table constitute a random sample of size $x_{i.}$ from a k-variate multinomial population. For each row then, one of the cell entries is determined by the constant total. One of the observable frequencies is redundant for each row, as is one of the probability parameters $P(A_i \cap B_j)$ for each i. Since

$$P(A_i \cap B_j) = P(A_i)P(B_j \mid A_i)$$

and $P(A_i) = x_{i.}/n$ is now fixed, we shall redefine the relevant parameters as $\theta_{ij} = P(B_j \mid A_i)$, where $\sum_{j=1}^{k} \theta_{ij} = 1$. The dimension of the parameter space is then reduced from $rk - 1$ to $r(k - 1)$. The B family is independent of the A_i category if $\theta_{ij} = P(B_j \mid A_i) = P(B_j) = \theta_j$ for all $j = 1, 2, \ldots, k$, where $\sum_{j=1}^{k} \theta_j = 1$. Under H_0 then, $E(X_{ij}) = x_{i.}\theta_j$, and if the θ_j were specified, the test statistic for independence between B and A_i would be

$$\sum_{j=1}^{k} \frac{(X_{ij} - x_{i.}\theta_j)^2}{x_{i.}\theta_j} \tag{5.3}$$

which is approximately chi square distributed with $k - 1$ degrees of freedom. The B family is completely independent of the A family if $\theta_{ij} = \theta_j, j = 1, 2, \ldots, k$, all $i = 1, 2, \ldots, r$, so that the null hypothesis can be written as

$$H_0: \theta_{1j} = \theta_{2j} = \cdots = \theta_{rj} = \theta_j \qquad \text{for } j = 1, 2, \ldots, k$$

The test statistic for complete independence then is the statistic in (5.3) summed over all $i = 1, 2, \ldots, r$,

$$\sum_{i=1}^{r} \sum_{j=1}^{k} \frac{(X_{ij} - x_{i.}\theta_j)^2}{x_{i.}\theta_j} \tag{5.4}$$

which under H_0 is the sum of r independent chi-square variables, each having $k - 1$ degrees of freedom, and therefore has $r(k - 1)$ degrees of freedom. Of course, in our case the θ_j are not specified, and so the test statistic is (5.4), with the θ_j replaced by their maximum-likelihood estimates and the degrees of freedom reduced by $k - 1$, the number of independent parameters estimated. The likelihood function under H_0 of all n observations with row totals fixed is

$$L(\theta_1, \theta_2, \ldots, \theta_k) = \prod_{i=1}^{r} \prod_{j=1}^{k} \theta_j^{x_{ij}} = \prod_{j=1}^{k} \theta_j^{x_{.j}}$$

so that $\hat{\theta}_j = X_{.j}/n$. Substituting this result in (5.4), we find the test criterion unchanged from the previous case with random totals given in (5.2), and the degrees of freedom are $r(k-1) - (k-1) = (r-1)(k-1)$, as before. By similar analysis, it can be shown that the same result holds for fixed column totals or both row and column totals fixed.

THE CONTINGENCY COEFFICIENT

As a measure of the degree of association between families in a contingency table classifying a total of n experimental units, Pearson (1904) proposed the contingency coefficient C, defined as

$$C = \left(\frac{Q}{Q+n}\right)^{\frac{1}{2}}$$

where Q is the appropriate test statistic for the hypothesis of independence. If the families are completely independent, the values of Q and C are both small. Further, increasing values of C imply an increasing degree of association since large values of Q are a result of more significant departures between the observed and expected cell frequencies. Although clearly the value of C cannot exceed 1 for any n, a disadvantage of C as a measure of association is that it cannot attain the value 1.

It is easily shown that for a two-way contingency table of dimension $r \times k$, the maximum value of C is

$$c_{\max} = \left(\frac{t-1}{t}\right)^{\frac{1}{2}} \qquad \text{where } t = \min(r,k)$$

Without loss of generality, we can assume $r \geq k$. Then n must be at least r so that there is one element in each row and each column to avoid any zero denominators in the test statistic. Consider n fixed at its smallest value r, so that $x_{i.} = 1$ for $i = 1, 2, \ldots, r$, and x_{ij} is 0 or 1 for all i, j. The number of cells for which $x_{ij} = 1$ for fixed j is $x_{.j}$, and the number for which $x_{ij} = 0$ is $r - x_{.j}$. The value of Q then is

$$\sum_{j=1}^{k} \frac{(r - x_{.j})(0 - x_{.j}/r)^2 + x_{.j}(1 - x_{.j}/r)^2}{x_{.j}/r}$$

$$= \sum_{j=1}^{k} \frac{x_{.j}(r - x_{.j})[x_{.j} + (r - x_{.j})]}{rx_{.j}} = r(k-1)$$

Then the contingency coefficient is

$$c = \left[\frac{r(k-1)}{rk - r + r}\right]^{\frac{1}{2}} = \left(\frac{k-1}{k}\right)^{\frac{1}{2}}$$

As a result of this property, contingency coefficients for two different sets of count data are not directly comparable to measure association unless $\min(r,k)$ is the same for both tables.

The sampling distribution of C is not known. However, this is of no consequence since C is a function of Q, and a test of significance based on Q would be equivalent to a test of significance based on C^2.

SOME SPECIAL RESULTS FOR A 2 × 2 CONTINGENCY TABLE

In a 2×2 contingency table, each family is simply a dichotomy, and it is a simple algebraic exercise to show that the test statistic for independence can be written in an equivalent form as

$$Q = \sum_{i=1}^{2} \sum_{j=1}^{2} \frac{(X_{ij} - X_{i.}X_{.j}/n)^2}{X_{i.}X_{.j}/n} = \sum_{i=1}^{2} \frac{(Y_i - n_i\hat{p})^2}{n_i\hat{p}(1 - \hat{p})} \tag{5.5}$$

or

$$Q = \frac{(Y_1/n_1 - Y_2/n_2)^2}{\hat{p}(1 - \hat{p})(1/n_1 + 1/n_2)} \tag{5.6}$$

where

$$Y_i = X_{i1} \qquad n_i - Y_i = X_{i2} \qquad \text{for } i = 1, 2 \qquad \hat{p} = \frac{Y_1 + Y_2}{n}$$

In these two expressions, if B_1 and B_2 are regarded as successes and failures, and A_1 and A_2 are termed sample 1 and sample 2, we see that:

1. In (5.5), the chi-square test statistic is the sum of the squares of two standardized binomial random variables with the parameter p estimated by its consistent estimator $\hat{p}$.
2. In (5.6), the chi-square test statistic is the square of the difference between two proportions divided by the estimated variance of their difference. In other words, Q is the square of the classical standard normal test statistic used for the hypothesis that two population proportions are equal.

Substituting the original X_{ij} notation in (5.6), a little algebraic manipulation gives another equivalent form for Q as

$$Q = \frac{n(X_{11}X_{22} - X_{12}X_{21})^2}{X_{.1}X_{.2}X_{1.}X_{2.}} \tag{5.7}$$

This expression is related to the sample Kendall tau coefficient of Chap. 12. Suppose that the two families A and B are factors or qualities, both dichotomized into categories which can be called presence and absence of the factor or possessing and not possessing the quality. Suppose further that we have a single sample of size n, and make two observations on each element in the sample, one for each of the two factors. We record the observations using the scheme 1 for presence and 2 for absence. The observations then consist of n sets of pairs, for which the Kendall tau coefficient T of Chap. 12 can be determined as a measure of association between the factors. The numerator of T is the number of sets of pairs of observations, say (a_i, b_i) and (a_j, b_j), whose differences $a_i - a_j$ and $b_i - b_j$ have the same sign but are not zero. The differences here are both positive or both negative only for a set (1,1) and (2,2), and of opposite signs for a set (1,2) and (2,1). If X_{ij} denotes the number of observations where factor A was recorded as i and factor B was recorded as j for $i, j = 1, 2$, the number of differences with the same sign is $X_{11}X_{22}$, the number of pairs which agreed in the sense that both factors were present or both were absent. The number of differences with opposite signs is $X_{12}X_{21}$, the number of pairs which disagreed. Since there are so many ties, it seems most appropriate to use the definition of T modified for ties, given in (12.2.37). Then the denominator of T is the square root of the product of the numbers of pairs with no ties for each factor, or $X_{.1}X_{.2}.X_{.1}X_{.2}$. Therefore the tau coefficient is

$$T = \frac{X_{11}X_{22} - X_{12}X_{21}}{(X_{.1}X_{.2}X_{1.}X_{2.})^{\frac{1}{2}}} = \left(\frac{Q}{n}\right)^{\frac{1}{2}} \tag{5.8}$$

and Q/n estimates τ^2, the parameter of association between factors A and B. For this type of data, the Kendall measure of association is sometimes called the phi coefficient.

13.6 KENDALL'S TAU COEFFICIENT FOR PARTIAL CORRELATION

Coefficients of partial correlation are useful measures in studying relationships between two random variables since they are ordinary correlations between two variables with the effects of some other variables eliminated because they are held constant. In other words, the coefficients measure association in the conditional probability distribution of two variables given one or more other variables. An advantage of Kendall's tau coefficient over Spearman's rank-correlation coefficient is that it can be extended to the theory of partial correlation.

Assume we are given m independent observations of triplets (X_i, Y_i, Z_i), $i = 1, 2, \ldots, m$, from a trivariate population where the marginal distributions of each variable are continuous. We wish to determine a sample measure of the association between X and Y when Z is held constant. Define the indicator variables

$$U_{ij} = \text{sgn}(X_j - X_i) \qquad V_{ij} = \text{sgn}(Y_j - Y_i) \qquad W_{ij} = \text{sgn}(Z_j - Z_i)$$

and for all $1 \leq i < j \leq m$, let $n(u,v,w)$ denote the number of values of (i,j) such that $u_{ij} = u$, $v_{ij} = v$, $w_{ij} = w$. Now we further define the count variables

$$X_{11} = n(1,1,1)$$
$$X_{22} = n(-1,-1,1)$$
$$X_{12} = n(-1,1,1)$$
$$X_{21} = n(1,-1,1)$$

Then X_{11} is the number of sets of (i,j) pairs, $1 \leq i < j \leq m$, of each variable such that X and Y are both concordant with Z, X_{22} is the number whereby X and Y are both discordant with Z, X_{12} is the number such that X is discordant with Z and Y is concordant with Z, and X_{21} is the number where X is concordant with Z and Y is discordant with Z. If these counts are presented in the form of a 2×2 contingency table and T calculated as in (5.8) of the last section, we obtain a measure $T_{XY.Z}$ of the association between the X and Y samples when Z is held constant,

$$T_{XY.Z} = \frac{X_{11}X_{22} - X_{12}X_{21}}{(X_{.1}X_{.2}X_{1.}X_{2.})^{1/2}} \tag{6.1}$$

The value of this coefficient ranges between -1 and $+1$. At either of these two extremes, we have

$$(x_{11} + x_{21})(x_{12} + x_{22})(x_{11} + x_{12})(x_{21} + x_{22}) - (x_{11}x_{22} - x_{12}x_{21})^2 = 0$$

a sum of products of three or more nonnegative numbers whose exponents total four equal to zero. This occurs only if at least two of the numbers are zero. If $x_{ij} = x_{hk} = 0$ for $i = h$ or $j = k$, either X or Y is in perfect concordance or discordance with Z. The nontrivial cases then are where both diagonal entries are zero. If $x_{12} = x_{21} = 0$, the X and Y sample values are always either both concordant or both discordant with Z and $t_{XY.Z} = 1$. If $x_{11} = x_{22} = 0$, they are never both in the same relation and $t_{XY.Z} = -1$. The sampling distribution of the partial tau coefficient is unknown, so that tests of significance cannot be performed. Even though as before $T_{XY.Z} = (Q/n)^{1/2}$, where $n = \binom{m}{2}$, a test of significance based on the random variable Q is not appropriate in this situation. The contingency-table entries are not independent even if X and Y are independent for fixed Z, since all categories involve pairings with the Z sample. However, $T_{XY.Z}$ provides a useful relative measure of the degree to which X and Y are concordant when their relation with Z is eliminated.

It is interesting to look at the partial tau coefficient in a different algebraic form. Using the X_{ij} notation above, the Kendall tau coefficients for the three different paired samples would be

$$\binom{m}{2} T_{XY} = (X_{11} + X_{22}) - (X_{12} + X_{21})$$

$$\binom{m}{2} T_{XZ} = (X_{11} + X_{21}) - (X_{22} + X_{12})$$

$$\binom{m}{2} T_{YZ} = (X_{11} + X_{12}) - (X_{22} + X_{21})$$

Since $\binom{m}{2} = X_{11} + X_{12} + X_{21} + X_{22} = n$, we have

$$1 - T_{XZ}^2 = \frac{4(X_{11} + X_{21})(X_{12} + X_{22})}{n^2} = \frac{4X_{.1}X_{.2}}{n^2}$$

$$1 - T_{YZ}^2 = \frac{4(X_{11} + X_{12})(X_{22} + X_{21})}{n^2} = \frac{4X_{1.}X_{2.}}{n^2}$$

and

$$n^2 T_{XY} = [(X_{11} + X_{22}) - (X_{12} + X_{21})][(X_{11} + X_{22}) + (X_{12} + X_{21})]$$
$$n^2(T_{XY} - T_{XZ}T_{YZ}) = 4(X_{11}X_{22} - X_{12}X_{21})$$

Therefore (6.1) can be written as

$$T_{XY.Z} = \frac{T_{XY} - T_{XZ}T_{YZ}}{[(1 - T_{XZ}^2)(1 - T_{YZ}^2)]^{\frac{1}{2}}} \tag{6.2}$$

Some other approaches to defining a measure of partial correlation have appeared in the journal literature. One of the more useful measures is the index of matched correlation proposed by Quade (1967). However, the partial tau defined here has a particularly appealing property in that it can be generalized to the case of more than three variables. Note that the form in (6.2), with each T replaced by its corresponding R, is identical to the expression for a Pearson product-moment partial correlation coefficient. This is because both are special cases of a generalized partial correlation coefficient which is discussed in Somers (1959). With his generalized form, extensions of the partial tau coefficient to higher orders are possible.

PROBLEMS

13.1. Four varieties of soybean are each planted in three blocks. The yields are:

	Variety of soybean			
Block	A	B	C	D
1	45	48	43	41
2	49	45	42	39
3	38	39	35	36

Use Friedman's analysis of variance by ranks to test the hypothesis that there is no difference in the average yields of the four varieties of soybean.

13.2. A beauty contest has eight contestants. The three judges are each asked to rank the contestants in a preferential order of pulchritude. The results are:

				Contestant				
Judge	A	B	C	D	E	F	G	H
1	2	1	3	5	4	8	7	6
2	1	2	4	5	7	6	8	3
3	3	2	1	4	5	8	7	6

(a) Calculate Kendall's coefficient of concordance between rankings.

(b) Calculate the coefficient of rank correlation for each of the three pairs of rankings and verify the relation between r_{av} and w given in (3.7).

(c) Estimate the true preferential order of pulchritude.

13.3. Derive by enumeration the exact null distribution of W for three sets of rankings of two objects.

13.4. Derive the maximum-likelihood estimators for the parameters in the likelihood function of (5.1).

13.5. Show that (5.2) is still the appropriate test statistic for independence in a two-way $r \times k$ contingency table when both the row and column totals are fixed.

13.6. Given the following triplets of rankings of six objects:

X:	1	3	5	6	4	2
Y:	1	2	6	4	3	5
Z:	2	1	5	4	6	3

(a) Calculate the Kendall coefficient of partial correlation between X and Y from (6.1).

(b) Calculate (6.2) for these same data to verify that it is an equivalent expression.

13.7. Verify the equivalences expressed in (5.5), (5.6), and (5.7).

14
Asymptotic Relative Efficiency

14.1 INTRODUCTION

In Chap. 1 the concept of Pitman efficiency was defined as a criterion for the comparison of any two test statistics. Many of the nonparametric tests covered in this book can be considered direct analogs to some classical test which is known to be most powerful under certain specific distribution assumptions. The asymptotic efficiencies of the nonparametric tests relative to a "best" test have been simply stated here without discussion. In this chapter we shall investigate the concept of efficiency more thoroughly and prove some theorems which simplify the calculation. The theory will then be illustrated by applying it to various tests covered in earlier chapters in order to derive numerical values of the ARE for some particular distributions. The theory presented here is attributed to Pitman and follows the presentation in Fraser (1957).

Suppose that we have two test statistics, denoted by T and T^*, which can be used for the same sampling situation and similar types of inferences regarding simple hypotheses. One method of comparing the performance of the two tests was described in Chap. 1 as relative power efficiency. The power efficiency of test T relative to test T^* is defined as the ratio n^*/n, where n^* is the sample size necessary to attain the power γ at significance level α when test T^* is used and n is the sample size required by test T to attain the same values γ and α.

As a simple numerical example, consider a comparison of the normal-theory test T^* and the ordinary sign test T for the respective hypothesis-testing situations

$$H_0: \mu = 0 \qquad \text{versus} \qquad H_1: \mu = 1$$

and

$$H_0: M = 0 \qquad \text{versus} \qquad H_1: M = 1$$

The inference is to be based on a single random sample from a population which is assumed to be normally distributed with known variance equal to 1. Then the hypothesis sets above are identical. Suppose we are interested in the relative sample sizes for a power of 0.90 and a significance level of 0.05. For the most powerful (normal-theory) test based on n^* observations, the hypothesis is rejected when $\sqrt{n^*}\,\bar{X} \geq 1.64$ for $\alpha = 0.05$. Setting the power γ equal to 0.90, n^* is found as follows:

$$Pw(1) = \gamma(1) = P(\sqrt{n^*}\,\bar{X} \geq 1.64 \mid \mu = 1)$$
$$= P[\sqrt{n^*}\,(\bar{X} - 1) \geq 1.64 - \sqrt{n^*}] = 0.90$$
$$\Phi(1.64 - \sqrt{n^*}) = 0.10 \qquad 1.64 - \sqrt{n^*} = -1.28 \qquad n^* = 9$$

The sign test T of Sec. 6.2 has rejection region $K \geq k_\alpha$, where K is the number of positive observations X_i and k_α is chosen so that

$$\sum_{k=k_\alpha}^{n} \binom{n}{k} 0.5^n = \alpha \tag{1.1}$$

The power of the test T then is

$$\sum_{k=k_\alpha}^{n} \binom{n}{k} \theta^k (1 - \theta)^{n-k} = \gamma(1) \tag{1.2}$$

where $\theta = P(X > 0 \mid M = 1) = 1 - \Phi(-1) = 0.8413$, since the mean and median coincide for a normal population. The numbers n and $k_{0.05}$ will be those values of n and k_α, respectively, which simultaneously satisfy (1.1) and (1.2) when $\alpha = 0.05$ and $\gamma = 0.90$. If θ is rounded off to 0.85, ordinary tables of the binomial distribution can be used instead of actual calculations. Some of the steps relevant to finding the simultaneous solution are shown in Table 1.1.

Table 1.1

n	k_α	α	$\gamma = 1 - \beta$	Randomized decision rule for exact $\alpha = 0.05$	
				Probability of rejection	$\gamma(1)$
17	13	0.0245	0.9013		
	12	0.0717	0.9681		
16	12	0.0383	0.9211	1	
					0.9308
	11	0.1050	0.9766	0.1754	
15	12	0.0176	0.8226	1	
					0.9151
	11	0.0593	0.9382	0.8010	
14	11	0.0288	0.8535	1	
					0.8881
	10	0.0899	0.9533	0.3470	
13	10	0.0461	0.8820		
	9	0.1334	0.9650		

If we do not wish to resort to the use of randomized decision rules, we can either (1) choose those values for n and k_α such that α and γ are both as close as possible to the preselected numbers or (2) choose the smallest value of n such that the smallest value of k_α gives α and $\beta = 1 - \gamma$ no larger than the preselected numbers. We obtain $n = 13$ and $k_{0.05} = 10$ using method 1, and $n = 16$, $k_{0.05} = 12$ with method 2. These methods are undesirable for a number of obvious reasons, but mainly because method 1 may not lead to a unique answer and method 2 may be too conservative with respect to both types of errors. A preferable approach for the purposes of comparison would be to use randomized decision rules. Then we can either make exact $\alpha = 0.05$ or exact $\gamma = 0.90$ but probably not both. When deciding to make exact $\alpha = 0.05$, the procedure is also illustrated in Table 1.1. Starting with the smallest n and corresponding smallest k_α for which simultaneously $\alpha \leq 0.05$ and $\gamma \geq 0.90$, the randomized decision rule is found by solving for p in the expression

$$\sum_{k=k_\alpha}^{n} \binom{n}{k} 0.5^n + p \binom{n}{k_\alpha - 1} 0.5^n = 0.05$$

Then the power for this exact 0.05 size test is

$$\sum_{k=k_\alpha}^{n} \binom{n}{k} (0.85^k)(0.15^{n-k}) + p \binom{n}{k_\alpha - 1} (0.85^k)(0.15^{n-k})$$

Do the same set of calculations for the next smaller n, etc., until $\gamma \leq 0.90$. The selected value of n may either be such that $\gamma \geq 0.90$ or γ is as close as possible to 0.90, as before, but at least here the choice is always between two consecutive numbers for n. From Table 1.1 the answers in these two cases are $n = 15$ and $n = 14$, respectively.

This example shows that the normal test here requires only nine observations to be as powerful as a sign test using 14 or 15, so that the power efficiency is around 0.60 or 0.64. This result applies only for the particular numbers α and β (or γ) selected and therefore is not in any sense a general comparison even though both the null and alternative hypotheses are simple.

Since fixing the value for α is a well-accepted procedure in statistical inference, we might perform calculations similar to those above for some additional and arbitrarily selected values of γ and plot the coordinates (γ,n) and (γ,n^*) on the same graph. From these points the curves $n(\gamma)$ and $n^*(\gamma)$ can be approximated. The numerical processes can be easily programmed for computer calculation. Some evaluation of general relative performance of two tests can therefore be made for the particular value of α selected. However, this power-efficiency approach is satisfactory only for a simple alternative. Especially in the case of nonparametric tests, the alternative of interest is usually composite. In the above example, if the alternative were $H_1: \mu > 0$ $(M > 0)$, curves for the functions $n[\gamma(\mu)]$ and $n^*[\gamma(\mu)]$ would have to be compared for all $\mu > 0$ and a preselected α. General conclusions for any μ and γ are certainly difficult if not impossible.

In many important cases the limit of the ratio n^*/n turns out not to be a function of α and γ, or even the parameter value when it is in the neighborhood of the hypothesized value. Therefore, even though it is a large-sample property for a limiting type of alternative, the asymptotic relative efficiency of two tests is a somewhat more satisfying criterion for comparison in the sense that it leads to a single number and consequently a well-defined conclusion for large samples. It is for this reason that the discussion here will be limited to comparisons of tests using this standard.

14.2 THEORETICAL BASES FOR CALCULATING THE ARE

Suppose that we have two test statistics T_n and T_n^*, for data consisting of n observations, both of which are consistent for a test of

$$H_0: \theta \in \omega \qquad \text{versus} \qquad H_1: \theta \in \Omega - \omega$$

In other words, for all $\theta \in \Omega - \omega$

$$\lim_{n \to \infty} Pw[T_n(\theta)] = 1 \qquad \text{and} \qquad \lim_{n \to \infty} Pw[T_n^*(\theta)] = 1$$

Suppose further that a subset of the space Ω can be indexed in terms of a sequence of parameters $\{\theta_0, \theta_1, \theta_2, \ldots, \theta_n, \ldots\}$ such that θ_0 specifies a value in ω and the remaining $\theta_1, \theta_2, \ldots$ are in $\Omega - \omega$ and that $\lim_{n \to \infty} \theta_n = \theta_0$. For example, in the case of a one-sided alternative $\theta > \theta_0$, we take a monotonic decreasing sequence of numbers $\theta_1, \theta_2, \ldots$ which converge to θ_0 from above. If each θ_i specifies a probability distribution for the test statistics, we might say that the alternative distribution is getting closer and closer to the null distribution as n approaches infinity. Under these conditions, a formal definition of the ARE of T relative to T^* can be given.

Definition   *Let T_n and T_n^* be two sequences of test statistics, all with the same significance level α. Let $\{n_i\}$ and $\{n_i^*\}$ be two monotonic increasing sequences of positive integers such that*

$$\lim_{i \to \infty} Pw(T_{n_i} \mid \theta = \theta_i) = \lim_{i \to \infty} Pw(T_{n_i*}^* \mid \theta = \theta_i) = \gamma$$

where γ is not equal to 0 or 1. Then the asymptotic relative efficiency of test T relative to test T^* is

$$\mathrm{ARE}(T, T^*) = \lim_{i \to \infty} \frac{n_i^*}{n_i}$$

if this limit exists and is constant for all sequences $\{n_i\}$ and $\{n_i^*\}$.[†]

In other words, the ARE is the inverse ratio of the sample sizes necessary to obtain any power γ for the tests T and T^*, respectively, while simultaneously the sample sizes approach infinity and the sequences of alternatives approach θ_0, and both tests have the same significance level. It is thus a measure of asymptotic and localized power efficiency. In the case of the more general tests of hypotheses concerning distributions like $F = F_\theta$, the same definition holds.

Now suppose that our consistent size α tests T_n and T_n^* are for the one-sided alternative

$$H_0: \theta = \theta_0 \qquad \text{versus} \qquad H_1: \theta > \theta_0$$

and have respective rejection regions of the form

$$T_n \in R \quad \text{for } t_n \geq t_{n,\alpha} \qquad \text{and} \qquad T_n^* \in R^* \quad \text{for } t_n^* \geq t_{n,\alpha}^*$$

where $t_{n,\alpha}$ and $t_{n,\alpha}^*$ are chosen such that

$$P(T_n \geq t_{n,\alpha} \mid \theta = \theta_0) = \alpha \qquad \text{and} \qquad P(T_n^* \geq t_{n,\alpha}^* \mid \theta = \theta_0) = \alpha$$

[†] Fraser (1957), p. 270.

We make the following regularity assumptions for the test T_n, and analogous ones for T_n^*.

1. $dE(T_n)/d\theta$ exists and is nonzero for $\theta = \theta_0$, and is continuous at θ_0.
2. There exists a positive constant c such that

$$\lim_{n \to \infty} \frac{dE(T_n)/d\theta \Big|_{\theta = \theta_0}}{\sqrt{n} \, \sigma(T_n) \Big|_{\theta = \theta_0}} = c$$

3. There exists a sequence of alternatives $\{\theta_n\}$ such that for some constant $d > 0$, we have

$$\theta_n = \theta_0 + \frac{d}{\sqrt{n}}$$

$$\lim_{n \to \infty} \frac{dE(T_n)/d\theta \Big|_{\theta = \theta_n}}{dE(T_n)/d\theta \Big|_{\theta = \theta_0}} = 1$$

$$\lim_{n \to \infty} \frac{\sigma(T_n) \Big|_{\theta = \theta_n}}{\sigma(T_n) \Big|_{\theta = \theta_0}} = 1$$

4. $\displaystyle \lim_{n \to \infty} P\left[\frac{T_n - E(T_n) \Big|_{\theta = \theta_n}}{\sigma(T_n) \Big|_{\theta = \theta_n}} \leq z \Big|_{\theta = \theta_n} \right] = \Phi(z)$

Theorem 2.1 *Under the four regularity conditions above, the limiting power of the test T_n is*

$$\lim_{n \to \infty} Pw(T_n \mid \theta = \theta_n) = 1 - \Phi(z_\alpha - dc)$$

where z_α is that number for which $1 - \Phi(z_\alpha) = \alpha$.

Proof The limiting power is

$$\lim_{n \to \infty} P(T_n \geq t_{n,\alpha} \mid \theta = \theta_n) = \lim_{n \to \infty} P\left[\frac{T_n - E(T_n) \Big|_{\theta = \theta_n}}{\sigma(T_n) \Big|_{\theta = \theta_n}} \right.$$

$$\left. \geq \frac{t_{n,\alpha} - E(T_n) \Big|_{\theta = \theta_n}}{\sigma(T_n) \Big|_{\theta = \theta_n}} \right]$$

$$= 1 - \Phi(z) \qquad \text{from regularity condition 4}$$

where

$$z = \lim_{n \to \infty} \frac{t_{n,\alpha} - E(T_n)\big|_{\theta = \theta_n}}{\sigma(T_n)\big|_{\theta = \theta_n}} = \lim_{n \to \infty} \left[\frac{t_{n,\alpha} - E(T_n)\big|_{\theta = \theta_n}}{\sigma(T_n)\big|_{\theta = \theta_0}} \frac{\sigma(T_n)\big|_{\theta = \theta_0}}{\sigma(T_n)\big|_{\theta = \theta_n}} \right]$$

$$= \lim_{n \to \infty} \frac{t_{n,\alpha} - E(T_n)\big|_{\theta = \theta_n}}{\sigma(T_n)\big|_{\theta = \theta_0}} \qquad \text{from condition 3}$$

Expanding $E(T_n)\big|_{\theta = \theta_n}$ in a Taylor's series about θ_0, we obtain

$$E(T_n)\big|_{\theta = \theta_n} = E(T_n)\big|_{\theta = \theta_0} + (\theta_n - \theta_0)\frac{dE(T_n)}{d\theta}\bigg|_{\theta = \theta_0} + \epsilon$$

where $\lim_{n \to \infty} \epsilon = 0$ because of regularity condition 3. Substituting this in the above expression for z, we obtain

$$z = \lim_{n \to \infty} \left\{ \frac{t_{n,\alpha} - E(T_n)\big|_{\theta = \theta_0}}{\sigma(T_n)\big|_{\theta = \theta_0}} - \frac{(\theta_n - \theta_0)[dE(T_n)/d\theta]\big|_{\theta = \theta_0} - \epsilon}{\sigma(T_n)\big|_{\theta = \theta_0}} \right\}$$

$$= \lim_{n \to \infty} \left[\frac{t_{n,\alpha} - E(T_n)\big|_{\theta = \theta_0}}{\sigma(T_n)\big|_{\theta = \theta_0}} \right] - dc \qquad \text{from conditions 2 and 3}$$

$$= z_\alpha - dc$$

since condition 4 also implies that if $t_{n,\alpha}$ is the size α critical value, then

$$\alpha = \lim_{n \to \infty} P(T_n \geq t_{n,\alpha} \mid \theta = \theta_0) = \lim_{n \to \infty} P\left[\frac{T_n - E(T_n)\big|_{\theta = \theta_0}}{\sigma(T_n)\big|_{\theta = \theta_0}} \right.$$

$$\left. \geq \frac{t_{n,\alpha} - E(T_n)\big|_{\theta = \theta_0}}{\sigma(T_n)\big|_{\theta = \theta_0}} \right] = 1 - \Phi(z_\alpha)$$

Theorem 2.2 *If T_n and T_n^* are two tests satisfying the four regularity conditions above, the ARE of T relative to T^* is*

$$\text{ARE } (T,T^*) = \lim_{n \to \infty} \left[\frac{dE(T_n)/d\theta \big|_{\theta = \theta_0}}{dE(T_n^*)/d\theta \big|_{\theta = \theta_0}} \right]^2 \frac{\sigma^2(T_n^*)\big|_{\theta = \theta_0}}{\sigma^2(T_n)\big|_{\theta = \theta_0}} \qquad (2.1)$$

Proof From Theorem 2.1, the limiting powers of tests T_n and T_n^*, respectively, are

$$1 - \Phi(z_\alpha - dc) \quad \text{and} \quad 1 - \Phi(z_\alpha - d^*c^*)$$

These quantities are equal if

$$z_\alpha - dc = z_\alpha - d^*c^*$$

or, equivalently, for

$$\frac{d^*}{d} = \frac{c}{c^*}$$

From regularity condition 3, the sequences of alternatives are the same if

$$\theta_n = \theta_0 + \frac{d}{\sqrt{n}} = \theta_n^* = \theta_0 + \frac{d^*}{\sqrt{n^*}}$$

or, equivalently, for

$$\frac{d}{\sqrt{n}} = \frac{d^*}{\sqrt{n^*}} \quad \text{or} \quad \frac{d^*}{d} = \left(\frac{n^*}{n}\right)^{\frac{1}{2}}$$

Since the ARE is the limit of the ratio of sample sizes when the limiting power and sequence of alternatives are the same for both tests, we have

$$
\begin{aligned}
\text{ARE } (T, T^*) &= \frac{n^*}{n} = \left(\frac{d^*}{d}\right)^2 = \left(\frac{c}{c^*}\right)^2 \\
&= \lim_{n \to \infty} \left[\frac{dE(T_n)/d\theta \Big|_{\theta = \theta_0}}{\sqrt{n}\, \sigma(T_n) \Big|_{\theta = \theta_0}} \frac{\sqrt{n}\, \sigma(T_n^*) \Big|_{\theta = \theta_0}}{dE(T_n^*)/d\theta \Big|_{\theta = \theta_0}} \right]^2 \\
&= \lim_{n \to \infty} \frac{\dfrac{[dE(T_n)/d\theta]^2}{\sigma^2(T_n)} \Big|_{\theta = \theta_0}}{\dfrac{[dE(T_n^*)/d\theta]^2}{\sigma^2(T_n^*)} \Big|_{\theta = \theta_0}} \quad (2.2)
\end{aligned}
$$

which is equivalent to (2.1).

From expression (2.2) we see that when these regularity conditions are satisfied, the ARE can be interpreted as the limit as n approaches infinity of the ratio of two quantities

$$\text{ARE}(T,T^*) = \lim_{n \to \infty} \frac{e(T_n)}{e(T_n^*)} \tag{2.3}$$

where $e(T_n)$ is called the *efficacy* of the test statistic T_n when used to test the hypothesis $\theta = \theta_0$ and

$$e(T_n) = \frac{[dE(T_n)/d\theta]^2}{\sigma^2(T_n)}\bigg|_{\theta = \theta_0} \tag{2.4}$$

Theorem 2.3 *The statement in Theorem 2.2 remains valid as stated if both tests are two-sided, with rejection region*

$$T_n \in R \qquad \text{for } t_n \geq t_{n,\alpha_1} \qquad \text{or} \qquad t_n \leq t_{n,\alpha_2}$$

where the size is still α, and a corresponding rejection region is defined for T_n^* with the same α_1 and α_2. Then the alternative is also two-sided, as $H_1\colon \theta \neq \theta_0$.

Note that the result for the ARE in Theorem 2.2 is independent of both the quantities α and γ. Therefore, when the regularity conditions are satisfied, the ARE does not suffer the disadvantages of the power-efficiency criterion. However, it is only an approximation to efficiency for any finite sample size and/or alternative not in the neighborhood of the null case.

In the two-sample cases, where the null hypothesis is equal distributions, if the hypothesis can be parameterized in terms of θ, the same theorems can be used for either one- or two-sided tests. The limiting process must be restricted by assuming that as m and n approach infinity, $m/n = \lambda$, a constant. When m is approximately a fixed proportion of n regardless of the total sample size $m + n$, the theory goes through as before for n approaching infinity. For two-sample linear rank tests, evaluation of the efficacies is simplified by using the general results for mean and variance given in Theorem 8.3.8.

14.3 EXAMPLES OF THE CALCULATION OF EFFICACY AND ARE

ONE–SAMPLE AND PAIRED–SAMPLE PROBLEMS

In the one-sample and paired-sample problems treated in Chap. 6, the null hypothesis concerned the value of the population median or median of the population of differences of pairs. This might be called a one-sample location problem with the distribution model

$$F_X(x) = F(x - \theta) \tag{3.1}$$

for some continuous distribution F with median zero. Since F_X then has median θ, the model implies the null hypothesis

$$H_0: \theta = 0$$

against one- or two-sided alternatives.

The nonparametric tests for this model can be considered analogs of the one-sample or paired-sample Student's t test for location of the mean or difference of means if F is any continuous distribution symmetric about zero since then θ is both the mean and median of F_X. For a single random sample of size N from a population F_X with mean μ and variance σ^2, the t test statistic of the null hypothesis

$$H_0: \mu = 0$$

is

$$T_N^* = \frac{\sqrt{N}\,\bar{X}_N}{S_N} = \left[\frac{\sqrt{N}(\bar{X}_N - \mu)}{\sigma} + \frac{\sqrt{N}\,\mu}{\sigma} \right] \frac{\sigma}{S_N}$$

where $S_N{}^2 = \sum_{i=1}^{N} (X_i - \bar{X})^2/(N-1)$. Since $\lim_{N\to\infty} (S_N/\sigma) = 1$, T_N^* is asymptotically equivalent to the Z (normal-theory) test for σ known. The moments of T_N^* for N large then are

$$E(T_N^*) = \frac{\sqrt{N}\,\mu}{\sigma} \qquad \text{and} \qquad \mathrm{var}(T_N^*) = \frac{N\,\mathrm{var}(\bar{X}_N)}{\sigma^2} = 1$$

and

$$\frac{d}{d\mu} E(T_N^*)\Big|_{\mu=0} = \frac{\sqrt{N}}{\sigma}$$

Using (2.4), the large-sample efficacy of Student's t test for observations from any population F_X with mean μ and variance σ^2 is

$$e(T_N^*) = \frac{N}{\sigma^2} \tag{3.2}$$

The ordinary sign-test statistic K_N of Sec. 6.2 is appropriate for the model (3.1) with

$$H_0: M = \theta = 0$$

Since K_N follows the binomial probability distribution, its moments are

$$E(K_N) = Np \qquad \text{and} \qquad \mathrm{var}(K_N) = Np(1-p)$$

where

$$p = P(X > 0) = 1 - F_X(0)$$

If θ is the median of the population F_X, $F_X(0)$ is a function of θ, and for the location model (3.1) we have

$$\frac{dp}{d\theta}\Big|_{\theta=0} = \frac{d}{d\theta}[1 - F_X(0)]\Big|_{\theta=0}$$

$$= \frac{d}{d\theta}[1 - F(-\theta)]\Big|_{\theta=0}$$

$$= f(\theta)\Big|_{\theta=0} = f(0) = f_X(\theta)$$

When $\theta = 0$, $p = \frac{1}{2}$, so that $\mathrm{var}(K_N)\Big|_{\theta=0} = N/4$. The efficacy of the ordinary sign test for N observations from any population F_X with median θ is therefore

$$e(K_N) = 4Nf_X^2(\theta) = 4Nf^2(0) \tag{3.3}$$

These efficacy results in (3.2) and (3.3) can be applied to any population, although the distribution must be symmetric for the hypotheses to be equivalent. Three examples are given below.

1. Normal distribution

$$F_X \text{ is } N(\theta,\sigma^2) \qquad F_X(x) = \Phi\left(\frac{x - \theta}{\sigma}\right) \qquad \text{or} \qquad F(x) = \Phi\left(\frac{x}{\sigma}\right)$$

$$f(0) = (2\pi\sigma^2)^{-\frac{1}{2}} \qquad e(K_N) = \frac{2N}{\pi\sigma^2}$$

$$\mathrm{ARE}\,(K,T^*) = \frac{2}{\pi}$$

2. Uniform distribution

$$f_X(x) = 1 \qquad \text{for } \theta - \frac{1}{2} < x < \theta + \frac{1}{2}$$

or

$$f(x) = 1 \qquad \text{for } -\frac{1}{2} < x < \frac{1}{2}$$
$$f(0) = 1 \qquad \mathrm{var}(X) = \frac{1}{12}$$
$$e(T_N^*) = 12N \qquad e(K_N) = 4N$$
$$\mathrm{ARE}\,(K,T^*) = \frac{1}{3}$$

3. Double exponential distribution

$$f_X(x) = \frac{1}{2}\lambda e^{-\lambda|x-\theta|} \qquad \text{or} \qquad f(x) = \frac{1}{2}\lambda e^{-\lambda|x|}$$

$$f(0) = \frac{1}{2}\lambda \qquad \mathrm{var}(X) = \frac{2}{\lambda^2}$$

$$e(T_N^*) = \frac{N\lambda^2}{2} \qquad e(K_N) = N\lambda^2$$

$$\mathrm{ARE}\,(K,T^*) = 2$$

For the sampling situation where we have paired-sample data and the hypotheses concern the median or mean difference, all results obtained above are directly applicable if the random variable X is replaced by the difference variable $D = X - Y$. It should be noted that the parameter σ^2 in (3.2) then denotes the variance of the population of differences,

$$\sigma_D{}^2 = \sigma_X{}^2 + \sigma_Y{}^2 - 2\operatorname{cov}(X,Y)$$

and the $f_X(\theta)$ in (3.3) now becomes $f_D(\theta)$.

TWO-SAMPLE LOCATION PROBLEMS

For the general location problem in the case of two independent random samples of sizes m and n, the distribution model is

$$F_Y(x) = F_X(x - \theta) \tag{3.4}$$

and the null hypothesis of identical distributions is

$$H_0\colon \theta = 0$$

The corresponding classical test statistic for populations with a common variance σ^2 is the two-sample Student's t test statistic

$$T^*_{m,n} = \sqrt{\frac{mn}{m+n}}\,\frac{\bar{Y}_n - \bar{X}_m}{S_{m+n}} = \sqrt{\frac{mn}{m+n}}\left(\frac{\bar{Y}_n - \bar{X}_m - \theta}{\sigma} + \frac{\theta}{\sigma}\right)\frac{\sigma}{S_{m+n}}$$

where

$$S^2_{m+n} = \frac{\displaystyle\sum_{i=1}^{m}(X_i - \bar{X})^2 + \sum_{i=1}^{n}(Y_i - \bar{Y})^2}{m + n - 2}$$

is the pooled estimate of σ^2. Since S_{m+n}/σ approaches 1 as $n \to \infty$, $m/n = \lambda$, the moments of $T^*_{m,n}$ for n large, $\theta = \mu_Y - \mu_X$, are

$$E(T^*_{m,n}) = \theta\,\frac{\sqrt{mn/(m+n)}}{\sigma} \qquad \operatorname{var}(T^*_{m,n}) = \frac{mn}{m+n}\,\frac{\sigma^2/m + \sigma^2/n}{\sigma^2} = 1$$

Therefore

$$\frac{d}{d\theta}E(T^*_{m,n}) = \frac{\sqrt{mn/(m+n)}}{\sigma}$$

and the efficacy of Student's t test for any population is

$$e(T^*_{m,n}) = \frac{mn}{\sigma^2(m+n)} \tag{3.5}$$

For the Mann-Whitney test statistic given in (7.5.2), the mean is

$$E(U_{m,n}) = mnP(Y < X) = mnP(Y - X < 0) = mn\pi$$

A general expression for π was given in (7.5.3) for any distributions. For the location model in (3.4), this integral becomes

$$\pi = \int_{-\infty}^{\infty} F_X(x - \theta) f_X(x) \, dx$$

so that

$$\frac{d}{d\theta} E(U_{m,n}) \Big|_{\theta=0} = mn \frac{d\pi}{d\theta} \Big|_{\theta=0} = -mn \int_{-\infty}^{\infty} f_X^2(x) \, dx$$

Since $\pi = \frac{1}{2}$ under H_0, the null variance is found from (7.5.15) to be

$$\text{var}(U_{m,n}) = \frac{mn(m + n + 1)}{12}$$

The efficacy then is

$$e(U_{m,n}) = \frac{12mn \left[\int_{-\infty}^{\infty} f_X^2(x) \, dx \right]^2}{m + n + 1} \tag{3.6}$$

The ARE of the Mann-Whitney test relative to Student's t test can be found by evaluating the efficacies in (3.5) and (3.6) for any population F_X with variance σ^2. Since Student's t test is the best test for normal distributions satisfying the general location model, we shall use this as an example. If F_X is $N(\mu_X, \sigma^2)$,

$$\int_{-\infty}^{\infty} f_X^2(x) \, dx = \int_{-\infty}^{\infty} (2\pi\sigma^2)^{-1} \exp\left[-\left(\frac{x - \mu_X}{\sigma}\right)^2 \right] dx$$

$$= \frac{1}{\sqrt{2\pi\sigma^2}} \sqrt{\frac{1}{2}} = (2\sqrt{\pi}\,\sigma)^{-1}$$

$$e(T_{m,n}^*) = \frac{mn}{\sigma^2(m + n)} \qquad e(U_{m,n}) = \frac{3mn}{\pi\sigma^2(m + n + 1)}$$

$$\text{ARE}\,(U_{m,n}, T_{m,n}^*) = \frac{3}{\pi}$$

For the uniform and double exponential distributions, the relative efficiencies are 1 and $\frac{3}{2}$, respectively (Prob. 14.1).

This evaluation of efficacy of the Mann-Whitney test does not make use of the fact that it can be written as a linear rank statistic. As an illustration of how the general results given in Theorem 8.3.8 simplify the calculation of efficiencies, we shall show that the Terry and van der Waerden tests discussed in Sec. 9.3 are asymptotically optimum rank tests for normal populations differing only in location.

The weights for the van der Waerden test of (9.3.2) in the notation of Theorem 8.3.8 are

$$a_i = \Phi^{-1}\left(\frac{i}{N+1}\right) = \Phi^{-1}\left(\frac{i}{N}\frac{N}{N+1}\right) = J_N\left(\frac{i}{N}\right)$$

The combined population cdf for the general location model of (3.4) is

$$H(x) = \lambda_N F_X(x) + (1 - \lambda_N)F_X(x - \theta)$$

so that

$$J[H(x)] = \lim_{N \to \infty} J_N(H) = \Phi^{-1}[\lambda_N F_X(x) + (1 - \lambda_N)F_X(x - \theta)]$$

Applying Theorem 8.3.8 now to this J function, the mean is

$$\mu_N = \int_{-\infty}^{\infty} \Phi^{-1}[\lambda_N F_X(x) + (1 - \lambda_N)F_X(x - \theta)]f_X(x)\,dx$$

To evaluate the derivative, we note that since

$$\Phi^{-1}[g(\theta)] = y \quad \text{if and only if} \quad g(\theta) = \Phi(y)$$

it follows that

$$\frac{d}{d\theta}g(\theta) = \varphi(y)\frac{dy}{d\theta} \quad \text{or} \quad \frac{dy}{d\theta} = \frac{d[g(\theta)]/d\theta}{\varphi(y)}$$

Therefore the derivative of μ_N above is

$$\frac{d}{d\theta}\mu_N\bigg|_{\theta=0} = \int_{-\infty}^{\infty} \frac{-(1 - \lambda_N)f_X^2(x)}{\varphi\{\Phi^{-1}[F_X(x)]\}}\,dx \tag{3.7}$$

Now to evaluate the variance when $\theta = 0$, we can use Corollary 8.3.8 to obtain

$$N\lambda_N\sigma_N^2 = (1 - \lambda_N)\left\{\int_0^1 [\Phi^{-1}(u)]^2\,du - \left[\int_0^1 \Phi^{-1}(u)\,du\right]^2\right\}$$

$$= (1 - \lambda_N)\left\{\int_{-\infty}^{\infty} x^2\varphi(x)\,dx - \left[\int_{-\infty}^{\infty} x\varphi(x)\,dx\right]^2\right\}$$

$$= 1 - \lambda_N$$

The integral in (3.7) reduces to a simple expression when $F_X(x)$ is $N(\mu_X, \sigma^2)$ since then

$$F_X(x) = \Phi\left(\frac{x - \mu_X}{\sigma}\right) \quad \text{and} \quad f_X(x) = \frac{1}{\sigma}\varphi\left(\frac{x - \mu_X}{\sigma}\right)$$

$$\frac{d}{d\theta}\mu_N\bigg|_{\theta=0} = -\frac{1 - \lambda_N}{\sigma^2}\int_{-\infty}^{\infty} \frac{\varphi^2[(x - \mu_X)/\sigma]}{\varphi[(x - \mu_X)/\sigma]}\,dx$$

$$= -\frac{1 - \lambda_N}{\sigma}\int_{-\infty}^{\infty} \frac{1}{\sigma}\varphi\left(\frac{x - \mu_X}{\sigma}\right)\,dx = -\frac{1 - \lambda_N}{\sigma}$$

The efficacy of the van der Waerden X_N test in this normal case is then

$$e(X_N) = \frac{N\lambda_N(1 - \lambda_N)}{\sigma^2} = \frac{mn}{N\sigma^2} \tag{3.8}$$

which equals the efficacy of the Student's t test $T^*_{m,n}$ given in (3.5).

TWO–SAMPLE SCALE PROBLEMS

The general distribution model of the scale problem for two independent random samples is

$$F_Y(x) = F_X(\theta x) \tag{3.9}$$

where we are assuming without loss of generality that the common location is zero. The null hypothesis of identical distributions then is

$$H_0: \theta = 1$$

against either one- or two-sided alternatives. Given two independent random samples of sizes m and n, the analogous parametric test for the scale problem is the statistic

$$T^*_{m,n} = \frac{(n - 1) \sum\limits_{i=1}^{m} (X_i - \bar{X})^2}{(m - 1) \sum\limits_{i=1}^{n} (Y_i - Y)^2}$$

Since X and Y are independent and in our model above

$$\mathrm{var}(X) = \theta^2 \, \mathrm{var}(Y)$$

the expected value is

$$E(T^*_{m,n}) = E\left[\frac{\sum\limits_{i=1}^{m} (X_i - \bar{X})^2}{m - 1}\right] E\left[\frac{n - 1}{\sum\limits_{i=1}^{n} (Y_i - \bar{Y})^2}\right]$$

$$= (n - 1) \, \mathrm{var}(X) \, E\left[\frac{1}{\sum\limits_{i=1}^{n} (Y_i - \bar{Y})^2}\right]$$

$$= (n - 1)\theta^2 \, E\left[\frac{\mathrm{var}(Y)}{\sum\limits_{i=1}^{n} (Y_i - \bar{Y})^2}\right]$$

$$= (n - 1)\theta^2 \, E\left(\frac{1}{V}\right)$$

The probability distribution of V depends on the particular distribution F_X, but in the normal-theory model, where $F_X(x) = \Phi(x)$, V has the chi-square distribution with $n - 1$ degrees of freedom. Therefore we can evaluate

$$E\left(\frac{1}{V}\right) = \frac{1}{\Gamma\left(\dfrac{n-1}{2}\right) 2^{(n-1)/2}} \int_0^\infty x^{-1} e^{-x/2} x^{[(n-1)/2]-1} \, dx$$

$$= \frac{\Gamma\left(\dfrac{n-3}{2}\right)}{2\Gamma\left(\dfrac{n-1}{2}\right)} = \frac{1}{n-3}$$

$$E(T^*_{m,n}) = \frac{(n-1)\theta^2}{n-3} \qquad \frac{d}{d\theta} E(T^*_{m,n}) \bigg|_{\theta=1} = \frac{2(n-1)}{n-3}$$

In this normal-theory model, under the null hypothesis the distribution of $T^*_{m,n}$ is Snedecor's F with $m - 1$ and $n - 1$ degrees of freedom. Since the variance of the F distribution with ν_1 and ν_2 degrees of freedom is

$$\frac{2\nu_2{}^2(\nu_1 + \nu_2 - 2)}{\nu_1(\nu_2 - 4)(\nu_2 - 2)^2}$$

we have here

$$\operatorname{var}(T^*_{m,n}) \bigg|_{\theta=1} = \frac{2(n-1)^2(N-4)}{(m-1)(n-5)(n-3)^2}$$

The statistic $T^*_{m,n}$ is the best test for the normal-theory model, and its efficacy for this distribution is

$$e(T^*_{m,n}) = \frac{2(m-1)(n-5)}{N-4} \approx \frac{2mn}{N} = 2N\lambda_N(1 - \lambda_N) \tag{3.10}$$

We shall now evaluate the efficacy of the Mood and Freund-Ansari-Bradley-Barton-David-Siegel-Tukey tests by applying the results of Theorem 8.3.8 to the two-sample scale model (3.9), for which

$$H(x) = \lambda_N F_X(x) + (1 - \lambda_N) F_X(\theta x)$$

For the Mood test statistic of Sec. 10.2, we write

$$M'_N = N^{-2} \sum_{i=1}^N \left(i - \frac{N+1}{2}\right)^2 Z_i = N^{-2} M_N$$

so that for M_N'

$$a_i = \left(\frac{i}{N} - \frac{N+1}{2N}\right)^2 = J_N\left(\frac{i}{N}\right)$$

$$\lim_{N\to\infty} J_N(H) = (H - \tfrac{1}{2})^2$$

The mean of M_N' then is

$$\mu_N = \int_{-\infty}^{\infty} [\lambda_N F_X(x) + (1 - \lambda_N)F_X(\theta x) - \tfrac{1}{2}]^2 f_X(x)\,dx$$

$$\frac{d\mu_N}{d\theta}\Big|_{\theta=1} = 2(1 - \lambda_N)\int_{-\infty}^{\infty} [F_X(x) - \tfrac{1}{2}]x f_X^2(x)\,dx$$

and the variance under the null hypothesis is

$$N\lambda_N\sigma_N^2 = (1 - \lambda_N)\left\{\int_0^1 (u - \tfrac{1}{2})^4\,du - \left[\int_0^1 (u - \tfrac{1}{2})^2\,du\right]^2\right\}$$

$$= \frac{1 - \lambda_N}{180}$$

so that the efficacy for any distribution F_X with median zero is

$$e(M_N) = 720 N\lambda_N(1 - \lambda_N)\left\{\int_{-\infty}^{\infty} [F_X(x) - \tfrac{1}{2}]x f_X^2(x)\,dx\right\}^2 \quad (3.11)$$

In order to compare the Mood statistic with the F test statistic, we shall calculate $e(M_N)$ for the normal-theory model, where $F_X(x) = \Phi(x)$. Then

$$\int_{-\infty}^{\infty} [\Phi(x) - \tfrac{1}{2}]x\varphi^2(x)\,dx = \int_{-\infty}^{\infty} x\Phi(x)\varphi^2(x)\,dx$$
$$- \frac{1}{2\sqrt{2\pi}}\int_{-\infty}^{\infty} x\varphi(x\sqrt{2})\,dx$$

$$= \int_{-\infty}^{\infty}\int_{-\infty}^{x} x\varphi(t)\varphi^2(x)\,dt\,dx$$

$$= \int_{-\infty}^{\infty} \varphi(t)\left(\int_t^{\infty} x\frac{1}{2\pi}e^{-x^2}\,dx\right)dt$$

$$= \frac{1}{4\pi}\int_{-\infty}^{\infty} \varphi(t)e^{-t^2}\,dt = \frac{1}{4\pi}\int_{-\infty}^{\infty} \frac{1}{\sqrt{2\pi}}e^{-3/2 t^2}\,dt = (4\pi\sqrt{3})^{-1}$$

For normal distributions, the result then is

$$e(M_N) = \frac{15 N\lambda_N(1 - \lambda_N)}{\pi^2} \qquad \text{ARE } (M_N, T_{m,n}^*) = \frac{15}{2\pi^2}$$

Using the same procedure for the tests of Sec. 10.3, we write the weights for the test A'_N so that $(N + 1)A'_N = A_N$, where A_N was given in (10.3.1). The result is

$$a_i = \left| \frac{i}{N + 1} - \frac{1}{2} \right| = \frac{N}{N + 1} \left| \frac{i}{N} - \frac{1}{2} - \frac{1}{2N} \right| = J_N\left(\frac{i}{N}\right)$$
$$J(H) = |H - \tfrac{1}{2}|$$

The mean of A'_N is

$$\mu_N = \int_{-\infty}^{\infty} |\lambda_N F_X(x) + (1 - \lambda_N)F_X(\theta x) - \tfrac{1}{2}|f_X(x)\, dx$$

$$\frac{d\mu_N}{d\theta}\bigg|_{\theta=1} = (1 - \lambda_N) \int_{-\infty}^{\infty} |x f_X(\theta x)| f_X(x)\, dx \bigg|_{\theta=1}$$

$$= (1 - \lambda_N) \int_{-\infty}^{\infty} |x| f_X{}^2(x)\, dx$$

If $f_X(x)$ is symmetric about its zero median, this reduces to

$$\frac{d\mu_N}{d\theta}\bigg|_{\theta=1} = 2(1 - \lambda_N) \int_0^{\infty} x f_X{}^2(x)\, dx \tag{3.12}$$

The variance when $\theta = 1$ is

$$N\lambda_N \sigma_N{}^2 = (1 - \lambda_N) \left[\int_0^1 |u - \tfrac{1}{2}|^2\, du - \left(\int_0^1 |u - \tfrac{1}{2}|\, du \right)^2 \right]$$

$$= \frac{1 - \lambda_N}{48}$$

For $f(x) = \varphi(x)$, the integral in (3.12) is easily evaluated, and the results are

$$e(A_N) = \frac{12N\lambda_N(1 - \lambda_N)}{\pi^2}$$

$$\text{ARE}\,(A_N, T^*_{m,n}) = \frac{6}{\pi^2} \qquad \text{ARE}\,(A_N, M_N) = \tfrac{4}{5}$$

TESTS OF ASSOCIATION

In the tests of association in Chap. 12, the null hypothesis was that the X and Y random variables are independent. In order to calculate efficiency, the model of relationship must be specified, and there are a wide variety of alternatives which may be appropriate to the null hypothesis of independence. Further, there is no general test for independence per se in classical statistics. However, if the joint distribution of X and Y is the bivariate normal, independence is equivalent to a zero product-moment correlation coefficient ρ and also to a zero slope β in the classical regression model. Therefore, we shall restrict our comparisons here to the bivariate normal distribution model, where

$$F_{Y|X} \text{ is } N(\alpha + \beta x, \sigma^2) \qquad \text{or} \qquad F_{Y|X}(y \mid x) = \Phi\left(\frac{y - \alpha - \beta x}{\sigma}\right)$$

(3.13)

and consider the tests of independence based on Kendall's tau coefficient and Spearman's rank-correlation coefficient as nonparametric analogs of

$$H_0: \beta = 0$$

The normal-theory test statistic for a random sample of n pairs (x_i, Y_i) is based on the sample estimated slope or

$$T_n^* = B = \frac{\sum\limits_{i=1}^{n} (x_i - \bar{x})(Y_i - \bar{Y})}{\sum\limits_{i=1}^{n} (x_i - \bar{x})^2}$$

It is well known that the sampling distribution of B is normal with moments

$$E(B) = \beta \qquad \text{and} \qquad \text{var}(B) = \frac{\sigma^2}{\sum\limits_{i=1}^{n} (x_i - \bar{x})^2}$$

The efficacy of this classical test statistic then is

$$e(T_n^*) = \frac{\sum\limits_{i=1}^{n} (x_i - \bar{x})^2}{\sigma^2}$$

(3.14)

If we assume that the numbers $x_1, x_2, \ldots, x_n$ are equally spaced, no generality is lost by replacing them by the first n positive integers, so that

$$\sum_{i=1}^{n} (x_i - \bar{x})^2 = \sum_{i=1}^{n} \left(i - \frac{n+1}{2}\right)^2 = \frac{n(n^2 - 1)}{12}$$

and the efficacy for this special case is

$$e(T_n^*) = \frac{n(n^2 - 1)}{12\sigma^2}$$

(3.15)

The Kendall tau statistic was given in (12.4.1) as

$$T_n = \frac{\sum\limits_{i=1}^{n} \sum\limits_{j=1}^{n} U_{ij} V_{ij}}{n(n - 1)}$$

where $U_{ij} = \text{sgn}(X_j - X_i)$ and $V_{ij} = \text{sgn}(Y_j - Y_i)$. If we assume without loss of generality that the subscripts are assigned such that $X_i < X_j$ for all $i < j$, T_n can be written as a function of V_{ij} only as

$$T_n = \frac{\displaystyle\sum\sum_{1 \le i < j \le n} V_{ij} - \sum\sum_{1 \le j < i \le n} V_{ij}}{n(n-1)} = \frac{2 \displaystyle\sum\sum_{1 \le i < j \le n} V_{ij}}{n(n-1)}$$

In the normal-theory model of (3.13), for any given x_i and x_j, the difference $Y_j - Y_i$ is normally distributed with moments

$$E(Y_j - Y_i \mid x_i, x_j) = \beta(x_j - x_i) \qquad \text{and} \qquad \text{var}(Y_j - Y_i \mid x_i, x_j) = 2\sigma^2$$

Therefore we can evaluate for any given x_i, x_j

$$E(V_{ij}) = P(Y_j - Y_i > 0) - P(Y_j - Y_i < 0)$$
$$= 1 - 2P(Y_j - Y_i < 0)$$
$$= 1 - 2\Phi\left(-\frac{\beta(x_j - x_i)}{\sqrt{2}\,\sigma}\right)$$

$$\frac{d}{d\beta} E(V_{ij})\Big|_{\beta=0} = 2\frac{x_j - x_i}{\sqrt{2}\,\sigma} \varphi(0) = \frac{x_j - x_i}{\sqrt{\pi}\,\sigma} \tag{3.16}$$

$$\frac{d}{d\beta} E(T_n)\Big|_{\beta=0} = \frac{2 \displaystyle\sum\sum_{1 \le i < j \le n} (x_j - x_i)}{\sqrt{\pi}\,\sigma n(n-1)}$$

Under the null hypothesis of independence, the variance was found in (12.2.29) to be

$$\text{var}(T_n) = \frac{2(2n+5)}{9n(n-1)}$$

so that the efficacy is

$$e(T_n) = \frac{18 \left[\displaystyle\sum\sum_{1 \le i < j \le n} (x_j - x_i) \right]^2}{n(n-1)(2n+5)\pi\sigma^2}$$

This efficacy can be expressed in a more convenient form for computation because of the algebraic identity

$$\sum\sum_{1 \le i < j \le n} (x_j - x_i) = \sum_{j=2}^{n} \sum_{i=1}^{j-1} x_j - \sum_{i=1}^{n-1} \sum_{j=i+1}^{n} x_i$$
$$= \sum_{j=1}^{n} (j-1)x_j - \sum_{i=1}^{n} (n-i)x_i$$
$$= 2\sum_{i=1}^{n} ix_i - n(n+1)\bar{x}$$

If $x_i = i$ for all $i = 1, 2, \ldots, n$, this reduces to

$$\sum_{1 \le i < j \le n} (x_j - x_i) = 2 \sum_{i=1}^{n} i^2 - \frac{n(n+1)^2}{2} = \frac{n(n^2-1)}{6}$$

and the efficacy is

$$e(T_n) = \frac{n(n-1)(n+1)^2}{2(2n+5)\pi\sigma^2} \tag{3.17}$$

Spearman's rank-correlation coefficient, as in (12.3.4) and (12.4.3), can be written

$$n(n^2 - 1)R_n = 12 \sum_{i=1}^{n} i(S_i - \bar{S}) = -6 \sum_{i=1}^{n} \sum_{j=1}^{n} iV_{ij}$$

with V_{ij} defined as before. The derivative then using the result in (3.16) when $x_i = i$ is

$$n(n^2 - 1) \frac{d}{d\beta} E(R_n) \Big|_{\beta=0} = -6 \sum_{i=1}^{n} \sum_{j=1}^{n} \frac{i(j-i)}{\sqrt{\pi}\,\sigma}$$

$$= \frac{-6 \left[\left(\sum_{i=1}^{n} i \right)^2 - n \sum_{i=1}^{n} i^2 \right]}{\sqrt{\pi}\,\sigma}$$

$$= \frac{n^2(n^2-1)}{2\sqrt{\pi}\,\sigma}$$

$$\frac{d}{d\beta} E(R_n) \Big|_{\beta=0} = \frac{n}{2\sqrt{\pi}\,\sigma}$$

In (12.3.13) the variance of R_n for independent random variables was found to be $\text{var}(R_n) = 1/(n-1)$, so that the efficacy is

$$e(R_n) = \frac{n^2(n-1)}{4\pi\sigma^2} \tag{3.18}$$

The relevant asymptotic relative efficiencies for the normal-theory regression model with equally spaced x values are found from (3.15), (3.17), and (3.18) to be

$$\text{ARE}(R,T^*) = \text{ARE}(T,T^*) = \frac{3}{\pi} \qquad \text{ARE}(R,T) = 1$$

PROBLEMS

14.1. Use the results of Theorem 8.3.8 to evaluate the efficacy of the two-sample Wilcoxon test statistic of (9.2.1) for the location model $F_Y(x) = F_X(x - \theta)$ where:

(a) F_X is $N(\mu_X, \sigma^2)$ or $F_X(x) = \Phi\left(\dfrac{x - \mu_X}{\sigma}\right)$

(b) F_X is uniform, with

$$F_X(x) = \begin{cases} 0 & x \leq -\frac{1}{2} \\ x + \frac{1}{2} & -\frac{1}{2} < x \leq \frac{1}{2} \\ 1 & x > \frac{1}{2} \end{cases}$$

(c) F_X is double exponential, with

$$F_X(x) = \begin{cases} \frac{1}{2}e^{\lambda x} & x \leq 0 \\ 1 - \frac{1}{2}e^{-\lambda x} & x > 0 \end{cases}$$

14.2. Calculate the efficacy of the two-sample Student's t test statistic in cases (b) and (c) of Prob. 14.1.

14.3. Use your answers to Probs. 14.1 and 14.2 to verify the following results for the ARE of the Wilcoxon (or Mann-Whitney) test to Student's t test:

Normal: $3/\pi$
Uniform: 1
Double exponential: $\frac{3}{2}$

14.4. Evaluate the efficiency of the Klotz normal-scores test of (10.5.1) relative to the F test statistic for the normal-theory scale model.

14.5. Evaluate the efficacies of the M_N and A_N test statistics and compare their efficiency for the scale model where, as in Prob. 14.1:

(a) F_X is uniform.
(b) F_X is double exponential.

14.6. Use (6.3.5) and (6.3.6) to evaluate the efficacy of the Wilcoxon signed-rank test T_N^+ statistic of Sec. 6.3 for the location model

$$F_X(x) = F(x - \theta)$$

where θ is the median of F_X, for the distributions (a), (b), and (c) of Prob. 14.1. Calculate the value of the ARE for these three populations for T_N^+ versus the ordinary sign-test statistic K_N, and T_N^+ versus Student's t test statistic T_N^*.

References

Alexander, D. A., and D. Quade (1968): On the Kruskal-Wallis Three Sample *H*-statistic, *Univ. North Carolina Dept. Biostatistics Inst. Statistics Mimeo Ser.* 602.

Ansari, A. R., and R. A. Bradley (1960): Rank Sum Tests for Dispersion, *Ann. Math. Statistics*, **31**: 1174–1189.

Auble, J. D. (1953): Extended Tables for the Mann-Whitney Statistic, *Bull. Inst. Educ. Res.*, **1**: i–iii and 1–39.

Bateman, G. (1948): On the Power Function of the Longest Run as a Test for Randomness in a Sequence of Alternatives, *Biometrika*, **35**: 97–112.

Birnbaum, Z. W. (1952): Numerical Tabulation of the Distribution of Kolmogorov's Statistic for Finite Sample Size, *J. Am. Statist. Assoc.*, **47**: 425–441.

―――― and F. H. Tingey (1951): One-sided Confidence Contours for Probability Distribution Functions, *Ann. Math. Statistics*, **22**: 592–596.

Bradley, J. V. (1968): "Distribution-free Statistical Tests," Prentice-Hall, Inc., Englewood Cliffs, N.J.

Capon, J. (1965): On the Asymptotic Efficiency of the Kolmogorov-Smirnov Test, *J. Am. Statist. Assoc.*, **60**: 843–853.

Chernoff, H., and E. L. Lehmann (1954): The Use of the Maximum Likelihood Estimate in χ^2 Tests for Goodness of Fit, *Ann. Math. Statistics*, **25**: 579–586.

―――― and I. R. Savage (1958): Asymptotic Normality and Efficiency of Certain Nonparametric Test Statistics, *Ann. Math. Statistics*, **29**: 972–994.

Cochran, W. G. (1952): The χ^2 Test of Goodness of Fit, *Ann. Math. Statistics*, **23**: 315–345.

―――― and G. M. Cox (1957): "Experimental Designs," John Wiley & Sons, Inc., New York, 520–544.

Cramer, H. (1928): On the Composition of Elementary Errors, *Skand. Aktuarietidssk.*, **11**: 13-74, 141–180.

―――― (1946): "Mathematical Methods of Statistics," Princeton University Press, Princeton, N.J.

Darling, D. A. (1957): The Kolmogorov-Smirnov, Cramer–von Mises Tests, *Ann. Math. Statistics*, **28**: 823–838.

David, F. N. (1947): A Power Function for Tests for Randomness in a Sequence of Alternatives, *Biometrika*, **34**: 335–339.

―――― and D. E. Barton (1958): A Test for Birth-order Effects, *Ann. Human Eugenics*, **22**: 250–257.

David, H. T. (1952): Discrete Populations and the Kolmogorov-Smirnov Tests, *Univ. Chicago Statist. Res. Center Rept.* SRC-21103D27.

Drion, E. F. (1952): Some Distribution Free Tests for the Difference between Two Empirical Cumulative Distributions, *Ann. Math. Statistics*, **23**: 563–574.

Edgington, E. S. (1961): Probability Table for Number of Runs of Signs of First Differences in Ordered Series, *J. Am. Statist. Assoc.*, **56**: 156–159.

Fieller, E. C., H. O. Hartley, and E. S. Pearson (1957): Tests for Rank Correlation Coefficients: I, *Biometrika*, **44**: 470–481.

—— and E. S. Pearson (1961): Tests for Rank Correlation Coefficients: II, *Biometrika*, **48**: 29–40.

Fisz, M. (1963): "Theory of Probability and Mathematical Statistics," John Wiley & Sons, Inc., New York.

Fraser, D. A. S. (1957): "Nonparametric Methods in Statistics," John Wiley & Sons, Inc., New York.

Freund, J. E., and A. R. Ansari (1957): Two-way Rank Sum Tests for Variance, *Virginia Polytech. Inst. Tech. Rept. to Office of Ordnance Res. and Natl. Sci. Foundation* 34.

Friedman, M. (1937): The Use of Ranks to Avoid the Assumption of Normality Implicit in the Analysis of Variance, *J. Am. Statist. Assoc.*, **32**: 675–701.

—— (1940): A Comparison of Alternative Tests of Significance for the Problem of *m* Rankings, *Ann. Math. Statistics*, **11**: 86–92.

Gastwirth, J. L. (1965): Percentile Modifications of Two Sample Rank Tests, *J. Am. Statist. Assoc.*, **60**: 1127–1141.

Gibbons, J. D., and J. L. Gastwirth (1966): Small Sample Properties of Percentile Modified Rank Tests, *Johns Hopkins Univ. Dept. Statistics Tech. Rept.* 60.

Gnedenko, B. V. (1954): Tests of Homogeneity of Probability Distributions in Two Independent Samples (in Russian), *Dokl. An SSSR*, **80**: 525–528.

Goodman, L. A. (1954): Kolmogorov-Smirnov Tests for Psychological Research, *Psychological Bull.*, **51**: 160–168.

—— and W. H. Kruskal (1954, 1959, 1963): Measures of Association for Cross-classifications, *J. Am. Statist. Assoc.*, **49**: 732–764; **54**: 123–163; **58**: 310–364.

Gumbel, E. J. (1944): Ranges and Midranges, *Ann. Math. Statistics*, **15**: 414–422.

—— (1958): "Statistics of Extremes," Columbia University Press, New York.

Hajek, J., and Z. Sidak (1967): "Theory of Rank Tests," Academic Press, Inc., New York.

Harter, H. L. (1961): Expected Values of Normal Order Statistics, *Biometrika*, **48**: 151–165.

Hartley, H. O. (1942): The Probability Integral of the Range in Samples of *n* Observations from the Normal Population, *Biometrika*, **32**: 301–308.

Hodges, J. L., Jr. (1958): The Significance Probability of the Smirnov Two-sample Test, *Arkiv Mat.*, **3**: 469–486.

Hoeffding, W. (1951): Optimum Nonparametric Tests, pp. 83–92, in J. Neyman (ed.), *Proc. Second Berkeley Symp. Math. Statistics Probability*.

Hoel, P. G. (1962): "Introduction to Mathematical Statistics," John Wiley & Sons, Inc., New York.

Hogg, R. V., and A. T. Craig (1965): "Introduction to Mathematical Statistics," The Macmillan Company, New York.

Kamat, A. R. (1956): A Two-sample Distribution-free Test, *Biometrika*, **43**: 377–387.

Kendall, M. G. (1962): "Rank Correlation Methods," Hafner Publishing Company, Inc., New York.

Klotz, J. H. (1962): Nonparametric Tests for Scale, *Ann. Math. Statistics*, **33**: 495–512.

—— (1964): On the Normal Scores Two-sample Rank Test, *J. Am. Statist. Assoc.*, **49**: 652–664.

Kolmogorov, A. (1933): Sulla determinazione empirica di una legge di distribuzione, *Giorn. Inst. Ital. Attuari*, **4**: 83–91.

—— (1941): Confidence Limits for an Unknown Distribution Function, *Ann. Math. Statistics*, **12**: 461–463.

Kruskal, W. H. (1952): A Nonparametric Test for the Several Sample Problem, *Ann. Math. Statist.*, **23**: 525–540.

—— and W. A. Wallis (1952): Use of Ranks in One-criterion Analysis of Variance, *J. Am. Statist. Assoc.*, **47**: 583–621; errata, *ibid.*, **48**: 907–911 (1953).

Lehmann, E. L. (1953): The Power of Rank Tests, *Ann. Math. Statistics*, **24**: 23–43.

Levene, H. (1952): On the Power Function of Tests of Randomness Based on Runs Up and Down, *Ann. Math. Statistics*, **23**: 34–56.

—— and J. Wolfowitz (1944): The Covariance Matrix of Runs Up and Down, *Ann. Math. Statistics*, **15**: 58–69.

Lieberman, G. J., and D. B. Owen (1961): "Tables of the Hypergeometric Probability Distribution," Stanford University Press, Stanford, Calif.

Lilliefors, H. W. (1967): On the Kolmogorov-Smirnov Test for Normality with Mean and Variance Unknown, *J. Am. Statist. Assoc.*, **62**: 399–402.

—— (1969): On the Kolmogorov-Smirnov Test for the Exponential Distribution with Mean Unknown, *J. Am. Statist. Assoc.*, **64**: 387–389.

Mann, H. B., and D. R. Whitney (1947): On a Test Whether One of Two Random Variables Is Stochastically Larger than the Other, *Ann. Math. Statistics*, **18**: 50–60.

Massey, F. J. (1950): A Note on the Estimation of a Distribution Function by Confidence Limits, *Ann. Math. Statistics*, **21**: 116–119; correction, *ibid.*, **23**: 637–638 (1952).

—— (1951a): The Distribution of the Maximum Deviation between Two Sample Cumulative Step Functions, *Ann. Math. Statistics*, **22**: 125–128.

—— (1951b): The Kolmogorov-Smirnov Test for Goodness of Fit, *J. Am. Statist. Assoc.*, **46**: 68–78.

—— (1952): Distribution Table for the Deviation between Two Sample Cumulatives, *Ann. Math. Statistics*, **23**: 435–441.

Mises, R. von (1931): "Wahrscheinlichkeitsrechnung und ihre Anwendung in der Statistik und theoretischen Physik," F. Deuticke, Leipzig-Wien, 574.

Mood, A. M. (1940): The Distribution Theory of Runs, *Ann. Math. Statistics*, **11**: 367–392.

—— (1950): "Introduction to the Theory of Statistics," McGraw-Hill Book Company, New York, 394–406.

—— (1954): On the Asymptotic Efficiency of Certain Nonparametric Two-sample Tests, *Ann. Math. Statistics*, **25**: 514–522.

—— and F. A. Graybill (1963): "Introduction to the Theory of Statistics," 2d ed., McGraw-Hill Book Company, New York.

Moses, L. E. (1963): Rank Tests for Dispersion, *Ann. Math. Statistics*, **34**: 973–983.

Mosteller, F. (1941): Note on an Application of Runs to Quality Control Charts, *Ann. Math. Statistics*, **12**: 228–232.

National Bureau of Standards (1949): Tables of the Binomial Probability Distribution, *Appl. Math. Ser.* 6.

Noether, G. E. (1967): "Elements of Nonparametric Statistics," John Wiley & Sons, Inc., New York.

Olmstead, P. S. (1946): Distribution of Sample Arrangements for Runs Up and Down, *Ann. Math. Statistics*, **17**: 24–33.

Owen, D. B. (1962): "Handbook of Statistical Tables," Addison-Wesley Publishing Company, Inc., Reading, Mass.

Parzen, E. (1960): "Modern Probability Theory and Its Applications," John Wiley & Sons, Inc., New York.

Pearson, K. (1900): On the Criterion That a Given System of Deviations from the Probable in the Case of a Correlated System of Variables Is Such That It Can Reasonably Be Supposed to Have Arisen in Random Sampling, *Phil. Mag.*, **50**: 157–175.

—— (1904): On the Theory of Contingency and Its Relation to Association and Normal Correlation, *Drapers' Company Res. Mem. Biometric Ser.* I.

Quade, D. (1967): Nonparametric Partial Correlation, *Univ. North Carolina Dept. Biostatistics Inst. Statistics Mimeo Ser.* 526.

Rijkoort, P. J. (1952): A Generalization of Wilcoxon's Test, *Proc. Koninkl. Ned. Akad. Wetenschap.*, **A55**: 394–404.

Rosenbaum, S. (1953): Tables for a Nonparametric Test of Dispersion, *Ann. Math. Statistics*, **24**: 663–668.

—— (1965): On Some Two-sample Non-parametric Tests, *J. Am. Statist. Assoc.*, **60**: 1118–1126.

Ruben, H. (1954): On the Moments of Order Statistics in Samples from Normal Populations, *Biometrika*, **41**: 200–227.

Sarhan, A. E., and B. G. Greenberg (1962): "Contributions to Order Statistics," John Wiley & Sons, Inc., New York.

Savage, I. R. (1962): "Bibliography of Nonparametric Statistics," Harvard University Press, Cambridge, Mass.

Siegel, S. (1956): "Non-parametric Statistics for the Behavioral Sciences," McGraw-Hill Book Company, New York.

—— and J. W. Tukey (1960): A Nonparametric Sum of Ranks Procedure for Relative Spread in Unpaired Samples, *J. Am. Statist. Assoc.*, **55**: 429–445; correction, *ibid.*, **56**: 1005 (1961).

Smirnov, N. V. (1935): Über die Verteilung des allgemeinen Gliedes in der Variationsreihe, *Metron*, **12**: 59–81.

—— (1936): Sur la distribution de ω^2, *Compt. Rend.*, **202**: 449–452.

—— (1939): Estimate of Deviation between Empirical Distribution Functions in Two Independent Samples (in Russian), *Bull. Moscow Univ.*, **2**: 3–16.

—— (1948): Table for Estimating the Goodness of Fit of Empirical Distributions, *Ann. Math. Statistics*, **19**: 279–281.

Somers, R. H. (1959): The Rank Analogue of Product-moment Partial Correlation and Regression, with Application to Manifold, Ordered Contingency Tables, *Biometrika*, **46**: 241–246.

Stuart, A. (1954): The Correlation between Variate-values and Ranks in Samples from a Continuous Distribution, *Brit. J. Statist. Psychol.*, **7**: 37–44.

Sukhatme, B. V. (1957): On Certain Two Sample Nonparametric Tests for Variances, *Ann. Math. Statistics*, **28**: 188–194.

Swed, F. S., and C. Eisenhart (1943): Tables for Testing Randomness of Grouping in a Sequence of Alternatives, *Ann. Math. Statistics*, **14**: 66–87.

Tate, M. W., and R. C. Clelland (1957): "Nonparametric and Shortcut Statistics," The Interstate Publishers & Printers, Danville, Ill.

Terry, M. E. (1952): Some Rank Order Tests Which Are Most Powerful against Specific Parametric Alternatives, *Ann. Math. Statistics*, **23**: 346–366.

Waerden, B. L. van der (1952, 1953): Order Tests for the Two-sample Problem and Their Power, I, II, III, *Indagationes Math.* **14** [*Proc. Koninkl. Ned. Akad. Wetenschap.* **55**]: 453–458; *Indag.* **15** [*Proc.* 56]: 303–310, 311–316; correction, *Indag.* **15** [*Proc.* 56]: 80.

――― and E. Nievergelt (1956): "Tafeln zum Vergleich Zweier Stichproben mittels X-test und Zeichentest," Springer-Verlag OHG, Berlin-Gottingen-Heidelberg.

Wald, A., and J. Wolfowitz (1940): On a Test Whether Two Samples Are from the Same Population, *Ann. Math. Statistics*, **11**: 147–162.

Walsh, J. E. (1949a): Applications of Some Significance Tests for the Median Which Are Valid under Very General Conditions, *J. Am. Statist. Assoc.*, **44**: 342–355.

――― (1949b): Some Significance Tests for the Median Which Are Valid under Very General Conditions, *Ann. Math. Statistics*, **20**: 64–81.

――― (1962, 1965): "Handbook of Nonparametric Statistics," Van Nostrand Company, Inc., Princeton, N.J.

Westenberg, J. (1948): Significance Test for Median and Interquartile Range in Samples from Continuous Populations of Any Form, *Proc. Koninkl. Ned. Akad. Wetenschap.*, **51**: 252–261.

Wilcoxon, F. (1945): Individual Comparisons by Ranking Methods, *Biometrics*, **1**: 80–83.

――― (1947): Probability Tables for Individual Comparisons by Ranking Methods, *Biometrics*, **3**: 119–122.

――― (1949): Some Rapid Approximate Statistical Procedures, American Cyanamid Company, Stanford Research Laboratories, Stanford, Calif.

Wilks, S. S. (1948): Order Statistics, *Bull. Am. Math. Soc.*, **54**: 6–50.

Wolfowitz, J. (1944a): Asymptotic Distribution of Runs Up and Down, *Ann. Math. Statistics*, **15**: 163–172.

――― (1944b): Note on Runs of Consecutive Elements, *Ann. Math. Statistics*, **15**: 97–98.

――― (1949): Non-parametric Statistical Inference, pp. 93–113, in J. Neyman (ed.), *Proc. Berkeley Symp. Math. Statistics Probability*.

Index